The Theo
Catering

Fifth Edition

Ronald Kinton
Garnett College, College for Teachers for Further and Higher Education

Victor Ceserani MBE
Former Head of School of Hotelkeeping and Catering
Ealing College of Higher Education

Edward Arnold

© Ronald Kinton and Victor Ceserani 1984

First Published 1964
by Edward Arnold (Publishers) Ltd.,
41 Bedford Square,
London WC1B 3DQ

Edward Arnold (Australia) Pty Ltd,
80 Waverley Road,
Caulfield East 3145,
PO Box 234,
Melbourne.

Reprinted 1965, 1967, 1968 (twice)
Second Edition 1970
Reprinted 1970, 1971
Third Edition 1973
Reprinted 1974
Fourth Edition 1978
Reprinted 1978, 1980, 1981, 1982
Fifth Edition, 1984

Reprinted 1984, 1985

British Library Cataloguing in Publication Data
Kinton, Ronald
The Theory of Catering.—5th ed.
1. Caterers and catering
I. Title II. Ceserani, Victor
642′.4 TX 943

ISBN 0-7131-0984 X
ISBN 0-7131-0985 8 Pbk

All rights reserved. No part of this publication may be reproduced, stored in a retrieval system, or transmitted in any form or by any means, electronic, mechanical, photocopying, recording or otherwise, without the prior permission of Edward Arnold (Publishers) Ltd.

Text set in 10/11 pt Times Compugraphic by Colset Private Ltd,
Printed and bound in Great Britain by Richard Clay
(The Chaucer Press) Ltd, Bungay, Suffolk.

Contents

Introduction to Fifth Edition

This book is designed to meet the needs of students studying for the Business and Technician Education Council (BTEC), City & Guilds of London Institute and HCIMA examinations.

This edition has been prepared taking into account developments which have occured and are now having considerable effect on the catering industry, to those employed in the industry and especially those preparing to make a career in catering.

These developments concern:

1 *legislation* affecting people at work; therefore basic information on the legal aspects regarding employees and employer's responsibilities and rights and essential points from the Health, Safety and Welfare at Work Act have been included.
2 *technological* changes and eating habits of consumers which are affecting the many types of establishments and their menus have been taken into account.
3 *educational* changes required by the examining bodies BTEC; and C & GLI.

Many courses have their requirements stated in objective terms to enable students and teachers to know what is expected of them. Objectives may be expressed generally or specifically; general objectives are those indicating understanding, knowing, appreciating or awareness; specific objectives are those which can be measured, where there is some tangible proof that the intention or objective has been realised.

Each chapter is preceded by relevant objectives, however it is important to *realise the relationship between the theoretical and practical knowledge and experience* as well as *the overlap of objectives between the chapters*. The true test of the achievement of the objectives is in the practical situation. For example it is important that students and employees know the legal requirements regarding hygiene, it is even more important that they have the right attitude to them and apply them in practice. To check if the learning objectives have been achieved it is recommended that students use *Questions on Theory of Catering* which is published in conjunction with this book.

As in previous editions we have not attempted to write a comprehensive book but rather to set out a simple outline as a basis for further study. In this way we hope that it may assist students studying at all levels and to help those who wish to study at greater depth further references are suggested where appropriate.

Acknowledgements

The authors gratefully acknowledge the contribution of the following people and organisations:

Mr Peter Bateman, Director of Rentokil Ltd, for scrutinising Chapter 2 Hygiene; Mr A Brown, North Thames Gas; Mr T Burgess, XI Data Systems; Mrs F Cable BSc, Ealing College of Higher Education; Elisabeth Studd for advice on the French in Chapters 7, 12 and 13; Mr J Watkinson, Electricity Council; Mr R Wellman, Vauxhall College of Building, for Chapter 4 Water.

The photographs on pages 118–25 were taken by John Crocker, 139 Drayton Gardens, West Drayton, Middlesex and the loan of the equipment for the photographs was courtesy of Ferrari Bros, 60 Wardour Street, London.

The Publishers would like to thank the following for permission to include copyright material:

The Hotel and Catering Industry Training Board for their cartoon on page 40, Trusthouse Forte for their menus on pages 369 and 372–3 and Walton's Restaurants Limited for their menu on pages 360–1.

The Publishers would also like to thank the following organisations for permission to include copyright photographs:

Catering Management: 8, 12, 297, 299, 359t, 404; *Caterer*: 19, 114, 285, 294, 300, 401; British Railways Board: 20; The Portman Intercontinental Hotel: 23, 98, 266; Marks & Spencer plc: 26; *Catering Management* & Dawsons: 32; *Catering Management* & Johnson & Johnson Ltd: 35; Eurocaddy Systems Ltd: 37; *Catering Management* & BA: 39, 419; Rentokil Ltd: 50; Darton Engineering Ltd: 52; Exel Equipment Limited: 85; Science Museum, London: 96; Rank Hotels: 97, 405; BOC Limited: 100; General Electric, Illinois: 104; Fox Photos & Electrolux: 105; Thomson Regional Newspapers: 107; Garland Catering Equipment Ltd: 108; Francis-Thompson Studios Ltd: 109; Hobart Manufacturing Co Ltd: 113; John Crocker: 118, 119, 121, 122, 123, 124, 125; Mareno Industriale SpA: 126; *Caterer & Hotelkeeper*: 154, 165t, 303, 357b, 359b, 403, 412; Meat & Livestock Commission: 165b; Camera Press: 186; Dexion Ltd: 280; Michael Balfré: 357t, 407b; X1 Data Systems/T Burgess: 393, 395; Wimpy International Ltd: 400; The Plessey Co Ltd: 407t; Exel Equipment Limited: 408; *Catering Management* & El Al: 420;

The Publishers have made every effort to trace the copyright of the remaining photographs but have been unsuccessful.

1
Health and safety

Having read the chapter these learning objectives, both general and specific, should be achieved.

General objectives Be aware of the main points of the Health and Safety at Work Act. Know of the accidents which may occur in catering establishments and understand why there is a need for care to be taken by people in working situations. Create a favourable attitude to safety so that in practice it is second nature.

Specific objectives Explain the responsibilities of the employee and employer and the function of the environmental health officer. Specify the rules to be observed when handling all tools and equipment, both large and small. Explain how to prevent burns, scalds, and cuts. Specify the correct procedure for basic first aid and be able to apply first aid. Distinguish the correct equipment for dealing with fires and be able to use the appropriate appliance correctly.

Health and Safety at Work Act

Every year in this country a thousand people are killed at work; a million people suffer injuries; 23 million working days are lost annually because of industrial injury and disease. As catering is one of the largest employers of labour the catering industry is substantially affected by accidents at work.

In 1974 the Health and Safety at Work Act of Parliament was passed with two main aims:

1 to extend the coverage and protection of the law to all employers and employees,
2 to increase awareness of safety amongst those at work, both employers and employees.

The law imposes a general duty on an employer 'to ensure so far as is reasonably practicable, the health, safety and welfare at work of all his employees'. The law also imposes a duty on every employee while at work to:

a) take reasonable care for the health and safety of himself or herself and of other persons who may be affected by his or her acts or omissions at work,

b) to co-operate with his or her employer so far as is necessary to meet or comply with any requirement concerning health and safety,

c) not to interfere with, or misuse, anything provided in the interests of health, safety or welfare.

It can be clearly seen that both health and safety at work is everybody's responsibility.

Furthermore the Act protects the members of the public who may be affected by the activities of those at work.

Penalties are provided by the Act which include improvement notices, prohibition notices and criminal prosecution, the Health and Safety Executive has been set up to enforce the law and the Health and Safety Commission will issue codes of conduct and act as advisers.

Responsibilities of the employer

1 Provide and maintain premises and equipment that are safe and without risk to health.
2 To provide supervision, information and training.
3 To issue a written statement of Safety Policy to employees to include:
   a) general policy with respect to health & safety at work of employees,
   b) the organisation to ensure the policy is carried out,
   c) how the policy will be made effective.
4 To consult with the employees safety representative and to establish a Safety Committee.

Responsibilities of the employee

1 Take reasonable care to avoid injury to themselves or to others by their work activities.
2 Co-operate with their employer and others so as to comply with the law.
3 To refrain from misusing or interfering with anything provided for health and safety.

Enforcement of the Act

Health and safety inspectors and local authority inspectors (environmental health officers) have the authority to enforce the requirements of the Act. They are empowered to:

1 issue a *prohibition notice* which immediately prevents further business until remedial action has been taken,
2 issue an *improvement notice* whereby action must be taken within a stated time, to an employee, employer or supplier,

3 *prosecute any* person breaking the Act. This can be instead of or in addition to serving a Notice and may lead to a substantial fine or prison,
4 seize, render harmless or destroy anything that the inspector considers to be the cause of imminent danger.

Function of the environmental health officer

One aspect of the environmental health officer's role is to enforce the law, the other aspect is to act as an adviser and educator in the areas of food hygiene and catering premises. Their function is to improve the existing standard of hygiene and to advise how this may be achieved. Frequently health education programmes are organised by environmental health officers which may include talks and free literature. If in doubt about any matter concerning food hygiene, pests, premises or legal aspects of the Act the environmental health officer is there to be consulted.

Accidents

It is essential that people working in the kitchen are capable of using the tools and equipment in a manner which will neither harm themselves nor those with whom they work. Moreover, they should be aware of the causes of accidents and be able to deal with any which occur.

Accidents are caused in various ways:

a) excessive haste,
b) distraction,
c) failure to apply safety rules,

It should be remembered that most accidents could be prevented.

a) Excessive haste – the golden rule of the kitchen is 'never run' and this may be difficult to observe during a very busy service. Excessive haste causes people to take chances which inevitably lead to mishaps.
b) Distraction – accidents may be caused by not concentrating on the job in hand, through lack of interest, personal worry or distraction from someone else. The mind must always be kept on the work so as to reduce the number of accidents.

Reporting accidents

Any accident occuring on the premises where the employee works must be reported to the employer and a record of the accident must be entered in the Accident Book.

Any accident causing death or major injury to an employee or member of the public must be reported by the employer to the Environmental Health Department. Also accidents involving dangerous equipment must be reported even if no one is injured.

Prevention of accidents

It is the responsibility of everyone to observe the safety rules; in this way a great deal of pain and loss of time can be avoided.

Prevention of cuts and scratches

Knives
These should never be misused and the following rules should always be observed:

1 The correct knife should be used for the appropriate job.
2 Knives must always be sharp and clean, a blunt knife is more likely to cause a cut owing to excessive pressure having to be used.
3 Handles should be free from grease.
4 When carrying knives, the points must be held downwards.
5 They should be placed flat on the board or table so that the blade is not exposed upwards.
6 Knives should be wiped clean with the edge away from the hands.
7 Do not put knives in a washing-up sink.

Choppers
These should be kept sharp and clean and when used care should be taken that no other knives, saws, hooks, etc, can be struck by the chopper, which could cause them to fly into the air. This applies also when using a large knife for chopping.

Cutting blades on machines
Guards should always be in place when the machine is in use; they should not be tampered with nor should hands or fingers be inserted past the guards. Before the guards are removed for cleaning, the blade or blades must have stopped revolving.

When the guard is removed for cleaning, the blade should not be left unattended, in case someone should put a hand on it by accident. If the machine is electrically operated the plug should, when possible, be removed.

Cuts from meat and fish bones
Jagged bones can cause cuts which may turn septic, particularly fish bones and the bones of a calf's head which has been opened to remove the brain. Cuts of this nature, however slight, should never be neglected. Frozen meat should not be boned out until it is completely thawed out, because it is difficult to handle, the hands become very cold and the knife slips easily.

Prevention of burns and scalds

A burn is caused by dry heat and a scald by wet heat, both can be very painful and have serious effects, and certain precautions should be taken to prevent them.

1 Sleeves of jackets and overalls should be rolled down and aprons worn at a sensible length so as to give adequate protection.
2 A good thick dry cloth (rubber) is most important for handling hot utensils. It should never be used wet on hot objects and is best folded to give greater protection. It should not be used if thin, torn or with holes.
3 Trays containing hot liquid, for example roast gravy, should be handled carefully, one hand on the side and the other on the end of the tray so as to balance it.
4 Hot pans brought out of the oven should have something white, eg a little flour, placed on the handle and lid as a warning that it is hot. This should be done as soon as the pan is taken out of the oven.
5 Handles of pans should not protrude over the edge of the stove as the pan may be knocked off the stove.
6 Large full pans should be carried correctly – that is to say, when there is only one handle the forearm should run along the full length of the handle and the other hand should be used to balance the pan where the handle joins the pan. This should prevent the contents from spilling.
7 Certain foods require extra care when heat is applied to them, as for example when a cold liquid is added to a hot roux or when adding cold water to boiling sugar when making caramel. Extra care should always be taken when boiling sugar.
8 Frying, especially deep frying, needs careful attention. When shallow or deep frying fish, for example, the fish should be put into the pan away from the person so that any splashes will do no harm. With deep frying, fritures should be moved with care and if possible only when the fat is cool. Fritures should not be more than two-thirds full. Wet foods should be drained and dried before being placed in the fat, and when foods are tipped out of the frying basket a spider should be at hand. Should the fat in the friture bubble over on to a gas stove then the gas taps should be turned off immediately. Fire blankets and fire extinguishers should be provided in every kitchen conveniently sited ready for use.
9 Steam causes scalds just as hot liquids do. It is important to be certain that before opening steamers the steam is turned off and when the steamer door is opened no one is in the way of the escaping steam. The steamer should be in proper working condition; the drain hole should always be clear. The door should not be opened immediately the steam is turned off; it is better to wait for about half a minute before doing so.

10 Scalds can also be caused by splashing when passing liquids through conical strainers; it is wise to keep the face well back so as to avoid getting splashed. This also applies when hot liquids are poured into containers.

Machinery

Accidents are easily caused by misuse of machines (refer to page 35 for further information). These rules should always be put into practice.

1 The machine should be in correct running order before use.
2 The controls of the machine should be operated by the person using the machine. If two people are involved there is the danger that a misunderstanding can occur and the machine be switched on when the other person does not expect it.
3 Machine attachments should be correctly assembled and only the correct tools used to force food through mincers.
4 When using mixing machines the hands should not be placed inside the bowl until the blades, whisk or hook have stopped revolving. Failure to observe this rule may result in a broken arm.
5 Plugs should be removed from electric machines when they are being cleaned so that they cannot be accidentally switched on.

Explosions

The risk of explosion from gas is considerable. To avoid this occurring it is necessary to ensure that the gas is properly lit. On ranges with a pilot on the oven it is important to see that the main jet has ignited from the pilot. If the regulo is low sometimes the gas does not light at once, the gas collects and an explosion occurs. When lighting the tops of solid-top ranges it is wise not to place the centre ring back for a few minutes after the stove is lit as the gas may go out – gas then collects and an explosion can occur.

Floors, etc

Accidents are also caused by grease and water being spilled on floors and not being cleaned up. It is most important that floors are always kept clean and clear; pots and pans, etc, should never be left on the floor, nor should oven doors be left open, because anyone carrying something large may not see the door, or anything on the floor and trip over.

Many people strain themselves by incorrectly lifting or attempting to lift items which are too heavy. Large stock pots, rondeaus, forequarters and hindquarters of beef, for example, should be lifted with care. Particular attention should be paid to the hooks in meat so that they do not injure anyone.

On no account should liquids be placed in containers on shelves above eye-level, especially when hot. They may be pulled down by someone else.

Safe kitchens are those which are well lit and well ventilated and where the staff take precautions to prevent accidents happening. When accidents do happen it is necessary to know something of first aid. **Further information:** Royal Society for the Prevention of Accidents, Cannon House, Priory Queensway, Birmingham.

First aid

As the term implies it is the immediate treatment on the spot to a person who has been injured or is ill. From 1982 it is a legal requirement that adequate first-aid equipment, facilities and personnel to give first aid are provided at work. If the injury is serious the injured person should be treated by a doctor or nurse as soon as possible.

Shock

The signs of shock are faintness, coldness, clammy skin and whiteness. Shock should be treated by keeping the person comfortable, laid down and warm. Cover the person with a blanket or clothing, but do not apply hot water bottles.

Fainting

Fainting may occur after a long period of standing in a hot badly ventilated kitchen. The signs of an impending faint are whiteness, giddiness and sweating. A faint should be treated by raising the legs slightly above the level of the head, and when the person recovers consciousness, put the person in the fresh air for a while and make sure that the person has not incurred any injury in fainting.

Cuts

All cuts should be covered immediately with a waterproof dressing, after the skin round the cut has been washed. When there is considerable bleeding it should be stopped as soon as possible. Bleeding may be controlled by direct pressure, by bandaging a dressing firmly on the cut. It may be possible to stop bleeding from a cut artery by pressing the artery with the thumb against the underlying bone, such pressure may be applied while a dressing or bandage is being prepared for application but not for more than 15 minutes.

Fractures

A person suffering from broken bones should not be moved until the injured part has been secured so that it cannot move.

Burns and scalds

Place the injured part gently under slowly running water or immerse the part in cool water keeping it there for at least 10 minutes or until the pain ceases. If serious, the burn or scald should then be covered with a clean cloth and the person sent immediately to hospital.

Electric shock
Switch off the current. If this is not possible, free the person by using a dry insulating material such as cloth, wood or rubber, taking care not to use the bare hands otherwise the electric shock may be transmitted. If breathing has stopped, give artificial respiration and send for a doctor. Treat any burns as above.

Damage by fire

Gassing
Do not let the gassed person walk, but carry him or her into the fresh air. If breathing has stopped apply artificial respiration and send for a doctor.

Artificial respiration
There are several methods of artificial respiration. The most effective is mouth-to-mouth (mouth-to-nose) and this method can be used by almost all age groups and in almost all circumstances.

Further information: St. John Ambulance Association, 1 Grosvenor Crescent, London SW1.

Fire precautions

Fires in hotel and catering establishments are common and all too often can result in injury to the employee and in serious cases either injury or loss of life to employees and customers.

Fire prevention
A basic knowledge regarding fire should assist in preventing fires and handling them if they do occur. Three components are necessary for a fire to start, if one of the three is not present, or is removed, then the fire does not happen or it is extinguished. The three parts are:

a) Fuel – something to burn,
b) Air – oxygen to sustain combustion (to keep the fire going),
c) Heat – gas, electricity etc.

Methods of extinguishing a fire
To extinguish a fire the three principal methods are:

a) starving – removing the fuel,
b) smothering – removing the air (oxygen),
c) cooling – removing the heat.

Therefore one of the sides of the triangle is removed.

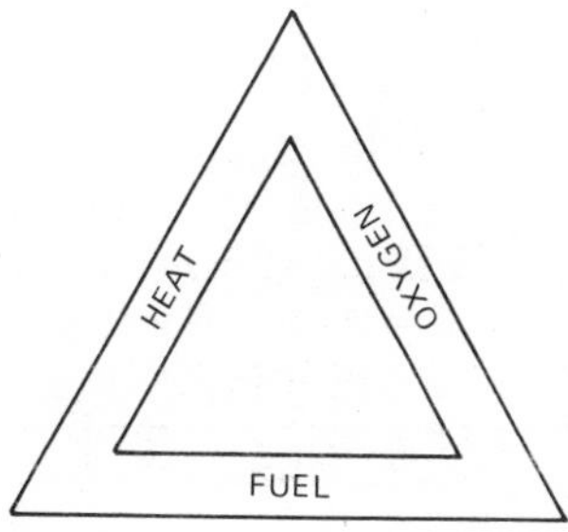

The fire triangle

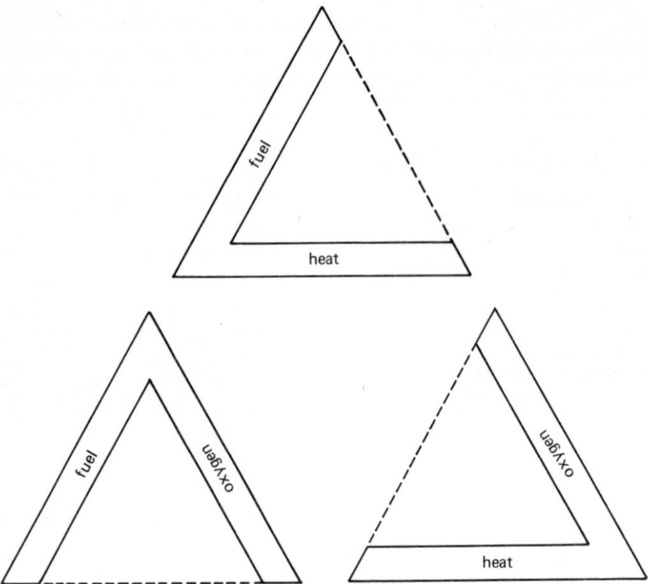

The fuel is that which burns, heat is that which sets the fuel alight and oxygen is needed for fire to burn. Eliminate one of these and the fire is put out. Oxygen is present in air so if air is excluded from the fuel and the heat then the fire goes out. For example, should the clothes of someone working in the kitchen catch alight then the action to be taken is to quickly wrap a fire blanket round the person and roll them on the floor. In so doing the flames have been cut off from the source of air. (The oxygen has been taken from the triangle.) In the event of a fire, windows and doors are to be closed so as to restrict the amount of air getting to the fire. Foam extinguishers work on the principle that the foam forms a 'blanket' thus excluding air from coming into contact with the fuel.

Should fat or oil in a pan ignite, then the pan should be quickly covered with a lid or other item or fire blanket so as to exclude air. It is also essential to turn off the source of heat, gas or electricity etc so that the heat is taken from the triangle.

Water extinguishes by dousing the flames thus taking the heat out of the triangle provided the fuel is material such as wood, paper etc. If fat or oil is alight water must *not* be used as it causes the ignited fat to spread thus increasing the heat. Water extinguishers must *not* be used on live electrical equipment because water is a conductor of electricity.

In the event of a small fire in a store it may be possible to remove items in the store to prevent the fire from spreading.

Fire doors are installed for the purpose of restricting an area so that in the event of a fire the fuel is limited.

FIRE NOTICE

Please read instructions carefully.

TELEPHONE: No matter how small the fire, immediately call the Hotel Telephone Exchange. State where the fire is in the Hotel.

DANGER: If the fire is already out of control, close your windows and doors and raise the alarm by ringing the bell situated next to the passenger lift opposite the Main Stairs.

DO NOT PANIC: Remember your calm, prompt action may save life.

EVACUATE: On notice of evacuation being made by the sounding of the fire-bell, leave the Hotel via the nearest fire exit and go directly to the assembly point at the Stratford Court Hotel in Marylebone Lane. This is on the other side of Oxford Street, 80 yard/75 metres to the right of Davies Street.

Do not attempt to struggle with your personal belongings and under no circumstances return for them.

Do not use lifts.

SIGNS OF FIRE: Please report any signs of fire. only good can come from the action. Thank you.

ZUR BEACHTUNG BEI FEUER

Bitte lesen Sie die folgenden Instruktionen sorgfältig durch.

TELEFON: Selbst wenn das Feuer nur sehr klein ist, rufen Sie sofort bei der Hotelzentrale an. Geben Sie die genaue Lage des Feuers im Hotel an.

GEFAHR: Wenn das Feuer nicht mehr unter Kontrolle zu bringen ist, schließen Sie die Fenster und Türen und schlagen Sie Alarm, indem Sie die Glocke, die sich neben dem Personenfahrstuhl gegenüber der Haupttreppe befindet, läuten.

GERATEN SIE NICHT IN PANIK: Denken Sie daran, daß durch Ihr ruhiges, schnelles Handeln Leben gerettet werden können.

EVAKUIERUNG: Wenn durch das Läuten der Feuerglocke das Zeichen zur Evakuierung gegeben wird, verlassen Sie das Hotel durch den nächstgelegenen Notausgang und gehen Sie direkt zum Sammelplatz im Stratford Court Hotel in Marylebone Lane. Dieses befindet sich auf der anderen Seite von Oxford Street, 75 Meter rechts von Davies Street.

Versuchen Sie nicht, Ihre persönlichen Sachen zu retten und kehren Sie auf keinen Fall danach zurück.

Benutzen Sie nicht die Fahrstühle.

ANZEICHEN VON FEUER: Bitte machen Sie auf jegliche Anzeichen von Feuer aufmerksam. Nur Gutes kann aus Ihrer Handlungsweise kommen. Danke schön!

INSTRUCTIONS EN CAS D'INCENDIE

Veuillez lire attentivement ces instructions.

TELEPHONE: Quelle que soit l'importance du feu, avertissez immédiatement le standard téléphonique de l'hôtel. Indiquez le lieu où s'est déclaré l'incendie dans l'hôtel.

DANGER: Si l'incendie est déjà hors de contrôle, fermez portes et fenêtres et donnez l'alarme en actionnant la sonnette située près de l'ascenseur, en face de l'escalier principal.

NE VOUS AFFOLEZ PAS: Gardéz votre calme, une prompte action peut sauver des vies humaines.

EVACUATION: Dès le signal d'évacuation annoncé par la sonnerie d'alarme, quittez immédiatement l'hôtel par la sortie de secours la plus proche et dirigez-vous directement au point de rassemblement au Stratford Court Hotel, dans Marylebone Lane, Cet hôtel se trouve de l'autre côté d'Oxford Street, à 75 mètres de Davies Street, sur la droite. N'essayez pas de vous encombrer de vos effets personnels et en aucun cas ne retournez les chercher.

Evitez d'utiliser les ascenseurs.

SIGNES DE DEBUT D'INCENDIE: Vous êtes prié de reporter tout signe de début d'incendie. Votre action ne peut être que profitable à tous. Merci.

火の用心

この次のお知らせをご注意深く調べて下さい。

電話. どんな小さい火事でも早速にホテルの電話交換局に連絡し火事場を告げて下さい

危険. 火事が消し止められないようでしたらまずご部屋の窓やドアを完全に閉じ そして 主階段の前にあるエレベーターの側の火警鐘を打鳴して下さい

慌てないよう. 万一火事があったらお客様のご落ち着きや良心のお陰で命が助かることもあります.

避難. 火警鐘が打鳴らされた場合早速にMARYLEBONE LANEのSTRATFORD COURT HOTELへ移して下さい、そのホテルはDAVIES STREETの右側から75メートルはなれたところでOXFORD STREETの向こう側にあります。

荷物をそのままにして置いてホテルから移して下さい。移した後決してお部屋へ戻らないで下さい。

エレベーターを利用しないで下さい。

火事の形跡. 火事の形跡を見付けるならば早速に電話交換局に知らせて下されば皆様のご安全性が心配ありません。

CUSTODY OF VALUABLES

We draw your attention to the notice displayed in the Main Hall regarding the liabilities of Hotel Proprietors. The proprietors cannot hold themselves responsible for any valuables that may be left unattended. Jewellery, money and other valuables should be deposited with the Cashier, for safe keeping.

GROSVENOR COURT

Fire notice

Fire blanket

Procedure in the event of a fire

1 Do not panic.
2 Warn other people in the vicinity.
3 Do not jeopardise your own safety or that of others.
4 Follow the fire instructions of the establishment.
5 If a small fire, use appropriate fire extinguisher.
6 Do close doors and windows, turn off gas, electricity and fans.
7 Do not wait for the fire to get out of control before calling the fire brigade.

It is important that in all catering establishments passageways are kept clear and that doors open outwards. Fire escape doors and windows should be clearly marked and fire fighting equipment must be readily available and in working order. Periodic fire drills should occur and be taken seriously since lives may be endangered if there was a fire. Fire alarm bells must be tested at least four times a year and staff should be instructed in the use of fire fighting equipment. All extinguishers should be refilled immediately after use.

All fire extinguishers should be manufactured in accordance with

British Standard specifications; they should be red with an additional colour code to indicate the type and with operating instructions on them.

Red – water
Cream – foam
Black – carbon dioxide
Blue – dry powder
Green – halon (vapourising liquid)

Use of portable fire extinguishers

1 Water (red)

Water is used for fire in ordinary combustible materials such as wood, paper etc. Water has better cooling properties than most other agents, therefore it is especially suitable for fires that may start up again if they are not cooled sufficiently. Most water extinguishers contain carbon dioxide gas which expels the water.

FIRE CLASS ACCORDING TO BS 4547, 1970	EXTINGUISHING PRINCIPLES	A.B.C. ALL PURPOSE POWDER (Blue)	CO_2 GAS Black	FOAM Cream	WATER Red	B.C.F. (halons) Green
CLASS A Fires involving solid materials usually of organic nature in which combustion normally takes place with the formation of glowing embers. Wood, paper, textiles etc.	Water Cooling or Combustion Inhibition	YES Excellent	NO	YES	YES Excellent	YES
CLASS B Fires involving liquids or liquefiable solids. Burning liquids, oil, fat, paint etc.	Flame Inhibiting or Surface blanketing and cooling	YES Excellent	YES	YES Excellent	NO	YES
CLASS C Fires involving gases		YES	YES	YES	NO	YES
FIRES INVOLVING ELECTRICAL HAZARDS	Flame Inhibiting	YES	YES Excellent	NO	NO	YES

Portable fire extinguishers

Disadvantages

a) Because water is a conductor of electricity it must never be used on live electrical equipment.
b) Water must never be used on fat fires because it may cause ignited fat to spread.

2 Foam (cream)

Foam puts out fires by forming a blanket of foam over the top of the fire. It is particularly good for putting out fat fires because the foam stays in position and so stops the fire re-igniting. Foam can also be used on fires of natural materials.

Disadvantages

a) Foam is a conductor of electricity and must not be used on live electrical equipment.
b) Foam is not effective on free flowing liquids.

3 Carbon dioxide (CO_2) (black)

Carbon dioxide gas is used on fires of inflammable liquids and has the advantage that it does not conduct electricity.

Disadvantage

CO_2 gas has limited cooling properties and therefore is not the most efficient way of putting out a fat fire.

4 Dry powder (blue)

Dry powder is commonly used for fat fires. It does not conduct electricity and some all-purpose powders can be used on fires in natural materials. Powders based on bicarbonate of soda are used in most extinguishers.

Disadvantage

Dry powders usually have limited cooling properties.

5 Halons (green)

These are also known as BCF which is short for bromochlorodifluoromethane. This is a gas which does not conduct electricity.

Disadvantage

If used in an enclosed situation halons give off a thick cloud which can irritate the user's throat and it should not be inhaled.

Other extinguishers

Fire hoses

Fire hoses are used for similar fires to those classified under water fire extinguishers. It is necessary to be familar with the instructions displayed by the fire hose before using it.

2 CATERING TIMES

NEWSFRONT

Fire risk times ten

For the average person the risk of dying in a fire is ten times greater when staying in a hotel than when he or she is in his or her own bed.

And over the last decade fire brigades in the United Kingdom have been called out to deal with, on average, nearly 20 fires in hotels every week.

Examples of fire extinguishers

Water sprinkler systems

These consist of sprinklers from the main water supply fitted in the ceiling. The system is designed to automatically spray water over the whole area when the temperature rises above a pre-set level eg 75°C (167°F).

Choice of fire extinguisher

	Type of fire risk	*Type of extinguisher*
1	Fires involving wood, paper, fabrics or similar materials requiring cooling or quenching.	*Water CO_2* (carbon dioxide) Operated by piercing a gas cylinder, the gas then forces the water out. When *hoses* are used for fighting a fire the hose should be connected to the mains water supply. *Soda-acid* Made in several sizes. Contains a canister of bicarbonate of soda and a bottle of dilute sulphuric acid. The acid is contained in a glass phial which when broken reacts with the bicarbonate of soda and forces water out of the nozzle.
2	Fire involving flammable liquids, petrols, oils, greases, fats requiring rapid action.	*Foam* Contains a small canister within a large one, both containing different chemicals which when mixed form foam. This is forced out of the canister and forms a blanket, so preventing air reaching the fire and thus causing it to go out. *Dry powder* *CO_2 (gas)*
3	Fires involving live electrical apparatus.	A non-conducting extinguishing agent must be used.

Each extinguisher should be fixed on a suitable bracket, be properly maintained and always available for use, and immediately refilled after use. It is important that staff learn how to use them.

Research and development by the manufacturers of fire-fighting equipment inevitably leads to changes and increased efficiency in the various appliances and as it is important that the best fire extinguishers are always available always consult the Fire Prevention Branch of the Fire Brigade and for a list of approved extinguishers apply to the Fire Officers' Committee. **Futher information**: Fire Protection Association, Alderney House, Queen Street, London EC4 NITJ.

Working methods

Having read this part of the chapter these learning objectives, both general and specific, should be achieved.

General objectives Be aware of the need for economy of time, energy and materials in the working situation. Develop an attitude to work which conserves energy and saves time and materials.

Specific objectives State examples of smooth work flow in catering situations. Relate the principles of work study so as to implement them by efficient practical application in the working environment.

A skilled craftsman is one who, among other things, completes the skill in the minimal time, to the highest standard and with the *least* effort. Effort requires expenditure of energy, and energy is the commodity which needs to be conserved, not wasted, in the kitchen. Any person working in a hot environment, with the stress of working against the clock, needs to get into the habit of working in such a way that energy is not wasted. To achieve this it is necessary to use common sense and to acquire some knowledge, to practise thinking about how to save energy so that the habit of working methodically and economically becomes second nature. This state of mind can be developed by students producing 12 items even though the real effects will only be evident when producing say 100 or 500 items.

The objective is to make work easier and this can be achieved by simplifying the operation, eliminating unnecessary movements, combining two operations into one or improving old methods. For example, if you are peeling potatoes and you allow the peelings to drop into the container in the first place, the action of moving the peelings into the bowl and the need to clean the table could have been eliminated. This operation is simplified if instead of a blunt knife a good hand potato peeler is used, because it is simple and safe to use, requires less effort, can be used more quickly and requires less skill to produce a better result. If the quantity of potatoes is sufficient, then a mechanical peeler could be used, but it would be necessary to remember that the electricity used would add to the cost and that the time needed to clean a mechanical aid may lessen its work-saving value. If it takes 25 minutes to clean a potato mashing machine which has been used to mash potatoes for 500 meals it could be time well spent in view of the time and energy saved mashing the potatoes. It may not be considered worth while using the machine to mash potatoes for 20 meals. Factors such as this need to be taken into account.

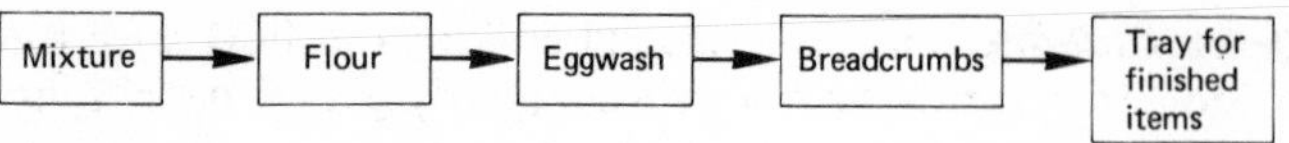

An example of correct sequence for working methods

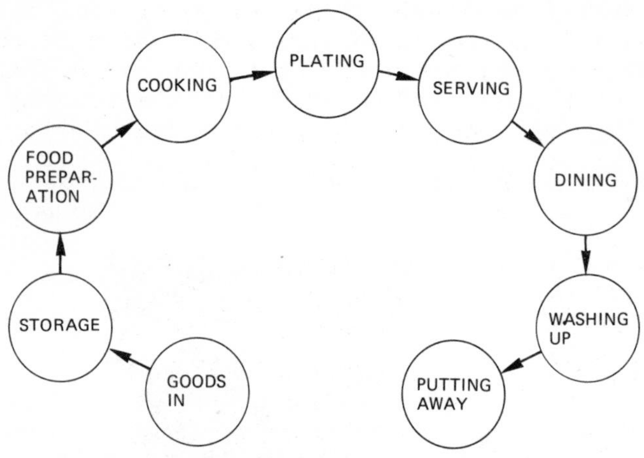

An example of planning the flow of work.

Working methods may be observed in catering at many different levels, from the experienced methodical chef wiping the knife after cutting a lemon, to the complexity of the Ganymede system in a large hospital or the carefully planned call-order unit in a fast food operation where, because of careful thought and study, wastage of time, money and materials is reduced to a minimum. Even with the aid of mechanical devices, labour-saving equipment and the extensive use of foods which have been partially or totally prepared, people at work still become fatigued. It is most important to stand correctly, well balanced with the weight of the body divided on to both legs with the feet sensibly spaced and the back reasonably straight when working for long periods in one place. Particular care is needed when lifting, it is desirable to stand with legs apart and bend the knees (not the back) and use the leg muscles to assist lifting. The object to be raised should be held close to the body.

It is possible to cultivate the right attitude to work as well as good working habits. Certain jobs are repetitive, some require considerable concentration, while others cause physical strain; not all work provides equal job satisfaction, therefore students need to develop the kind of attitude which is helpful. If 500 fish cakes have to be shaped it is worth while setting targets to complete a certain number in a certain time. Such simple things as not counting the completed items but counting those still to be done motivates some people to greater effort. Some circumstances do not lend themselves to overcoming the physical pressures, for example, if 150 people require 150 omelets, then, provided the eggs are broken and seasoned and kept in bulk with the correct sized ladle for portioning, the attitude to adopt may be to try to do each omelet better and quicker than the last.

If careful thought and study are given to all practical jobs wastage of time, labour and materials can often be eliminated.

Properly planned layout with adequate equipment, tools and materials to do the job are essential if practical work is to be carried out efficiently. If equipment is correctly placed then work will proceed smoothly in proper sequence without back-tracking or criss-crossing. Work tables, sinks and stores and refrigerators should be within easy reach in order to eliminate unnecessary walking. Equipment should be easily available during all working times.

A systematic working method

The storage, handling of foods, tools and utensils and the movement of food in various stages of production needs careful study. Many people carry out practical work by instinct and often evolve the most efficient method instinctively. Nevertheless careful observation of numerous practical workers will show a great deal of time and effort wasted through bad working methods. It is necessary to arrange work so that the shortest possible distance exists between storage and the place where the items are to be used.

When arranging storage see that the most frequently used items are nearest to hand. Place heavy items where the minimum of body strain is required to move them. Keep all items in established places so that time is not lost in hunting and searching. Adjustable shelving can be a help in organising different storage requirements.

After all the pre-planning to the job is complete then comes the actual work itself.

The kitchen on a high-speed train

Careful preparation of foods and equipment (a good mise-en-place) is essential if a busy service is to follow and is to be operated efficiently so that orders move out methodically without confusion.

The work to be done must be carefully planned so that the items requiring long preparation or cooking are started first. Where a fast production is required lining up will assist efficiency. Work carried out haphazardly without plan or organisation obviously takes longer to do than work done according to plan. There is a sequence to work that leads to high productivity and an efficient worker should learn this sequence quickly.

When preparing French beans they should all be topped and tailed, then cut - not topping, tailing the cutting each bean separately. When cutting food, articles to be cut should be on the left of the chopping board (for right-handed people), drawn with the left hand to the centre of the board, cut and pushed to the right. This should be a continuous, smoothly flowing process.

Food is often wasted by using bad working methods. For example: when preparing spinach, to tip a whole box of spinach into the sink of water, then pick off the stalks so that they drop back on to the unpicked spinach will always result in waste. This is a bad practice used by careless cooks because three-quarters of the way through the job the contents of the sink (including an amount of good spinach) are thrown away. This can happen in the preparation of other vegetables such as sprouts, potatoes, carrots.

These are just a few examples of how planning and working methodically can save time, energy and materials.

Further information: *Work Study*, Currie (Pitman); *Fast Food Operations*, Jenkins (Barrie and Jenkins).

2
Hygiene

Having read the chapter these learning objectives, both general and specific, should be achieved.

General objectives Understand why hygienic practice is essential and know the causes of ill health resulting from failure to exercise sound hygienic principles. Be aware of the need to have a healthy positive attitude and to practise high standards to the benefit of customers, employees, and employers.

Personal hygiene

General objectives Appreciate the need for personal hygiene and know how to maintain good health. Understand why those employed in the catering industry should acquire good hygienic habits. Develop a responsible attitude to hygienic practices.

Specific objectives State why personal hygiene is essential and describe how hygienic standards are achieved. Explain how germs may be transferred from the food handler to equipment, food and to other people in practical situations.

Germs or bacteria are to be found in and on the body and they can be transferred on to anything with which the body comes in contact. Personal cleanliness is essential to prevent germs getting on to food.

Personal cleanliness

Self-respect is necessary in every food-handler because a pride in one's appearance promotes a high standard of cleanliness and physical fitness. Persons suffering from ill-health or who are not clean about themselves should not handle food.

Bathing

Regular bathing at least once a week is essential, otherwise germs can be transferred on to the clothes and so on to food. If possible a daily bath or shower is ideal, but if this is not possible a thorough wash is satisfactory.

Hands

Hands must be thoroughly washed frequently; particularly after using the toilet, before commencing work and during the handling of food.

They should be washed in hot water with the aid of a brush and soap, rinsed, and dried on a *clean* towel, suitable paper towel or by hand hot-air drier. Hands and finger-nails if not kept clean can be a great source of danger as they can so easily transfer harmful bacteria on to the food.

Rings, watches and jewellery should not be worn where food is handled. Particles of food may be caught under a ring, and germs could multiply there until they are transferred into food.

Watches (apart from the fact that steam ruins them) should not be worn, because foodstuffs, eg salads and cabbage, which have to be plunged into plenty of water may not be properly washed because a watch is worn.

Jewellery should not be worn, since it may fall off into food, unknown to the wearer.

Handwashing point at the Portman Hotel

Finger-nails

These should always be kept clean and short as dirt can easily lodge under the nail and be dislodged when, for example, making pastry, so introducing bacteria into food. Nails should be cleaned with a nailbrush and nail varnish should not be worn.

Hair

Hair should be washed regularly and kept covered where food is being handled. Hair which is not cared for is likely to come out or shed dandruff which may fall into food. Men's hair should be kept short as it is easier to keep clean; it also looks neater. Women's hair should be covered as much as possible. The hair should never be scratched, combed or touched in the kitchen, as germs could be transferred via the hands to the food.

Nose

The nose should not be touched when food is being handled. If a handkerchief is used, the hands should be washed afterwards. Ideally, paper handkerchiefs should be used and then destroyed, the hands being washed afterwards. The nose is an area where there are vast numbers of harmful bacteria; it is therefore very important that neither food, people nor working surfaces are sneezed over, so spreading germs.

Mouth

There are many germs in the area of the mouth, therefore the mouth or lips should not be touched by the hands or utensils which may come into contact with food. No cooking utensils should be used for tasting food, nor should fingers be used for this purpose as germs may be transferred to food. A clean teaspoon should be used for tasting, and washed well afterwards.

Coughing over foods and working areas should be avoided as germs are spread long distances if not trapped in a handkerchief.

Ears

The ear-holes should not be handled while in the kitchen as, again, germs can be transferred.

Teeth

Sound teeth are essential to good health. They should be kept clean and visits to the dentist should be regular so that teeth can be kept in good repair.

Feet

As food-handlers are standing for many hours, care of the feet is important. They should be washed regularly and the toe-nails kept short and clean. Tired feet can cause general tiredness which leads to carelessness, and this results in a lowering of the standards of hygiene.

Cuts, burns, sores, etc

It is particularly important to keep all cuts, burns, scratches and similar openings of the skin covered with a waterproof dressing. Where the skin is septic, as with certain cuts, spots, sores, carbuncles, there are vast numbers of harmful bacteria which must not be permitted to get on food; in most cases people suffering in this way should not handle food.

Cosmetics

Cosmetics, if used by food-handlers, should be used in moderation, they should not be put on in the kitchen and the hands should be washed well afterwards. Cosmetics should be put on a clean skin, not used to cover up dirt.

Smoking

Smoking must never take place where there is food, because when a cigarette is taken from the mouth, germs from the mouth can be transferred to the fingers and so on to food. When the cigarette is put down the end which has been in the mouth can transfer germs on to working surfaces. Ash on food is most objectionable and it should be remembered that smoking where there is food is an offence against the law.

Spitting

Spitting should never occur, because germs can be spread by this objectionable habit.

Clothing and cloths (rubbers) – see also page 27

Clean whites (protective clothing) and clean underclothes should be worn at all times. Dirty clothes enable germs to multiply and if dirty clothing comes into contact with food the food may be contaminated. Cloths (rubbers) used for holding hot dishes should also be kept clean as the cloths are used in many ways such as wiping knives, wiping dishes and pans. All these uses could convey germs on to food.

Outdoor clothing, and other clothing which has been taken off before wearing whites, should be kept in a locker away from the kitchen.

General health and fitness

The maintenance of good health is essential to prevent the introduction of germs into the kitchen. To keep physically fit, adequate rest, exercise, fresh air and a wholesome diet are essential.

Sleep and relaxation

Persons employed in the kitchen require adequate sleep and relaxation as they are on the move all the time, often in a hot atmosphere where the tempo of work may be very fast. Frequently, the hours are long or extended over a long period of time, as with split duty, or they may extend into the night. In off-duty periods it may be wise to obtain some relaxation and rest rather than spend all the time energetically. The amount of sleep and rest required depends on each person's needs and the variation between one person and the next is considerable.

Exercise and fresh air

People working in conditions of nervous tension, rush, heat and odd hours need a change of environment and particularly fresh air. Swimming, walking or cycling in the country may be suitable ways of obtaining both exercise and fresh air.

Wholesome food and pure water

A well-balanced diet, correctly cooked, and pure water will assist in keeping kitchen personnel fit. The habit of 'picking' (eating small pieces of food while working) is bad; it spoils the appetite and does not allow the stomach to rest.

Meals should be taken regularly; long periods without food are also bad for the stomach. Pure water is ideal for replacing liquid lost by perspiring in a hot kitchen, or a soft drink to replace some of the salt as well as the fluid lost in sweating.

Good restroom facilities for employees

Worry

If possible it is best not to worry, as worrying causes a great deal of ill-health. The catering industry, like most others, has its worrying times and it is well to remember that not only does the job suffer, but health also, if this state of mind is not kept under control.

Clothing in the kitchen

It is of considerable importance that people working in the kitchen should wear suitable clothing and footwear. Suitable clothing must be:

1 Protective,
2 Washable,
3 Suitable colour,
4 Light in weight and comfortable,
5 Strong,
6 Absorbent.

1 *Protective*

Clothes worn in the kitchen must protect the body from excessive heat. For this reason chefs' jackets are double-breasted and have long sleeves; they are to protect the chest and arms from the heat of the stove and to prevent hot foods or liquids burning or scalding the body.

Aprons: These are designed to protect the body from being scalded or burned and particularly to protect the legs from any liquids which may be spilled; for this reason the apron should be of sufficient length to protect the legs.

Chef's hat: This is designed to enable air to circulate on top of the head, and can help to prevent baldness. The main purpose of the hat is to prevent loose hairs from dropping into food and to absorb perspiration on the forehead. The use of lightweight disposable hats is both acceptable and suitable.

Footwear: This should be stout and kept in good repair so as to protect and support the feet. As the kitchen staff are on their feet for many hours, boots for men give added support and will be found most satisfactory. Footwear such as sandals, training shoes etc are insufficient protection from spillage of hot liquids.

2 *Washable*

The clothing should be of a washable material as many changes of clothing are required.

3 *Colour*

Clothing which is white is readily seen when soiled and needs to be changed, and there is a tendency to work more cleanly when wearing 'whites'. Chef's trousers of blue and white check are a practical colour but require frequent changing.

4 *Light and comfortable*
Clothing must be light in weight and comfortable, not tight. Heavy clothing would be uncomfortable and a heavy hat in the heat of the kitchen would cause headaches.

5 *Strong*
Clothes worn in the kitchen must be strong to withstand hard wear and frequent washing.

6 *Absorbent*
Working over a hot stove causes people to perspire; this perspiration should be absorbed, and for this reason underclothes should be worn. The hat absorbs perspiration and the neckerchief is used to prevent perspiration from running down the body, for wiping the face and also to protect the neck, which is easily affected by draughts.

Summary of personal hygiene
The practice of clean habits in the kitchen is the only way to achieve a satisfactory standard of hygiene. These habits are as follows:

Hands must be washed frequently and always after using the toilet. Food should be handled as little as possible.

Bathing must occur frequently.

Hair must be kept clean and covered in the kitchen; it should not be combed or handled near food.

Nose and mouth should not be touched with the hands.

Cough and sneeze in a handkerchief not over food.

Jewellery, rings and watches should not be worn.

Smoking, spitting and snuff-taking must not occur where there is food.

Cuts and burns should be covered with a waterproof dressing.

Clean clothing should be worn and only clean cloths used.

Foods should be tasted with a clean teaspoon.

Tables should not be sat on.

Only healthy people should handle food.

Kitchen hygiene

General objectives Know why premises must be kept clean and understand the need for premises and equipment to be designed for ease of cleaning.

Specific objectives Specify the characteristics of well designed premises which facilitate cleaning. State the methods for cleaning the various materials used for kitchen equipment and explain how the equipment is cleaned. List the procedure for hygienic and safe cleaning of electrically powered equipment.

Neglect in the care and cleaning of any part of the premises and equipment could lead to a risk of food infection. Kitchen hygiene is of very great importance:

a) to those who work in the kitchen, because clean working conditions are more agreeable to work in than dirty conditions;
b) to the owners, because custom should increase when the public know the kitchen is clean;
c) to the customer - no one should want to eat food prepared in a dirty kitchen.

Cleaning materials and equipment

To maintain a hygienic working environment a wide range of materials and equipment are needed. These are some of the items which need to be budgeted for, ordered, stored and issued:

Brooms	Ammonia
Brushes	Detergent
Buckets	Disinfectant
Cloths	Dustbin powder
Dusters	Floor cleaner
Dustpans	Flyspray
Dustbins	Oven cleaner
Mops	Plastic sacks
Sponges	Scouring powder
Squeegee	Soap
Scrubbing machine	Soda
Wet suction cleaner	Steel wool
Dry suction cleaner	Washing powder

Kitchen premises

Ventilation

Adequate ventilation must be provided so that fumes from stoves are taken out of the kitchen, and stale air in the stores, larder, still-room, etc, is extracted. This is usually effected by erecting hoods over stoves and using extractor fans.

Hoods and fans must be kept clean; grease and dirt are drawn up by the fan and, if they accumulate, can drop on to food. Windows used for ventilation should be screened to prevent entry of dust, insects and birds. Good ventilation facilitates the evaporation of sweat from the body, which keeps one cool.

Lighting

Good lighting is necessary so that people working in the kitchen do not strain their eyes. Natural lighting is preferable to artificial lighting. Good lighting is also necessary to enable staff to see into corners so that the kitchen can be properly cleaned.

Plumbing

Adequate supplies of hot and cold water must be available for keeping the kitchen clean, cleaning equipment and for staff use. For certain cleaning hot water is essential, and the means of heating water must be capable of meeting the requirements of the establishment.

There must be hand washing and drying facilities and suitable provision of toilets, which must not be in direct contact with any rooms in which food is prepared.

Hand-washing facilities (separate from food preparation sinks) must also be available in the kitchen with a suitable means of drying the hands, eg hot air or paper towels.

Cleaning of toilets and sinks

Toilets must never be cleaned by food-handlers. Sinks and hand basins should be cleaned and thoroughly rinsed.

Floors

Kitchen floors have to withstand a considerable amount of wear and tear, therefore they must be:

a) capable of being easily cleaned,
b) smooth, but not slippery,
c) even,
d) without cracks or open joints,
e) impervious (non-absorbent).

Quarry tile floors, properly laid, are suitable for kitchens, since they fulfil the above requirements.

Cleaning – floors are swept, washed with hot detergent water and then dried. This can be done by machine or by hand, and should be carried out at least once a day.

Walls

Walls should be strong, smooth, impervious, washable and light in colour. The joint between the wall and floor should be rounded for ease of cleaning. Tiling is the best wall surface because it is easily cleaned and requires no further maintenance.

Cleaning – clean with hot detergent water and dry. This will probably be done monthly, but frequency will depend on circumstances.

Ceilings

Ceilings must be free from cracks and flaking. They should not be able to harbour dirt.

Doors and windows

Doors and windows should fit correctly and be clean. The glass should be clean inside and out so as to admit maximum light.

Food lifts

Lifts should be kept very clean and no particles of food should be allowed to accumulate as lift shafts are ideal places for rats, mice and insects to gain access into kitchens.

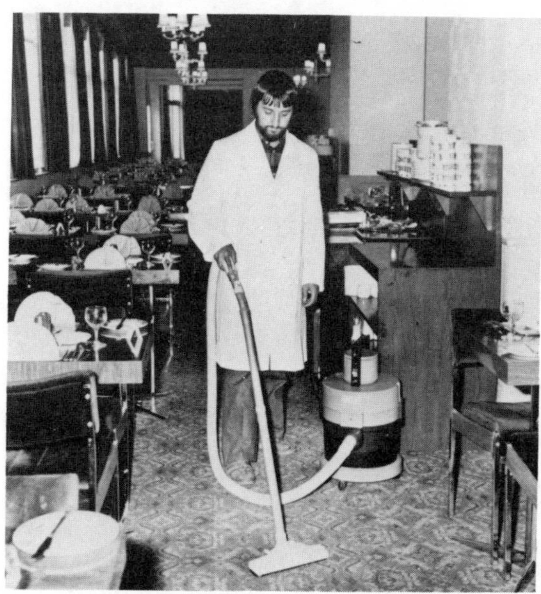

Equipment for cleaning walls and floors

Hygiene of kitchen equipment (also see Chapter 5)
Kitchen equipment should be so designed that it can be:

a) cleaned easily,
b) readily inspected to see that it is clean.

Failure to maintain equipment and utensils hygienically and in good repair may cause food poisoning.

Material used in the construction of equipment must be

hard so that it does not absorb food materials,
smooth so as to be easily cleaned,
resistant to rust,
resistant to chipping.

A pot and pan washer, specially designed for the proper cleansing of large equipment

Equipment must not be made from toxic materials, for example lead, or allowed to wear excessively, for example copper pans that need retinning on the inside so exposing harmful copper to food. Food must be protected from lubricants.

Easily cleaned equipment is free from unnecessary ridges, screws, ornamentation, dents, crevices, inside square corners, and has large smooth areas. Articles of equipment which are difficult to clean – for example mincers, sieves and strainers – are items where particles of food can lodge so allowing germs so multiply and contaminate food when the utensil is next used.

Normal cleaning of materials

Metals: As a rule all metal equipment should be cleaned immediately after use.

a) *Portable items:* Remove food particles and grease. Wash by immersion in hot detergent water. Thoroughly clean with a hard bristle brush or soak till this is possible. Rinse in water 77°C, by immersing in the water in wire racks.

b) *Fixed items:* Remove all food and grease with a stiff brush or soak with a wet cloth, using hot detergent water. Thoroughly clean with hot detergent water. Rinse with clean water. Dry with a clean cloth.

Abrasives should only be used in moderation as their constant scratching of the surface makes it more difficult to clean the article next time.

Marble: Scrub with a bristle brush and hot water and then dry.

Wood: Scrub with a bristle brush and hot detergent water, rinse and dry.

Plastic: Wash in reasonably hot water.

China, earthenware: Avoid extremes of heat and do not clean with an abrasive. Wash in hot water and rinse in very hot water.

Copper: Remove as much food as possible. Soak. Wash in hot detergent water with the aid of a brush. Clean the outside with a paste made of sand, vinegar and flour. Wash well. Rinse and dry.

Aluminium: Do not wash in water containing soda as the protective film which prevents corrosion may be damaged. To clean, remove food particles. Soak. Wash in hot detergent water. Clean with steel wool or abrasive. Rinse and dry.

Stainless steel: Stainless steel is easy to clean. Soak in hot detergent water. Clean with a brush. Rinse and dry.

Tin: Tin which is used to line pots and pans should be soaked, washed in detergent water, rinsed and dried. Tinned utensils where thin sheet steel has a thin coating of tin must be thoroughly dried, otherwise they are likely to rust.

Zinc: This is used to coat storage bins of galvanised iron and it should not be cleaned with a harsh abrasive.

Hygienic storage of equipment

Vitreous enamel: Clean with a damp cloth and dry. Avoid using abrasives.

Equipment requiring particular care in cleaning (sieves, conical strainers, mincers, graters). Extra attention must be paid to these items, because food particles clog the holes. The holes can be cleaned by using the force of water from the tap, by using a bristle brush and by moving the article, particularly a sieve, up and down in the sink, so causing water to pass through the mesh.

Whisks must be thoroughly cleaned where the wires cross at the end opposite the handle as food can lodge between the wires. The handle of the whisk must also be kept clean.

Saws and choppers, mandolins: These items should be cleaned in hot detergent water, dried and greased slightly.

Tammy cloths, muslins and piping bags: After use they should be emptied, food particles scraped out, scrubbed carefully and boiled. They should then be rinsed and allowed to dry.

Certain piping bags made of plastic should be washed in very hot water and dried. Nylon piping bags should be boiled.

Cleaning of large electrical equipment (mincers, mixers, choppers, slicers, etc)

1 Switch off the machine and remove the electric plug.
2 Remove particles of food with a cloth, palette knife, needle or brush as appropriate.
3 Thoroughly clean with hand hot detergent water all removable and fixed parts. Pay particular attention to threads and plates with holes on mincers.
4 Rinse thoroughly.
5 Dry and reassemble.
6 While cleaning see that exposed blades are not left uncovered or unguarded and that the guards are replaced when cleaning is completed.
7 Any specific maker's instructions should be observed.
8 Test that the machine is properly assembled by plugging in and switching on.

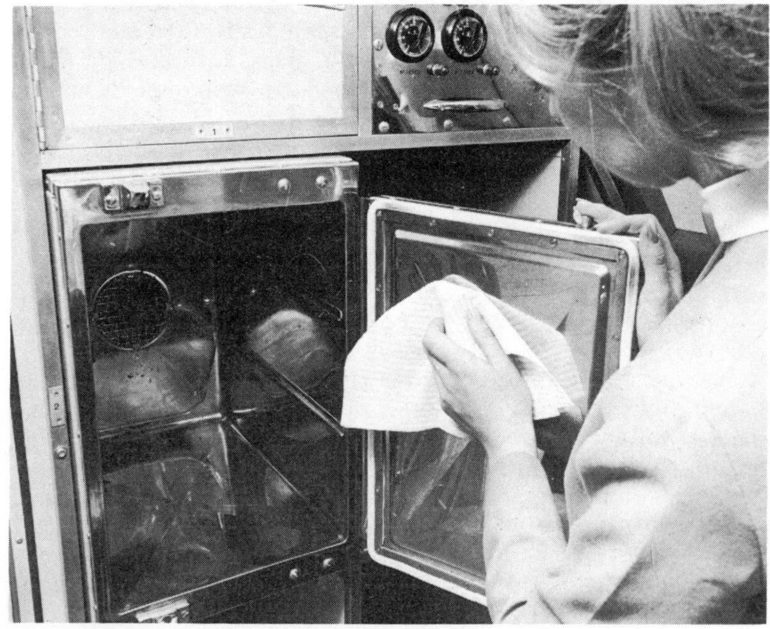

Disposable cloth being used for cleaning a convection oven

Newspaper extract

'A catering firm which supplies 20 foreign airlines with food for 2500 passengers daily is being taken to court for alleged contraventions of the Food and Drugs Act and the Food Hygiene Regulations.

'When council health inspectors visited the premises they found serious contraventions in all 11 main areas of the unit.

'In his report to the council's finance and general purposes committee, the chief environmental health officer said: "The risks of this type of operation are probably greater than any other, as each item of food is handled several times in several areas on a production line system.

' "No one person has control of the food from start to finish. The foods used are of the high risk variety - meats, fish and composite cream products - and a further risk is introduced by the time scales involved.

' "The food, of necessity, has to be prepared several hours before the departure of the relevant aircraft, and in addition may not be consumed until several hours into flight, giving adequate time for bacterial growth."

'It was imperative for the fabric of the building, together with the equipment used to prepare and handle the food, to be kept as clean as possible to minimise contamination, he said.

'This is what the inspectors allege they found:

'*Storeroom*: Walls, floors and ceiling not kept clean; wash basin being used for purposes other than personal hygiene; dirty equipment.

'*Meat cold room:* Smell of decomposition; walls, floor and ceilings dirty, with decomposed blood on the floor and stale food on the walls; racks and shelves covered with dried blood and meat; mouldy meat in trays full of stale blood.

'*Vegetable cold room:* Walls, floor and ceilings dirty; mouldy eggs and cheese stored with other fit items of food; greasy dust on equipment.

'*Deep freeze:* Dirty walls and floor; all 11 racks and 11 metal pallets dirty.

'*Cold kitchen:* Floor covered in food spillages; every item of equipment (including meat slicers) dirty; wash-hand basin not properly fixed to the wall; no soap or nailbrush.

'*Canteen:* No nailbrush or towel at hand basin; ceiling coated in brown, greasy film; air conditioning covered with a quarter-inch layer of greasy dust.

'*Staff toilets:* All toilet bowls stained and scaled; rooms dirty.

'*Bakery:* Equipment, including refrigerators, food mixers and pans, dirty; all racks dirty; food trolley caked in food debris; no nailbrush for hand basin.

'*Pot wash:* Roasting pans had extra rim of caked food; racks, food whisks and baking trays all dirty.

Mobile equipment for easy cleaning

'*Hot kitchen:* Drainage channels blocked with waste food residues; mobile hot cupboard soiled inside and out; all equipment, including choppers and slicers and ovens, caked in grease and food debris.

'The area used for bringing together individual food items to form complete meals prior to despatch was covered in greasy film, and equipment was dirty.

'The inspectors concluded that some areas were so dirty that they were incapable of being cleaned.'

Kitchen energy distribution systems

This concept operates from stainless steel housings (known as 'raceways') which are fastened to walls, floors, ceilings or may be island mounted. Inside the raceways are runs of electrical bus-bars or bus-wires and plumbing pipes. At intervals appropriate for the kitchen equipment served, are switch or valve sockets, electrical, gas, water, steam etc.

Connecting flexible cords and pipes from the kitchen equipment plug into the sockets and are designed to hang clear of the floor and are smooth plastic coated for easy cleaning.

For maximum advantage from this idea: the hygiene, safety, flexibility, ease of cleaning and maintenance etc most of the kitchen equipment is mounted on castors.

Periodic cleaning is carried out by pulling the equipment out from the wall or island, unplugging all the services then moving the equipment away on its castors giving free access to all wall and floor surfaces as well as backs and sides of equipment.

Further information: Eurocaddy Systems Ltd, Powder Mill Lane, Dartford, Kent DA1 1NN

Food hygiene

General objectives Understand the need to know how food may become contaminated and how to prevent this happening.

Specific objectives State the common methods whereby food may become a source of danger to health. List the ways food poisoning can be prevented. Specify the bacterial sources of contamination and the symptoms associated with each. Explain how bacterial growth is dependent on the factors of temperature, time, moisture and suitable food. Identify the foods most likely to need special care.

The most succulent mouth-watering dish into which has gone all the skill and art of the world's best chefs, using the finest possible ingredients, may look, taste and smell superb, yet be unsafe, even dangerous to eat because of harmful bacteria.

Grim toll of food poisoning

Food poisoning causes more than 23 million lost working days a year, a new report claims.

Twice as many cases are being reported now than 10 years ago, says the survey by two health officers.

Despite greater public awareness and higher hygiene standards, food poisoning continues to increase, Richard Foulgar and Edward Routledge say in "The Food Poisoning Handbook."

The authors, who work for the London Borough of Bromley add: "The unfortunate fact is that standards of food care and treatment in many establishments are still below safe limits".

There are more prosecutions too, not just because of increased vigilance by some local authorities, but by other factors which have multiplied risks of food contamination.

Plastic gloves used by food-handlers, British Airways

It is of the utmost importance that everyone who handles food, or who works in a place where food is handled, should know that food must be both clean and safe. Hygiene is the study of health and the prevention of disease, and because of the dangers of food poisoning, hygiene requires particular attention from everyone in the catering industry.

There are germs everywhere, particularly in and on our bodies; some of these germs if transferred to food can cause illness and in some cases death. These germs are so small they cannot be seen by the naked eye, yet food which looks clean and does not smell or taste bad may be dangerous to eat if harmful germs have contaminated it and multiplied.

The duty of every person concerned with food is to prevent contamination of food by germs and to prevent these germs or bacteria from multiplying.

Food-handlers must know the Food Hygiene Regulations, but no matter how much is written or read about food hygiene the practice of hygienic habits by people who handle food is the only way to safe food.

Food poisoning

Fourteen thousand people each year have been found by doctors to be suffering from food poisoning. This is the average number of notified cases for the last ten years, and there are thousands more who have not notified their doctor, but have suffered from food poisoning. This appalling amount of ill-health could be prevented. Failure to prevent it may be due to:

a) ignorance of the rules of hygiene,
b) carelessness, thoughtlessness or neglect,
c) poor standards of equipment or facilities to maintain hygienic standards,
d) accident.

Food poisoning can be prevented by:

1 High standards of personal hygiene.
2 Attention to physical fitness.

18 TAKEN ILL AT WEDDING

EIGHTEEN wedding guests taken to hospital last night suffering from suspected food poisoning after reception.

3 Maintaining good working conditions.
4 Maintaining equipment in good repair and in clean condition.
5 Adequate provision of cleaning facilities and cleaning equipment.
6 Correct storage of foodstuffs at the right temperature.
7 Correct reheating of food.
8 Quick cooling of foods prior to storage.
9 Protection of foods from vermin and insects.
10 Hygienic washing up procedure.
11 Food-handlers knowing how food poisoning is caused.
12 Food-handlers not only knowing but carrying out procedures to prevent food poisoning.

Food poisoning – what it is

Food poisoning can be defined as an illness characterised by stomach pains and diarrhoea and sometimes vomiting, developing within 1–36 hours after eating the affected food.

Causes of food poisoning

Food poisoning results when harmful foods are eaten. They may be harmful because:

a) *chemicals* have entered foods accidentally during the growth, preparation or cooking of the food;
b) *germs* (harmful bacteria) have entered the food from humans, animals or other sources and the bacteria themselves, or the toxins (poisons) produced in the food by certain bacteria, have caused the foods to be harmful. By far the greatest number of cases of food poisoning is caused by harmful bacteria.

Chemical food poisoning

Certain chemicals may accidentally enter food and cause food poisoning.

Arsenic is used to spray fruit during growth, and occasionally fruit has been affected by this poison.

Lead poisoning can occur from using water that has been in contact with lead pipes and then drunk or used for cooking.

Antimony or zinc: Acid foods if stored or cooked in poor quality enamelled or galvanised containers can also cause poisoning.

Copper pans should be correctly tinned and never used for storing foods, particularly acid foods, as the food could dissolve harmful amounts of copper.

Certain plants are poisonous – for example, poisonous mushrooms or fungi. Rhubarb leaves and the parts of potatoes which are exposed to the sun above the surface of the soil are also poisonous.

Rat poison may accidentally contaminate food.

Six die as poison bug hits hospital

By CLARE DOVEY

Medical Reporter

SIX patients have died since an outbreak of salmonella food poisoning at a hospital.

The source of the disease is still a mystery and hospital staff are taking massive precautions to stop the outbreak spreading.

The first death in Essex, was last Monday.

Outlining the battle against the disease, which can be carried by people who are not ill themselves, district administrator Mr John Hebb said all the patients had been screened.

Positive

"At one time 48 patients were positive and all 14 staff were positive for salmonella," he added.

"The staff were all sent home and replacements found. Four have since been cleared.

"We are testing food. The water supply has been sampled and tested. We have stopped visitors.

"We are sterilising crockery and barrier nursing patients to do all we can to stop the infection".

The patients who died were aged 49, 53, 69, 88 and two were 91.

Tests showed that salmonella was not involved in one death, was "a contributory factory"in another and results on the other four are still awaited.

Mr. Hebb added: "We badly need to know the cause of the outbreak . . . then we can rest more easily."

Prevention of chemical food poisoning

Chemical food poisoning can be prevented by

a) using correctly maintained and suitable kitchen utensils,
b) obtaining foodstuffs from reliable sources,
c) care in the use of rat poison, etc.

Bacterial food poisoning

Food contaminated by bacteria (germs) is by far the most common cause of food poisoning.

Bacteria

Bacteria are minute, single-celled organisms which can only be seen under a microscope. They are everywhere in our surroundings, and as most bacteria cannot move by themselves they are transferred to something by coming into direct contact with it.

Some bacteria form spores which can withstand high temperatures for long periods of time (even 6 hours) and on return to favourable conditions become normal bacteria again which then multiply.

Some bacteria produce toxins outside their bodies so that they mix with the food; the food itself is then poisonous and symptoms of food poisoning follow within a few hours.

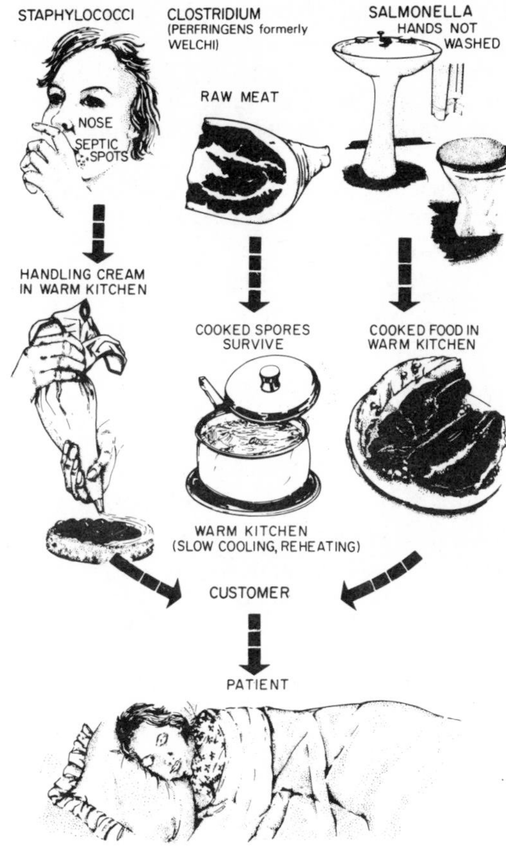

How food poisoning may be caused

Other bacteria cause food poisoning by virtue of large numbers of bacteria in food entering the digestive system, multiplying further and setting up an infection.

Certain bacteria produce toxins which are resistant to heat; foods in which this toxin has been produced may still cause illness, even though the food is heated to boiling-point and boiled for half an hour. Some bacteria will grow in the absence of air (anaerobes), others need it (aerobes).

Bacteria multiply by dividing in two, under suitable conditions, once every 20 minutes. Therefore one bacterium could multiply in 10–12 hours to between 500 million and 1000 million bacteria.

Not all bacteria are harmful, some are useful – for example, those used in cheese production. Some cause food spoilage – for example, souring of milk.

Some bacteria which are conveyed by food cause diseases other than food poisoning. These include typhoid, paratyphoid, dysentery and scarlet fever. In these cases the bacteria do not multiply in the food, they are only carried by it and the disease is known as a food-borne disease. With bacterial food poisoning the bacteria multiply in the food.

The time between eating the contaminated food (ingestion) to the beginning of the symptoms of the illness (onset) depends on the type of bacteria which have caused the illness.

Conditions favourable to bacterial growth

For the multiplication of bacteria certain conditions are necessary:

1 *Food* of the right kind.
2 *Temperature* must be suitable.
3 *Moisture* must be adequate.
4 *Time* must elapse.

Food

Most foods are easily contaminated; those less likely to cause food poisoning have a high concentration of vinegar, sugar or salt, or are preserved in some special way (see chapter on Food Preservation).

Foods most easily contaminated: The following foods are particularly susceptible to the growth of bacteria because of their composition. Extra care must be taken to prevent them from being contaminated.

1 Stock, sauces, gravies, soups.
2 Meat and meat products (sausages, pies, cold meats).
3 Milk and milk products.
4 Egg and egg products.
5 All foods which are handled.
6 All foods which are reheated.

Temperature

Food poisoning bacteria multiply rapidly at body temperature, 37°C. They grow between temperatures of 10°C and 63°C. This is a similar heat to a badly ventilated kitchen and for this reason foods should not be kept in the kitchen. They should be kept in the larder or refrigerator. Lukewarm water is an ideal heat for bacteria to grow in. Washing-up must not take place in warm water as bacteria are not killed and the conditions are ideal for their growth, therefore pots and pans, crockery and cutlery may become contaminated. Hot water must be used for washing up.

Boiling will kill bacteria in a few minutes, but to destroy toxins boiling for a half-hour is necessary. To kill heat-resistant spores, 4–5 hours' boiling is required. It is important to remember that it is necessary not only to heat foods to a sufficiently high temperature but also for a sufficient length of time to be sure of safe food. Extra care should be taken in warm weather to store foods at low temperatures and to

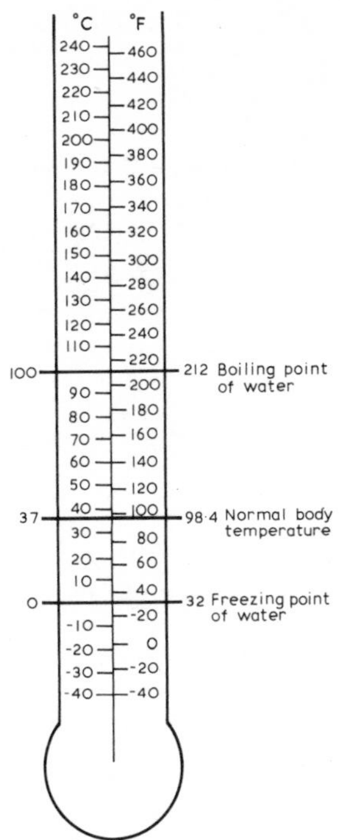

Fahrenheit and Celsius temperature scales

reheat thoroughly foods which cannot be boiled.

Bacteria are not killed by cold, although they do not multiply at very low temperatures, for example in a deep freeze they lie dormant for long periods. If foods have been contaminated before being made cold, on raising the temperature the bacteria will multiply. Foods which have been taken out of the refrigerator, kept in a warm kitchen and returned to the refrigerator for use later on may well be contaminated.

Moisture

Bacteria require moisture for growth, they cannot multiply on dry food. Ideal foods for their growth are jellies, custards, creams, sauces, etc.

Time

Under ideal conditions one bacterium divides into two every 20 minutes; in 5–6 hours millions of bacteria will have been produced. Small numbers of bacteria may have little effect, but in a comparatively short time sufficient numbers can be produced to cause food poisoning. Particular care therefore is required with foods stored overnight, especially if adequate refrigerated space is not available.

Germs multiply on or in moist foods in a warm temperature if left for a period of time

Types of food poisoning bacteria

The commonest food poisoning bacteria are:

1 the Salmonella group (cause food poisoning because of large numbers of bacteria in the food);
2 *Staphylococcus aureus* (causes food poisoning due to poison (toxin) production in the food);
3 *Clostridium perfringens* (formerly *welchii*) (causes food poisoning due to large numbers of bacteria producing toxins in the intestines).

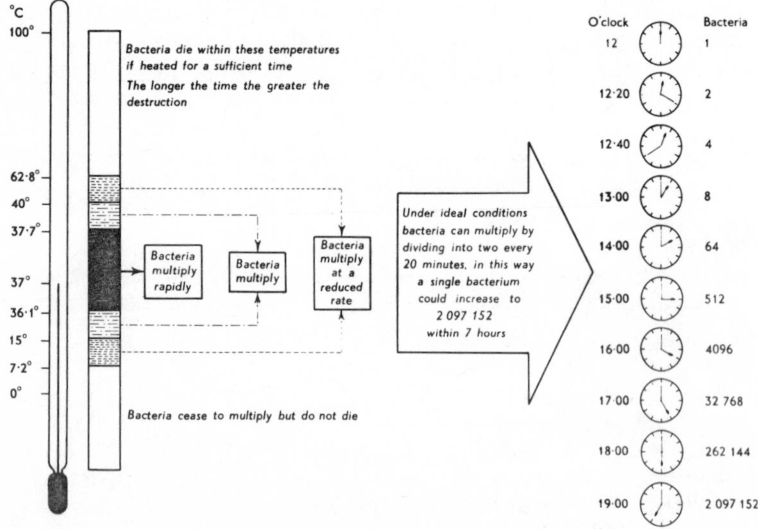

Danger temperature at which germs can multiply

Salmonella group

These bacteria are present in the intestines; they are excreted, and anything coming into contact directly or indirectly with the excreta may be contaminated. Infected excreta from human beings or animals may contaminate rivers and water supplies which may cause further infection. Salmonella infection is the result of human beings or animals eating food contaminated by salmonella-infected excreta originating from human beings or animals, so completing a chain of infection. For example, when flies land on the excreta of a dog which has eaten infected dog-meat and the flies then go on to food, the people who eat the contaminated food suffer from food poisoning.

Foods affected by the Salmonella group: Those most affected are eggs, meat, poultry. Contamination can be caused by:

1 Insects and vermin, because salmonellae are spread by droppings, feet, hairs, etc.
2 The food itself may be infected (as with duck eggs).
3 By cross contamination – for example, if a chicken is eviscerated on a board and the board is not properly cleaned before another food (for example, cold meat), is cut on the board.
4 The food could be infected by a human being who has the disease or who is a carrier (a person who does not suffer from food poisoning, but who carries and passes on the germs to others).

Staphylococci

These germs are present on human hands and other parts of the skin, or sores, spots, etc, and in the nose and throat.

Foods affected by staphylococci: Foods which have been handled are often contaminated because the hands have been infected from the nose or throat, cuts, etc. Brawn, pressed beef, pies, custards are foods frequently contaminated because they are ideal foods for the multiplication of bacteria.

Clostridium perfringens

These bacteria are distributed from the intestines of humans and animals and are found in the soil.

Foods affected by Clostridium perfringens: Raw meat is the main source of these bacteria, the spores of which survive light cooking.

Clostridium botulinum is another type of bacterium which causes food poisoning, but is rare in this country.

Sources of infection

Food-poisoning bacteria live in:

a) the soil;
b) humans – intestines, nose, throat, skin, cuts, sores, spots, etc;
c) animals, insects and birds – intestines and skin, etc.

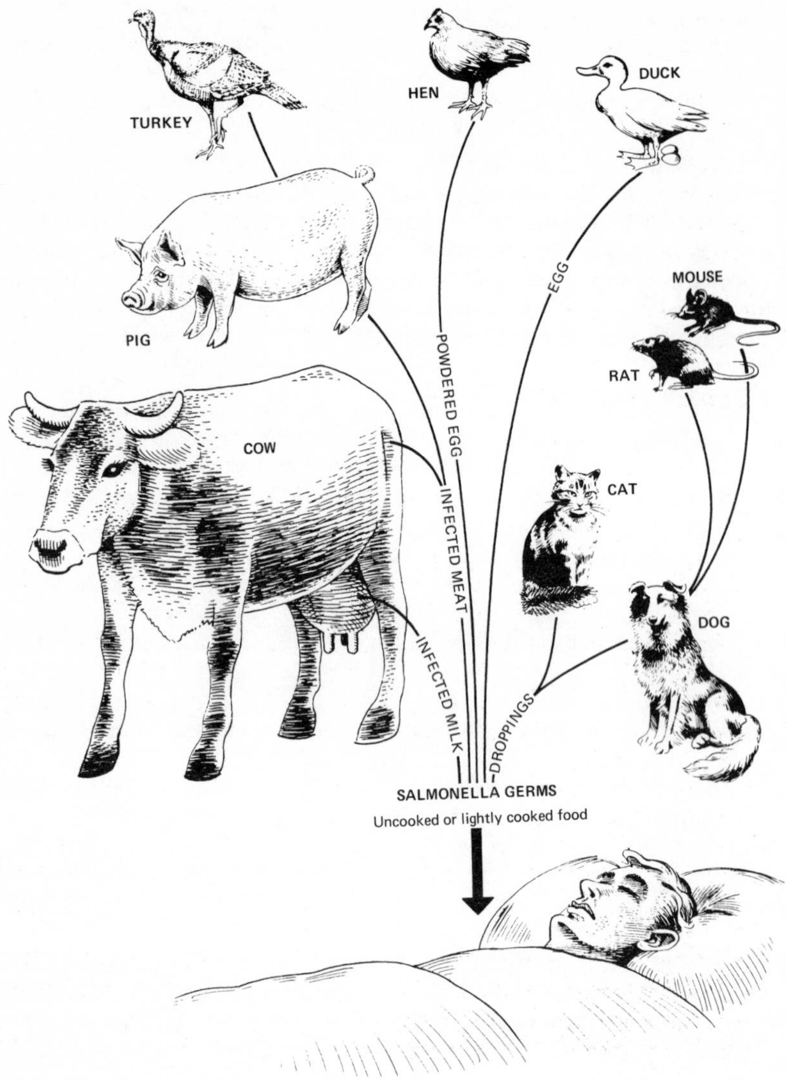

> Poultry served at a Christmas dinner caused 51 guests to be taken ill with salmonella poisoning because the birds were frozen and not fresh and needed longer cooking.

Foods contaminated by salmonella germs if uncooked or lightly cooked may result in food poisoning

Prevention of food poisoning from bacteria
To prevent food poisoning everyone concerned with food must:

a) prevent bacteria from multiplying;
b) prevent bacteria from spreading from place to place.

This means harmful bacteria must be isolated, the route of infection must be broken and conditions favourable to their growth eliminated. The conditions favourable to their growth – heat, time, moisture and a suitable food on which to grow – have been explained. It is also necessary to prevent harmful bacteria being brought into premises or getting on to food. This is achieved by a high standard of hygiene of personnel, premises, equipment and food-handling.

Spread of infection

a) Human – coughing, sneezing, by the hands.
b) Animal, insects, birds – droppings, hair, etc.
c) Inanimate objects – towels, dishcloths, knives, boards.

Human
People who are feeling ill, suffering from vomiting, sore throat or head cold must not handle food.

As soon as a person becomes aware that he or she is suffering from, or is a carrier of, typhoid or paratyphoid fever or salmonella or staphylococcal infection likely to cause food poisoning or dysentery, the person responsible for the premises must be informed. He or she must then inform the Medical Officer for Health.

Standards of personal hygiene should be high (see section on personal hygiene, page 22).

Animal
Vermin, insects, domestic animals and birds can bring infection into food premises.

Rats and mice are a dangerous source of food infection because they carry harmful bacteria on themselves and in their droppings. Rats infest sewers and drains, and since excreta is a main source of food-poisoning bacteria, it is therefore possible for any surface touched by rats to be contaminated.

Rats and mice frequent warm dark corners and are found in lift shafts, meter cupboards, lofts, openings in walls where pipes enter, under low shelves and on high shelves. They enter premises through any holes, defective drains, open doorways and in sacks of food-stuffs.

Signs to look for are droppings, smears, holes, runways, gnawing marks, grease marks on skirting boards and above pipes, paw foot-marks, damage to stock and also rat odour.

Rats spoil ten times as much food as they eat and there are at least as many rats as human beings. They are very prolific, averaging ten babies per litter and six litters per year, so that under ideal conditions it is theoretically possible for one pair of rats to increase to 350 million in three years. To prevent infestation from rats and mice the following measures should be taken:

a) Food stocks should be moved and examined to see that no rats or mice have entered the store-room.
b) No scraps of food should be left lying about.
c) Dustbins and swill-bins should be covered with tight-fitting lids.
d) No rubbish should be allowed to accumulate outside the building.
e) Buildings must be kept in good repair.
f) Premises must be kept clean.

Brown rat

Cockroach damage to ice cream wafers

Cockroach smears around defective pipe lagging

Common cockroach

Food poisoning

If the premises have become infested with rats or mice the environmental health inspector or a pest control contractor should be contacted.

Insect infection

House flies are the foremost of the insects which spread infection. Flies alight on filth and contaminate their legs, wings and bodies with harmful bacteria, and deposit these on the next object on which they settle; this may well be food. They also contaminate food with their excreta and saliva.

To control flies, the best way is to eliminate their breeding place. As they breed in rubbish and in warm, moist places, dustbins in summer are ideal breeding grounds, therefore:

a) Dustbins and swill-bins must be kept covered at all times with tight-fitting lids and the surrounding area kept clean.
b) The bins must be kept clean and sprayed with insecticide.
c) Ideally, dustbins and swill-bins should not be kept. Rubbish should be burnt immediately if it cannot be disposed of through a waste-master.
d) Paper or plastic lined bins which are destroyed with the rubbish are preferable to other types of bin.

Other ways to control flies are:

a) Screen windows to keep flies out of kitchens.
b) Install ultra-violet electrical fly killers.
c) Use sprays to kill flies (only where there is no food).
d) Employ a pest control contractor.

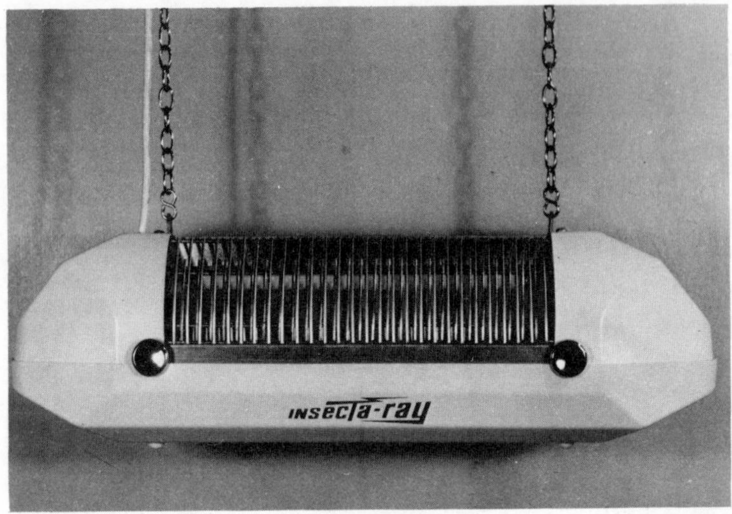

The Insecta-Ray consists of an ultra-violet lamp – strongly attractive to insects – set within an electrified grid which immediately kills any insect flying through it

Cockroaches: Cockroaches like warm, moist, dark places. They leave their droppings and a liquid which gives off a nauseating odour. They can carry harmful bacteria on their bodies and deposit them on anything with which they come into contact.

Silverfish: These small silver-coloured insects feed on, among other things, starchy foods and are found on moist surfaces.

Beetles: Beetles are found in warm places and can also carry harmful germs from place to place.

Insects are destroyed by using an insecticide, and it is usual to employ people familiar with this work. The British Pest Control Association has a list of member companies.

Cats and dogs

Domestic pets should not be permitted in kitchens or on food premises as they carry harmful bacteria on their coats and are not always clean in their habits. Cats also introduce fleas and should not be allowed to go into places where food is prepared.

Birds

Entry of birds through windows should be prevented as food and surfaces on which food is prepared may be contaminated by droppings.

Dust

Dust contains bacteria, therefore it should not be allowed to settle on food or surfaces used for food. Kitchen premises should be kept clean, then no dust will accumulate. Hands should be cleaned after handling dirty vegetables.

Washing up

The correct cleaning of all the equipment used for the serving and cooking of food is of vital importance to prevent multiplication of bacteria. This cleaning may be divided into the pan wash (plonge) or scullery and the china wash-up.

Scullery

For the effective washing up of pots and pans and other kitchen equipment the following method of work should be observed:

1 Pans should be scraped and all food particles placed in a bin.
2 Hot pans should be allowed to cool before being plunged into water.
3 Pans which have food stuck to them, should be allowed to soak (starchy foods, such as porridge and potatoes, are best soaked in cold water).
4 Frying-pans should be thoroughly wiped with a clean cloth, they should not be washed unless absolutely necessary.
5 Trays and tins used for pastry work should be thoroughly cleaned with a dry cloth, while warm.
6 Pots, pans and other equipment should be washed and cleaned with a stiff brush, steel wool or similar article, in hot detergent water.
7 The washing-up water must be changed frequently; it must be kept both clean and hot.
8 The cleaned items should be rinsed in very hot clean water to sterilise.
9 Pans, etc, which have been sterilised (minimum temperature 77°C) dry quickly; if it has not been possible to rinse in very hot water they should be dried with a clean cloth.
10 Equipment should be stored on clean racks, pans should be stacked upside down.

China wash-up

The washing up of crockery and cutlery may be by hand or machine.
Handwashing

1 Remove scraps from plates with a scraper or by hand.
2 Wash in water at 60°C containing a detergent.
3 Place in wire baskets and immerse them into water at 82°C for at least 2 minutes.

4 The hot utensils will air-dry without the use of a drying cloth.
5 Both the washing and sterilising water must be kept clean and at the correct temperature.

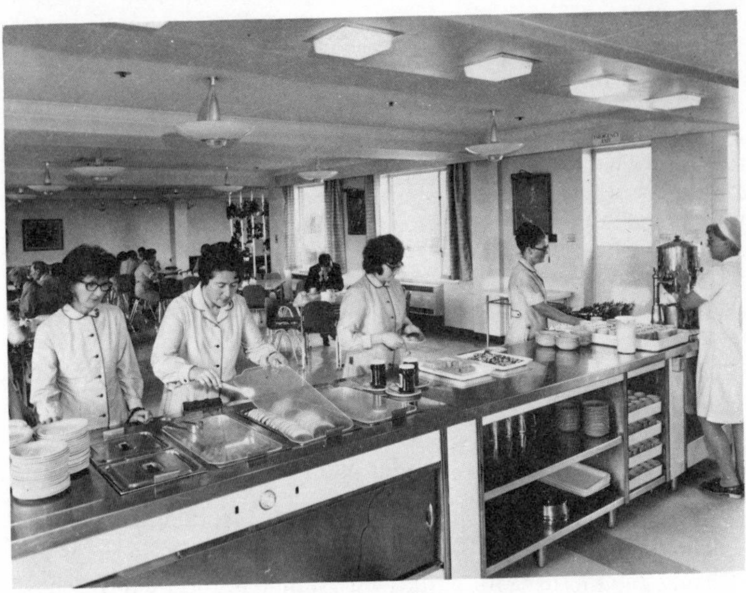

Hygienic service of food for employees in a staff restaurant

Machine washing-up. There are several types of machines which wash and sterilise crockery. In the more modern machines the detergent is automatically fed into the machine, which has continuous operation. To be effective the temperature of the water must be high enough to kill any harmful bacteria and the articles passing through the machine must be subjected to the water for sufficient time to enable the detergent water to thoroughly cleanse all the items. The detergent used must be of the correct amount and strength to be effective. Alternatively low temperature equipment is available which sterilises by means of a chemical, sodium hypochlorite (bleach). **Further information**: Lever Industrial.

Where brushes are used they must be kept free from food particles.

Hygienic storage

One of the most important ways to prevent contamination of food is the correct storage of food (see also chapter on Storekeeping, page 269). Foodstuffs of all kinds should be kept covered as much as possible to prevent infection from dust and flies. Foods should be kept in a refrigerated cold room or refrigerator where possible.

Hot foods which have to go into a refrigerator must be cooled quickly. This can be done by dividing large quantities of food into smaller containers, by cooling in a draught of air using fans or by raising the container and placing an article underneath, for example, a triangle or weight, so that air can circulate, or by placing the container in a sink with running cold water. If large quantities of food, for example minced beef, are left in one container the outside cools but the centre is still warm. When reheated the time taken to bring such a large quantity to the boil is sufficient to allow the bacteria to continue to multiply. If the food is not boiled long enough food poisoning can occur (see Temperature, page 44).

Particular care must be taken to store foods correctly in the warmer months; food not refrigerated in hot weather does not cool completely and, furthermore, flies and bluebottles are numerous in the summer.

Foods which require special attention

Meat

1 All made-up dishes, such as cottage pie, need extra care. They must be very thoroughly cooked.
2 Reheated meat dishes must be thoroughly reheated.
3 Pork must be well cooked (this is because pork may be affected by trichinosis, which is a disease caused by a worm).
4 Poultry which is drawn in the kitchen should be cleaned carefully; boards, tables and knives must be thoroughly cleaned afterwards, otherwise there is a danger of contamination from excreta.
5 Meat should be handled as little as possible. Minced and cut-up meats are more likely to become contaminated because of infection from the food-handler. Boned and rolled joints require extra care in cooking as inside surfaces may have been contaminated.
6 Sausages should be cooked right through.
7 Tinned hams are lightly cooked, therefore they must be stored in a refrigerator.

Fish

Fish is usually washed, cooked and eaten fresh and is not often a cause of food poisoning, except in reheated fish dishes. Care must be taken to reheat thoroughly such dishes as fish cakes, fish pie, coquilles de poisson, etc.

Some fish, such as oysters and mussels, have caused food poisoning because they have been bred in water which has been polluted by sewage. They are today purified before being sold. All shellfish should be used fresh. If bought alive, there is no doubt as to their freshness.

Eggs

Unless dried, hens' eggs do not cause food poisoning, but ducks' eggs should be used with caution and must be thoroughly cooked. Dried eggs if used should be reconstituted and used right away, not left in this condition in a warm kitchen as they may have been contaminated in processing. Bulk liquid egg undergoes pasteurisation but may be contaminated after the container is opened. Hollandaise sauce which is made with eggs is an example of a food which should not be kept in a warm kitchen for long. If not used in the morning it should not be used in the evening.

Milk

Milk when used in custards, trifles, puddings, etc, unless eaten soon after preparation should be treated with care. If required for the following day these dishes must be refrigerated.

Watercress and other green salad

Watercress must be thoroughly washed, as it grows in water which could be contaminated by animals. All green salads and other foods eaten raw should be well washed.

Synthetic cream

Synthetic cream can be a cause of food poisoning if allowed to remain in warm conditions for long periods. It is easily contaminated by handling.

Coconut

Desiccated coconut has been a cause of food poisoning, although nowadays production methods are much more hygienic so that contamination is unlikely.

Particular care is required in the handling of soups, sauces and gravies because bacteria multiply more rapidly in these foods.

Reheated foods — *Rechauffé*

In the interests of economy a sound knowledge of handling left-over food is necessary. Many tasty dishes can be prepared, but care must always be taken to see that the food is thoroughly and carefully reheated. If care is not taken then food poisoning can result (see page 45). Only sound food should be used ('if in doubt, throw it out').

After each meal service all unserved food should be cleared away in clean dishes, cooled quickly in the larder and placed in a cold room or refrigerator.

A good hors d'œuvrier can make interesting dishes out of left-over meats, poultry, fish and certain vegetables by mixing them with foods such as rice, gherkins, tomatoes, chives, parsley and a well-seasoned dressing such as mayonnaise or vinaigrette.

Trimmings and bones of meat, game and poultry should be used for stock. Trimmings of meat fat cooked or uncooked should be minced and rendered down for dripping.

Fish: Cooked kippers and haddock may be turned into savouries if freed from skin and bone, and finely minced or pounded with anchovy essence and a little butter and used as a spread on toast.

Cold fish may be used in many interesting dishes. For example:[1]

Coquilles de poisson Mornay
Curried fish
Fish cakes
Fish cutlets or croquettes
Fish pie
Fish kedgeree
Fish salad

Vegetables: Cold left-over cooked vegetables such as peas, cauliflower, haricot beans, potatoes, may be mixed with vinaigrette or mayonnaise and used for salads. Cold boiled potatoes can be used for potato salad or for sauté potatoes. Cold mashed potatoes may be used for fish cakes or potato cakes.

Meat: Left-over cooked items such as bacon, ham, tongue, kidneys or liver may be mixed with mince of any meat and used to give extra flavour to croquettes[1] and rissoles.[1]

Cold meats can be used for a number of dishes. For example:[1]

Minced for cottage pie
Minced lamb or mutton
Cornish pasties
Miroton of beef
Durham cutlet
Kromeski
Salad

Left-over poultry, such as chicken, if cut into joints can be reheated carefully in a curry sauce.

If the skin and bone are removed the poultry can be used for:

Salad[1]
Mayonnaise[1]
Cutlets or croquettes[1]
Vol-au-vent or bouchées[1]

Rice, spaghetti, macaroni can be turned into mixtures for hors d'œuvre with items such as chopped onion, chives, tomatoes, beetroot, cooked meat, haricot or French beans and a dressing of vinaigrette or mayonnaise.

Bread: Trimmings of crusts, etc, should be kept until dry, lightly browned in the oven, then passed through a mincer to make browned bread crumbs (chapelure) which may be used for crumbing cutlets, croquettes of fish, etc.

Stale bread can also be used for bread pudding. Stale sponge cake can be used for:

Trifles[1]
Cabinet puddings[1]
Queen of puddings[1]

Cheese: Left-overs of Cheddar cheese can be grated or chopped and used for Welsh rarebit.[1]

[1]The recipes may be found in the authors' *Practical Cookery Fifth Edition* (Arnold).

Food-borne diseases

Typhoid and paratyphoid are diseases caused by harmful bacteria carried in food or water. Scarlet fever, tuberculosis and dysentery may be caused by drinking milk which has not been pasteurised.

To prevent diseases being spread by food and water the following measures should be taken:

1 Water supplies must be purified.
2 Milk and milk products should be pasteurised.
3 Carriers should be excluded from food preparation rooms.

The Food Hygiene Regulations

These regulations should be known and complied with by all people involved in the handling of food. A copy of the full regulations can be obtained from HM Stationery Office and an abstract can be obtained which gives the main points of the full regulations.

These points are as follows:

Equipment

This must be kept clean and in good condition.

Personal requirements

1 All parts of the person liable to come into contact with food must be kept as clean as possible.
2 All clothing must be kept as clean as possible.
3 All cuts or abrasions must be covered with a waterproof dressing.
4 Spitting is forbidden.
5 Smoking and the use of snuff are forbidden in a food room or where there is food.
6 As soon as a person is aware that he is suffering from or is a carrier of such infections as typhoid, paratyphoid, dysentery, salmonella or staphylococcal infection he must notify his employer, who must notify the Medical Office of Health.

Requirements for food premises
Toilets

1 These must be clean, well lighted and ventilated
2 No food room shall contain or directly communicate with a toilet.
3 A notice requesting people to wash their hands after using the toilet must be displayed in a prominent place.
4 The ventilation of the soil drainage must not be in a food room.
5 The water supply to a food room and toilet is only permitted through an efficient flushing cistern.

Washing facilities

1 Hand basins and an adequate supply of hot water must be provided.

2 Supplies of soap, nail-brushes and clean towels or warm air machines must be available by the hand basins.

First aid: Bandages, waterproof dressings and antiseptics must be provided in a readily accessible position.

Lockers: Sufficient lockers must be available for outdoor clothes.

Lighting and ventilation: Food rooms must be suitably lighted and ventilated.

Sleeping room: Rooms in which food is prepared must not be slept in. Sleeping rooms must not be adjacent to a food room.

Refuse: Refuse must not be allowed to accumulate in a food room.

Buildings: The structure of food rooms must be kept in good repair to enable them to be cleaned and to prevent entry of rats, mice, etc.

Food temperatures: Certain foods must be kept at temperatures below 10°C or at not less than 62.8°C. These foods include meat, fish, gravy, imitation cream, egg products, milk and cream.

Storage: Foods should not be placed in a yard, etc, lower than 0.5 metre (18 inches) unless properly protected.

Penalties

Any person guilty of an offence shall be liable to a heavy fine and/or a term of imprisonment. Under the Food and Drugs (Control of Food Premises) Act, unhygienic premises can be closed down by a local authority in 72 hours.

Summary of food hygiene

Dangers to food

Chemical (copper, lead, etc).
Plant (toadstools).
Bacteria (cause of most cases of food poisoning).

Bacteria

Almost everywhere. Not all are harmful.
Must be magnified 500–1000 times to be seen.
Under ideal conditions, multiply by dividing in two every 20 minutes.

Sources of food-poisoning bacteria

Human – nose, throat, excreta, spots, cuts, etc.
Animal – excreta.
Foodstuffs – meat, eggs, milk, from animal carriers.

Method of spread of bacteria

Human – cough, sneezes, hands.
Animals – excreta (rats, mice, cows, pets, etc), infected carcasses.
Insects – flies, cockroaches.
Other means – equipment, china, towels.

Factors essential for bacterial growth
Suitable temperature.
Time.
Enough moisture.
Suitable food.

Method of control of bacterial growth

Heat – Sterilisation – using high temperatures to kill all micro-organisms.
Pasteurisation using lower temperatures to kill harmful bacteria only.
Cooking.

Cold – Refrigeration (3–5°C) stops growth of food poisoning bacteria and retards growth of other micro-organisms.
Deep freeze (–18°C) stops growth of all micro-organisms.

Foods commonly causing food poisoning
Made-up meat dishes.
Trifles, custards, synthetic cream.
Sauces.
Left-over foods.

Prevention of food poisoning
Care of person

1 Washing of hands.
2 Handle food as little as possible.
3 Cover cuts and burns with waterproof dressing.
4 Clean clothes and clean habits.

Care of food

1 Keep food cold during storage.
2 Cook meat and duck eggs thoroughly.
3 Cook and eat foods same day. 'Warmed-up' foods must be thoroughly reheated.
4 Protect foods from flies, rats, mice, etc.

Care of environment

1 Provide spacious well-lighted and well-ventilated premises.
2 Adequate wash and cleaning facilities.
3 Ample cold storage.
4 Suitable washing-up facilities.

Further information:

The Royal Society for the Promotion of Health, 13 Grosvenor Place, London, SW1X 7EN.
Royal Institute of Public Health and Hygiene, 28 Portland Place, London, W1N 4DE.
Local Environmental Health Departments.
Health and Safety Executive, Baynards House, 1 Chepstow Place, London W2.

Food Poisoning and Food Hygiene, Betty C. Hobbs (Arnold).
Food Hygiene Regulations (HM Stationery Office).
Food Hygiene Codes of Practice (HM Stationery Office).
Hygienic Food Handling (St John Ambulance Association).
Clean Catering (HM Stationery Office).
A guide to improving Food Hygiene, Aston and Tiffney, (Northwood Publication).
Food Hygiene and Food Hazards, Christie and Christie, (Faber).
Hygiene in Buildings, C. Lucas (Rentokil).

3
Gas and electricity

Gas

The studies of gas, water and electricity are highly technical. A simple study can be made, however, which is a help to the student in understanding how important a part all three play in the catering industry.

Having read this part of the chapter these learning objectives, both general and specific, should be achieved.

General objectives Know the part played by gas in the industry, and understand how gas is used for cooking, heating and refrigerating. Be aware of how to use gas safely.

Specific objectives State the basic principles of conduction, radiation and convection and give an example of each. Demonstrate how to read a gas meter and calculate gas bills. Explain the principles of thermostats, pressure governors, and gas refrigeration.

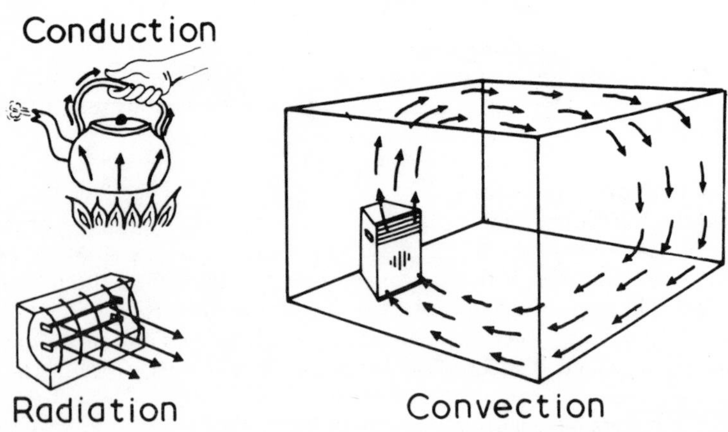

Three ways in which heat is transferred

Transference of heat

This is carried out by one of three methods: Conduction, Convection, Radiation.

Conduction is the travelling of heat through a solid, or from one solid to another, provided they are in contact, eg a poker in a fire or a saucepanful of water on a hot-plate.

Good conductors of heat are all metals.

Bad conductors of heat are cork, plastics, china, wood, string, cotton, and these are used in the manufacture of certain articles or parts of articles which require to be kept cool, eg pot triangles, dry oven cloths.

Convection is the passage of heat through liquids and gases, eg a saucepanful of water being heated, the water at the bottom becomes heated first and then because it is lighter rises to the top and the heavier cold water takes its place at the bottom, eg an Ideal boiler.

Radiation is the method by which heat travels from a hot object in straight rays as does light. Any object in the path of the rays becomes heated, eg grilling, electric fire.

The gas used today is almost entirely natural gas which is obtained from underground sources. A major contribution is produced from deep wells beneath the North Sea, some of which are many hundreds of feet below the surface. The discovery and exploitation of this vast underground source of energy has cost many millions of pounds and has brought about great advances in technology.

Natural gas is non-toxic, although in its natural state it has no smell, an artificial scent is introduced to give the characteristic smell, so that escapes can be easily detected. Natural gas is lighter than air. Oxygen is necessary for the combustion of natural gas. One volume of natural gas requires two volumes to oxygen.

Air contains four parts of nitrogen to one part of oxygen, therefore one volume of natural gas requires *ten* volumes of air.

Measuring gas consumption

Gas passes along mains into houses and other establishments through a gas meter which at present records the amount used in cubic feet. In the furture this will probably be changed to a metric unit.

How to read a gas meter

Read the four lower dials only. Copy down the readings in the order they appear. Where the hands are between two figures, put down the lower one. However, if the hands are between 9 and 0 you should put down 9. Then take the figures for the previous reading (shown on your gas bill) and deduct them from your new reading. You will then know how much gas has been used (in hundreds of cubic feet) since the last reading.

Some meters have a direct reading index in which the actual figures are all in a line.

Read only the white figures (not the red ones) for hundreds of cubic feet.

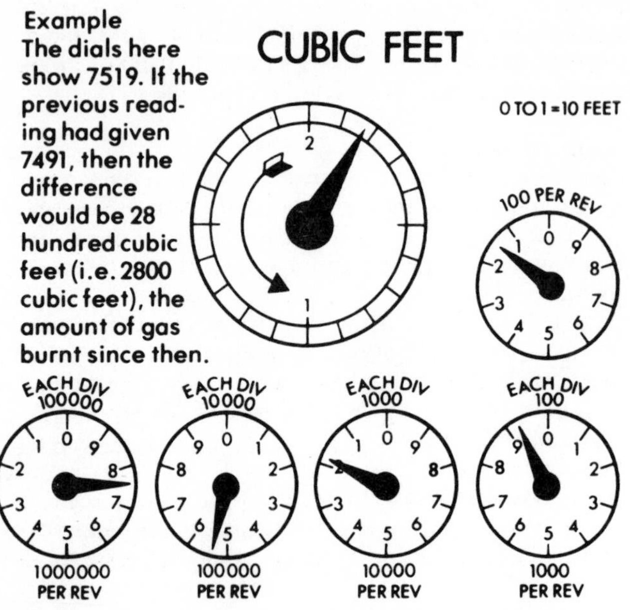

A gas meter

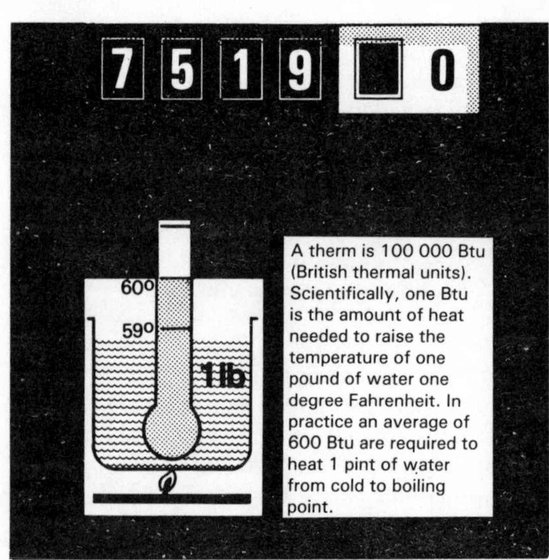

Gas is always measured in cubic feet, but is charged for in therms. To convert cubic feet into therms requires only a simple calculation.

A therm is 100 000 Btu (British thermal units). Scientifically, one Btu is the amount of heat needed to raise the temperature of one pound of water one degree Fahrenheit. In practice an average of 600 Btu are required to heat 1 pint of water from cold to boiling-point.

How to calculate a gas bill

To change cubic feet (cu ft) into therms you must know the heating power or calorific value (CV) of the gas. This will be shown on the gas bill and the simple calculation for working out the amount of heat supplied is:

$$\frac{\text{calorific value (CV)} \times \text{hundreds of cubic feet}}{1000} = \text{therms}$$

1 British Gas states the calorific value (CV) of the gas in British thermal units per cubic foot, ie 1035 Btu's/ft^3.

   To find how many Btu's have been used, multiply the consumption in ft^3 by the CV.

2 Gas is charged per therm. 1 therm = 100 000 Btu's, therefore divide the Btu's consumed by 100 000 to find the number of therms.

3 Supposing 1 therm costs 35p the calculation of the bill is as follows:

$$\begin{aligned} \text{previous reading} &= 42\,600\ ft^3 \\ \text{present reading} &= 44\,600\ ft^3 \\ \text{consumption} &= 2000\ ft^3 \\ \text{CV} &= 1035 \\ \text{price per therm} &= 35p \end{aligned}$$

$$\text{Therefore cost} = \frac{2000 \times 1035}{100\,000} \times 35p = £7.24$$

The burner

The burner is the source of heat in all gas appliances and it is essential that it is correctly adjusted.

Flash back or lighting back

This applies only to aerated-type burners and is caused by the gas being ignited at the injector of the burner, causing the flame at the burner head to be luminous with disturbed combustion.

Causes of lighting back

1 Incorrect lighting by placing igniter flame (match, taper, gas pilot or lighter) too close to the burner.
2 Incorrect gas/air adjustment.
3 Dirty burners causing gas starvation at gas/air adjustment.
4 Rough interior of burner tube.

The items 2, 3 and 4 can be corrected by the local Gas Region staff or manufacturers' staff.

Thermostats

The thermostat is a device for controlling the temperature in some ovens and fryers. There are two types:

a) The rod type – the action of which depends upon the fact that some metals expand more than others when heated.
b) Mercury vapour or liquid type – the action of a liquid thermostat depends upon the fact that a vapour expands when heated.

British Standard specification for gas-heated catering equipment

This British Standard was prepared under the authority of the Gas Industry Standards Committee, as a result of requests received from the Interdepartmental Committee of the Ministry of Fuel and Power, and was originally published in 1954. It has now been completely revised.

The standards of performance laid down will help to ensure that the appliances are safe and will give good service, although it is necessary to point out that compliance with this British Standard does not of itself guarantee that satisfactory service will be attained. Conditions of use vary greatly and it is necessary to relate the standards of performance to the actual use to which the appliance will be subjected during its life. Experience has shown that even detailed testing of samples by laboratories equipped for the purpose and staffed by personnel in touch with user requirements must be supplemented by practical experience. The local gas undertaking can supply details of gas quality, will have access to facilities for carrying out all tests specified in this standard and will frequently be concerned with the fitting and maintenance of the equipment.

Particular attention is drawn to the necessity for regular inspection and maintenance of catering equipment, particularly of safety devices, in order that they may be relied upon to give adequate service to the user. This specification covers:

1 Ovens
2 Boiling burners
3 Grills
4 Deep fryers
5 Steamers
6 Boiling pans (bulk liquid heaters)
7 Water boilers
8 Griddle plates
9 Hot cupboards (hot plates)
10 Bain-maries

The manufacturing requirements of equipment are laid down and in addition combustion, fire safety, stability and safety in use, and condition applicable to the use of governors, thermostats, flame failure devices.

British Gas issue an approved list of catering equipment covering some 700 appliances all of which have satisfied the requirements of the British Standard Specification.

Pressure governors

The purpose of the constant pressure governor is to maintain a constant pressure at its outlet, irrespective of any normal pressure fluctuating at its inlet.

Pressure governors are fitted to most catering appliances, particularly to those that have fixed injectors and those that have burners that are enclosed and concealed from visual observation.

Constant pressure governors may be fitted to the outlet of the gas meter to give a constant pressure to all gas-heated equipment. Individual control is to be preferred. The governors are made in a variety of designs and gas ratings to suit all equipment.

Solid top range

The drawing shows a side view of a popular-type gas cooker, with gas pipes omitted. When burner A is lighted the heat rises (as shown by arrows), causes thermostat B to expand (automatically controlling the gas supply) and then flows round the oven and out of vent C.

With the thermostat at medium setting, the temperature in the upper part of the oven in about 225°C; in the middle of the oven it is 220°C, but about 195°C at the bottom (temperatures measured with special

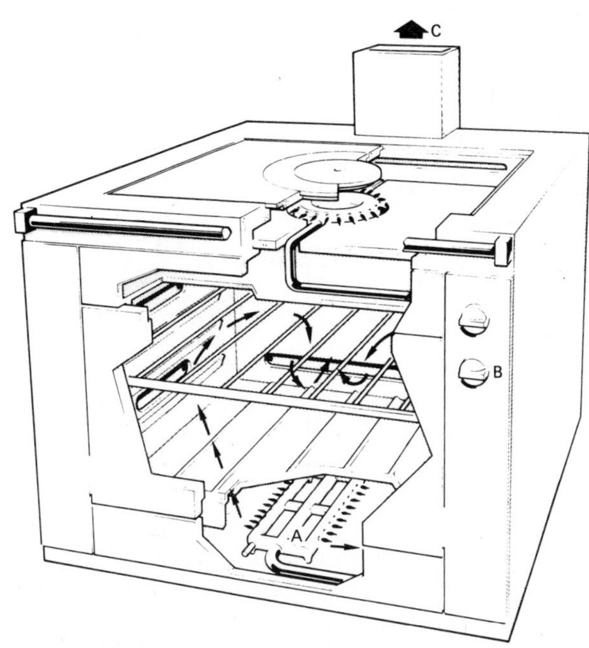

Solid top range

thermometers). The upper two-thirds of the lighted oven are therefore hotter than the third at the bottom. This enables dishes requiring different temperatures to be cooked in the oven at the same time.

The gas refrigerator

The simple principle on which gas refrigeration is based is that when a liquid evaporates it draws heat from its surroundings. A solution of ammonia in water is heated in the boiler by a small gas flame and the ammonia gas driven off is condensed to liquid ammonia in the air-cooled condenser. This is led into the evaporator with some hydrogen and begins to evaporate. In evaporating, heat is absorbed and refrigeration produced.

The gases produced are then led to the absorber and the ammonia absorbed by some weak liquid trickling down the absorber. The strong ammonia solution produced is then driven back into the boiler, while the hydrogen gas which is not absorbed is led into the evaporator.

The weak liquid trickling down the absorber is provided from the boiler.

In this way a complete cycle is obtained and refrigeration produced by heating only. There are no moving parts at all. The amount of heating is automatically controlled by a thermostat inside the refrigerator.

Further information: *A Student's Guide to Gas Catering* (British Gas Marketing Division, 326 High Holborn, London WCIV 7PT).

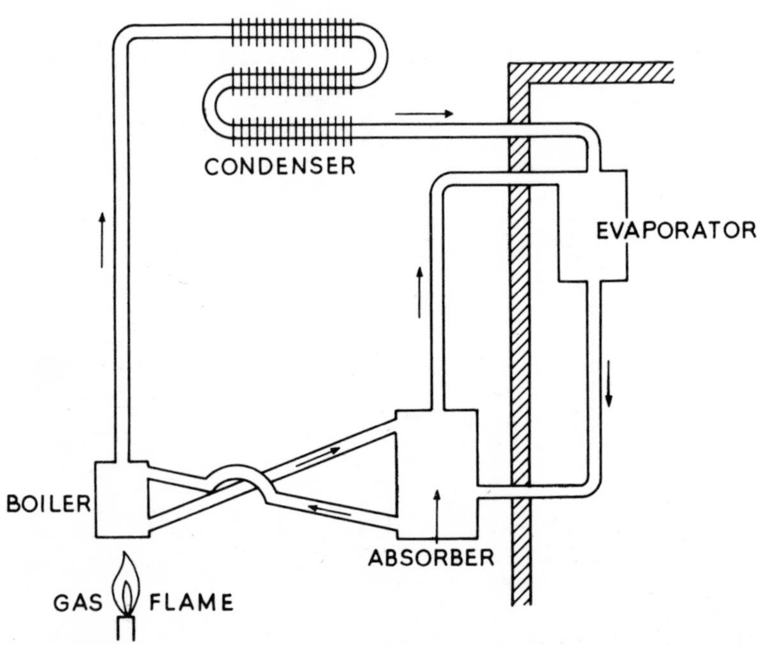

Gas refrigeration

Electricity

Having read this part of the chapter these learning objectives, both general and specific, should be achieved.

General Objectives Know the part played by electricity in the catering industry and it is used for cooking, heating and refrigerating. Appreciate the need to handle wires, plugs, appliances etc with care.

Specific objectives Make a comparison between gas and electricity. Read an electricity meter and calculate electricity bills. Explain the terms watt, volt, ampere and ohm. Identify which fuses to use with which appliance. Specify the safety measures necessary when dealing with electricity. Compare the costs of the various fuels available and evaluate them.

Generation of electricity

If a coil of wire is joined at both ends to another length of wire and a magnet is passed rapidly backwards and forwards through the coil a current of electricity is produced. In the electricity generating stations the magnets may be moved by turbines driven either by steam pressure or by water power and harnessed to drive the generators.

The most common electricity supply system now in use is A.C. (alternate current). Electricity is carried by cables called the grid and

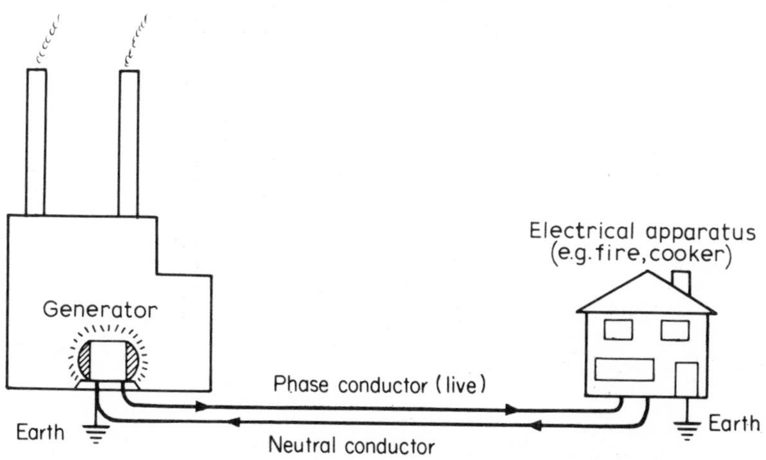

Electricity supply from producer to consumer

transformed at local substations. The consumer takes the supply of electricity from the substation.

Electricity must have a complete circuit from the source of supply through the load, eg an appliance back to the source of supply or it will cease to flow. Some substances are conductors of electricity – some are insulators.

All substances allow electricity to pass through them. Those which allow electricity to flow freely are known as *conductors*, eg metals, carbon, tap water, damp earth. Those which do not allow electricity to flow freely are known as *insulators*, eg glass, porcelain, wood, rubber, leather, plastic, stone. A substance's ability to be a good insulator depends on many things, such as its working temperature, the dampness of the surrounding air and its age. So when equipment is designed, an insulator is chosen so that it does not deteriorate after prolonged use under the conditions in which it is used.

Anyone who has received an electric shock will know that the human body also conducts electricity.

Electrical terms

Watts – measure power – that is, the rate at which any electrical appliance is using electric current for a given pressure (voltage).

Volts – measure pressure of flow. Comparing electricity to water, 'voltage' corresponds to pounds per square inch of a water supply. Before electricity can flow through a wire the electrical pressure at one end of the wire must be greater than at the other end. 240 volts is a common measure in domestic use. Knowledge of the voltage is essential.

Amperes – measure the rate of flow of a current, and can be obtained by dividing the 'watts' by the 'volts'.

Ohms – measure the resistance of the wires to the passage of electricity and is comparable with the friction offered by a water pipe to water flowing through it.

'Ohm's Law' on which the science of electricity is founded can be stated as:

$$\text{Amperes} = \frac{\text{Volts}}{\text{Ohms}} \quad \text{Volts} = \text{Amperes} \times \text{Ohms} \quad \text{Ohms} = \frac{\text{Volts}}{\text{Amperes}}$$

Examples:

If the voltage = 250
amperes = 5
wattage = V × A = 250 × 5 = 1250 watts
If the voltage = 110
amperes = 10
wattage = 1100 watts

To find the voltage, divide the watts by amperes:

$$V = \frac{W}{A}$$

$$\text{wattage} = 2000$$
$$\text{amperes} = 10$$
$$\therefore \text{voltage} = \frac{W}{A} = \frac{2000}{10} = 200 \text{ volts}$$
$$\text{wattage} = 10$$
$$\text{amperes} = \tfrac{1}{2}$$
$$\therefore \text{voltage} = \frac{10}{\frac{1}{2}} = \frac{10}{1} \times \frac{2}{1} = 20 \text{ volts}$$
$$\text{Amperes} = \frac{\text{Watts}}{\text{Volts}}$$
$$\text{wattage} = 1000$$
$$\text{volts} = 5$$
$$\therefore \text{amperes} = \frac{W}{V} = \frac{1000}{5} = 200 \text{ amperes}$$

A building has a load of 200 60-watt lamps. What current is required to supply the building if it is connected to a 240-volt supply?

$$W = 200 \times 60 = 12\,000$$
$$V = 240$$
$$A = \frac{W}{V} = \frac{12\,000}{240} = 50 \text{ amperes}$$

Find the current flowing through a 100-watt lamp connected to a 250-volt supply.

$$A = \frac{W}{V} = \frac{100}{250} = \frac{2}{5} = 0.4 \text{ ampere}$$

Therefore a 2 A plug could be safely used.

What ampere plug would be necessary for four 100-watt lamps using 240 volts?

$$A = \frac{W}{V} = \frac{400}{240} = 1\tfrac{2}{3}\text{A}$$

Therefore a 2 A plug could be safely used.

$$\text{Amperes} = \frac{\text{Volts}}{\text{Ohms}}$$

If the voltage is 230 and the resistance *R* (ohms) is 10 the socket outlet is able to supply 23 amperes.

$$\frac{V}{R} = A = \frac{230}{10} = 23 \text{ A}$$

If the voltage is 240 and the resistance is 24 ohms the socket outlet is able to supply 10 amperes.

$$\frac{V}{R} = A = \frac{240}{24} = 10\ A$$

If the resistance (R) is 25 ohms and the amperage is 5 the voltage is 125.

$$R \times A = V = 25 \times 5 = 125 \text{ volts}$$

If the resistance is 10 ohms and the amperage is 50 the voltage is 500.

$$R \times A = V = 10 \times 50 = 500 \text{ volts}$$

The number of watts is obtained by multiplying the amperage by the voltage. An electric fire rated 5 amps at 200 V would take $5 \times 200 = 1000$ watts of power to make it work.

Most electrical appliances have a label indicating their wattage or loading.

Calculate the loading of an electric iron connected to a 240-volt supply when a current of 2 amps is flowing.

$$W = A \times V = 2 \times 240 = 480 \text{ watts}$$

The kilowatt hour (kWh) or *unit* is the term used to indicate the amount of energy used and is the measure used on electricity bills.

It represents 1000 watts used for 1 hour.

eg A 1-kW fire in continuous use for 1 hour uses 1 kWh or unit. A 100-watt lamp in continuous use for 10 hours uses 1 kWh or unit.

If the price of a unit of electricity is 5p then a 100-watt lamp switched on for 1 hour costs 0.5p.

To find the resistance of, and the current flowing through, a 60-watt lamp connected to a 240-volt supply:

$$A = \frac{W}{V} = \frac{60}{240} = \frac{6}{24}\ A$$

$$R = \frac{V}{A} = \frac{240 \times 24}{6} = \frac{5760}{6} = 960 \text{ ohms}$$

What is the current passing through a 100-watt lamp *a*) on a 100-volt supply; *b*) on a 200-volt supply?

a) $A = \frac{W}{V} = \frac{100}{100} = 1\ A$

b) $A = \frac{W}{V} = \frac{100}{200} = \frac{1}{2}\ A$

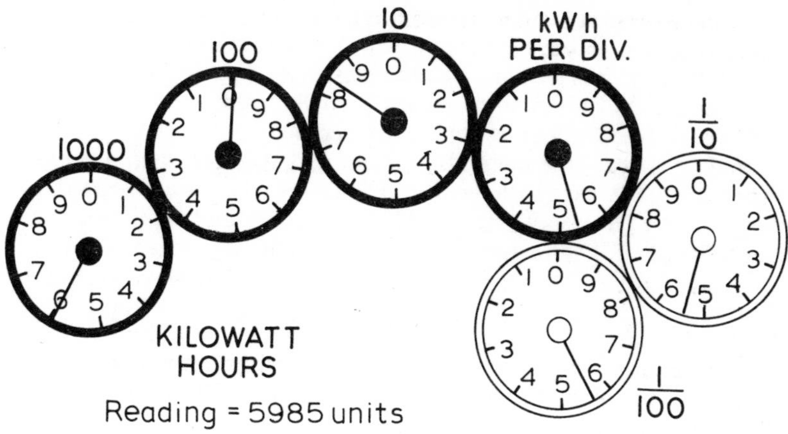

The electricity meter

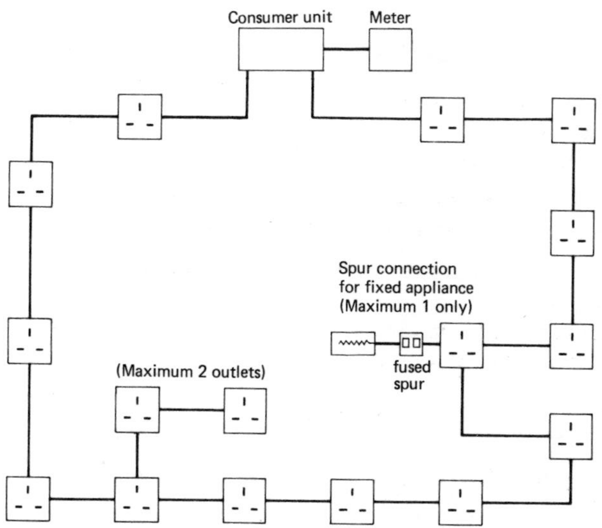

Ring circuit system of wiring

Electrical installation

House installation – the supply authority's fuse is sealed so that the householder cannot open it and it belongs to the local electricity board. It is designed to limit the total current taken for the whole house.

Meter – records the amount of electricity used in kilowatt hours. If off-peak energy of a cheaper rate is used then a two dial meter and time switch are used.

Main switch – to cut off entire lighting or power circuit (this is the customer's property and responsibility).

Distribution board (Consumer Unit) – divided into sub-circuits, each sub-circuit has its own fuse. In many cases, as a safety device, the main switch and distribution board are combined. It is best not to remove the fuse without switching off electricity.

Meter reading – the method for reading is the same as for the gas meter (see p. 64), except on the fourth dial where there is no need to add any noughts. The last two dials are 1/10 and 1/100 and are shown after a decimal point. Digital meters similar to gas meters are also used.

Fuses

The fuse in an electrical circuit acts as a safety device. Fuse wire is obtainable in varying thicknesses, usually 5, 10, 15 and 30 amperes, and for general purposes should be used as follows:

Lighting circuits	maximum of 5 amperes
Radial circuits	maximum of 15 or 20 amperes
Cookers or ring circuits	30 amperes.

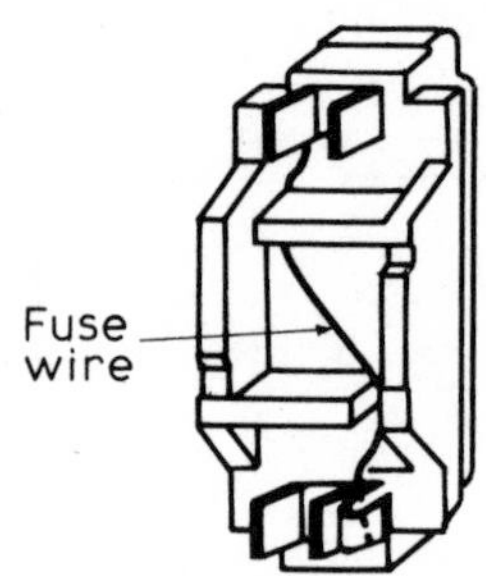

wire fuses

Cartridge fuses and miniature circuit breakers

Always fit fuses of the correct rating for the circuit. *Never* use a larger cartridge fuse or a thicker wire in a rewireable fuse, even as a temporary measure. Don't rely on the blown fuse as a guide, it may have been the wrong one in the first place. Instead of fuses you may have miniature circuit breakers. Once the fault has been fixed, you simply press the button or switch on to restore power.

Remember: always **switch off** before changing fuses.

If more than the maximum safe current passes through the circuit the wires will get hot, the insulation may burn, and there may be a fire.

The fuse consists of a short length of wire within the circuit and is of such a thickness that it will melt if more than the maximum safe current flows through it, so breaking the circuit. This is known as a blown fuse.

The fuse should always be connected in the live side of the supply. No fuse should ever be connected in the neutral side.

Plug fuses (usually 2 sizes as below)

3 amp red
Most appliances up to 720 watts (look for the rating plate on the appliance, usually on the base or back).

13 amp brown
Appliances rated over 700 watts. Also some appliances with motors – such as vacuum cleaners and spin dryers. (See manufacturers' instructions.)

Main fuses

5 amp white
Lighting circuit

15 amp blue
20 amp yellow
Immersion heater

30 amp red
Ring main circuit and average cookers

45 amp green
Larger cookers

Plug fuses

Causes of a blown fuse

A. Too many appliances plugged into a circuit: eg A 3-kW fire; 750-watt kettle; 750-watt iron plugged into a power circuit 15 A; voltage 240.

To find the current flowing through the three appliances:

$$W = V \times A$$

$$\therefore A = \frac{W}{V}$$

$$= \frac{4500}{240} = 18\tfrac{3}{4}\ A$$

therefore Current $= 18\frac{3}{4}$ amperes.

Therefore a greater amount of current than 15 A is flowing, and consequently the fuse will blow and the circuit will be broken.

B. Plugging a power appliance into a lighting circuit: eg A 2-kW fire plugged into a circuit designed to carry 5 amperes, voltage 240.

To find the current flow:

$$\text{Current flow} = \frac{W}{V} \text{ amperes}$$

$$= \frac{2000}{240} = 8\tfrac{1}{3} \text{ amperes}$$

therefore Current $= 8\frac{1}{3}$ amperes.

Therefore the fuse will blow. There is no earth connection.

C. Short circuit due to insulation failure. Means that lead and return wires touch and therefore current does not reach the appliance. Often due to wear of wire insulation.

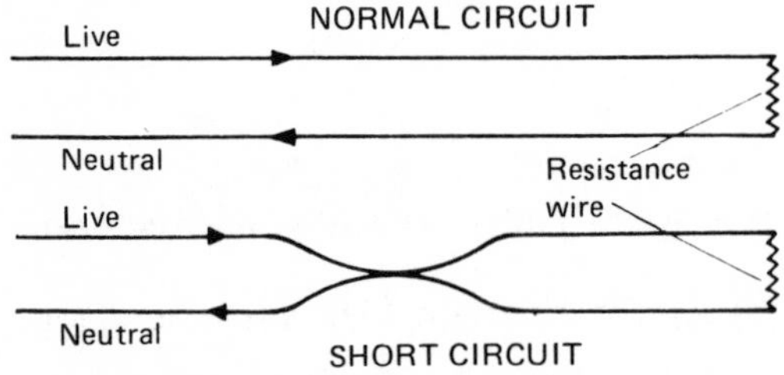

Repair of fuses

1 Turn off the main switch.
2 Find the fuse that has blown – this may be known beforehand; if not, try each one in turn.
3 Remove broken fuse wire.
4 Replace with new fuse wire of the correct size (5, 10, 15 or 30 amperes).
5 Before replacing the repaired fuse and before switching on, endeavour to trace fault and repair.

The wiring of plugs

1 Remove the outer sheath for about 3.8 cm – 5cm, thus leaving three coloured wires exposed.
2 If the plug has a bridge, clamp the flex sheath lightly under the cord grip.
3 Measure each wire – cut it off at a length which will allow it to go round the terminal. The last portion of the wire only is cleared of insulation.
4 Wind uncovered wire in clockwise direction around terminal or insert in terminal and tighten screw.
5 Replace the casing of the plug.

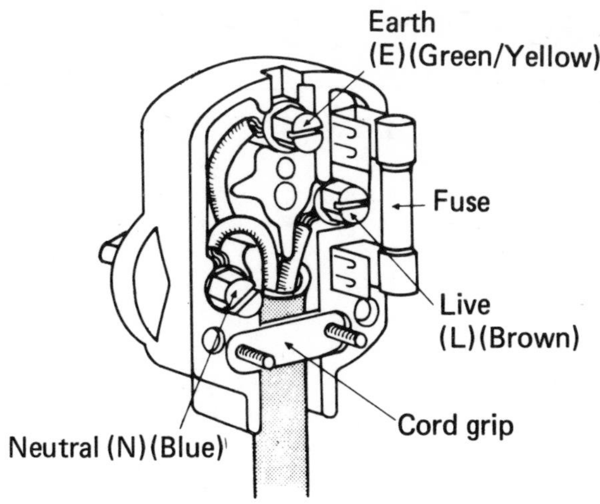

3 pin cartridge fused plug

Electric cookers

The oven. As for a gas cooker it has double walls, the space between being lagged.

There are a variety of positions for the heating elements – sides, top and bottom.

The boiling plates. The totally enclosed type is insulated with magnesium oxide, a heat conductor. An average loading is approximately 1800 W.

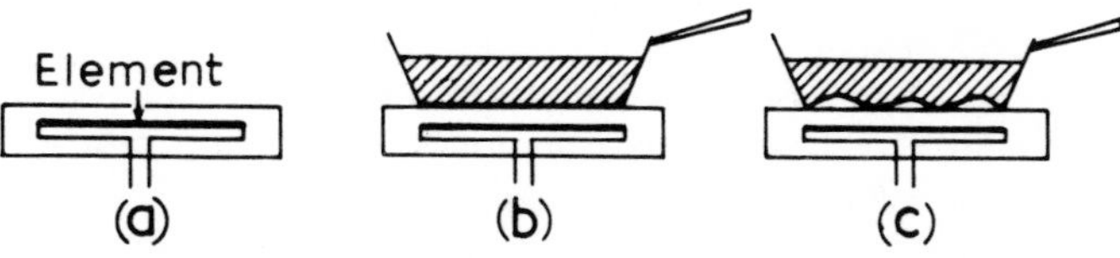

Electric boiling plates: a) Element inside plate b) Correct type of pan – bottom has perfect fit with stove c) Incorrect type of pan – bottom has bad fit with stove, giving inefficient conduction of heat

Loading: eg Boiling plate 1800 W
Grill 1800 W
Oven 2000 W

Total = 11 kW

Voltage is 240 V

$$\text{Therefore } \frac{W}{V} = \frac{11\,000}{240} = 45.8 \text{ amps}$$

As the grill, ovens and top plates are seldom all used together and because of the function of the thermostats, there are usually less than 30 amps of current flowing at any one time.

The main switch is a double pole switch – that is, it breaks the live and neutral conductors simultaneously and it often has a light when it is turned on. The box cannot be opened without first turning off the main switch. Very often there is a 13-amp socket outlet for an electric kettle.

Heat control – oven thermostatic control is similar to that in use in gas cookers. See page 66. Boiling plates are controlled by time control heat devices known as simmerstat heat regulators.

Refrigeration

How refrigeration works

Cold is the absence of heat and to make things cold it is necessary to remove the heat that things have absorbed. This is the function of a refrigerator. To raise water to boiling-point (100°C) from tap water temperature (10°C) heat must be added. Add further heat and the boiling water turns to water vapour or steam. A substance such as ammonia which boils at – 30°C would absorb heat at even lower temperatures. Ammonia is one of a number of liquids called refrigerants which are used to absorb heat from the interior of well-insulated containers or cabinets known as refrigerators. Water from which heat is removed cools down and forms ice. Ice blocks can be used to cool a refrigerator interior. The heat from the food stored in the refrigerator is absorbed by the ice and turned back into water as the interior cools down. Further ice blocks are then required to prolong the cooling process.

By using a refrigerant like liquid ammonia circulating through the pipes in the refrigerated compartment, heat is taken out of the food and the liquid ammonia boils, becoming a vapour or 'gas'. By removing or sucking out this vapour and replacing it with more liquid the cooling

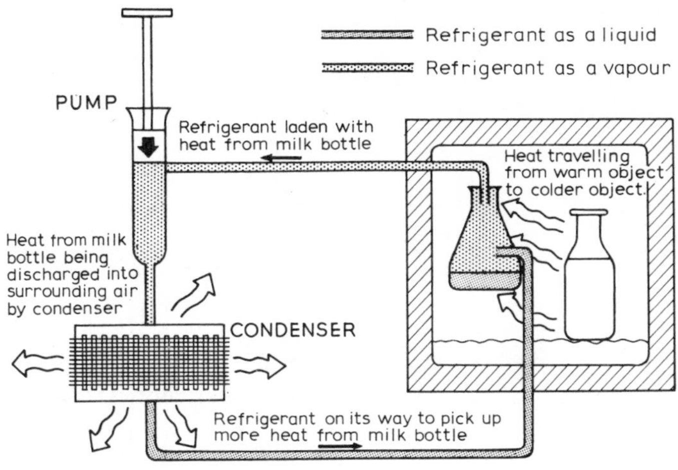

How refrigeration works

process can be maintained as long as required. In practice the vapour is not lost but compressed back into a liquid, cooled by giving up its heat to the air outside the refrigerator and then pumped back into the refrigerator cooling units for re-use.

Compression-type refrigerator (diagram, page 80)
The compression-type refrigerator works on the principle described. The three main parts are:

1 *The evaporator* inside the cabinet where the refrigerant boils and changes into a vapour and absorbs the heat.
2 *The compressor* which puts pressure on the refrigerant so that it can get rid of its heat. To enable the compressor to work, it is necessary to have an electric motor to operate the compressor or pump.
3 *The condenser* which helps to discharge the heat.

Absorption-type refrigerator (diagram, page 81)
This type of refrigerator does not have any moving parts as no compressor is used. A solution of ammonia gas in water is the refrigerant. The gas is given off from the solution when it is heated by an electrical element or gas flame. The ammonia gas passes to the condenser where it is liquefied. The liquefied ammonia passes to the evaporator with some hydrogen, where it expands and draws the heat from the cabinet. The gases produced pass to the absorber, where they mix with water and become a solution again. The solution returns to be heated again while the hydrogen goes to the evaporator.

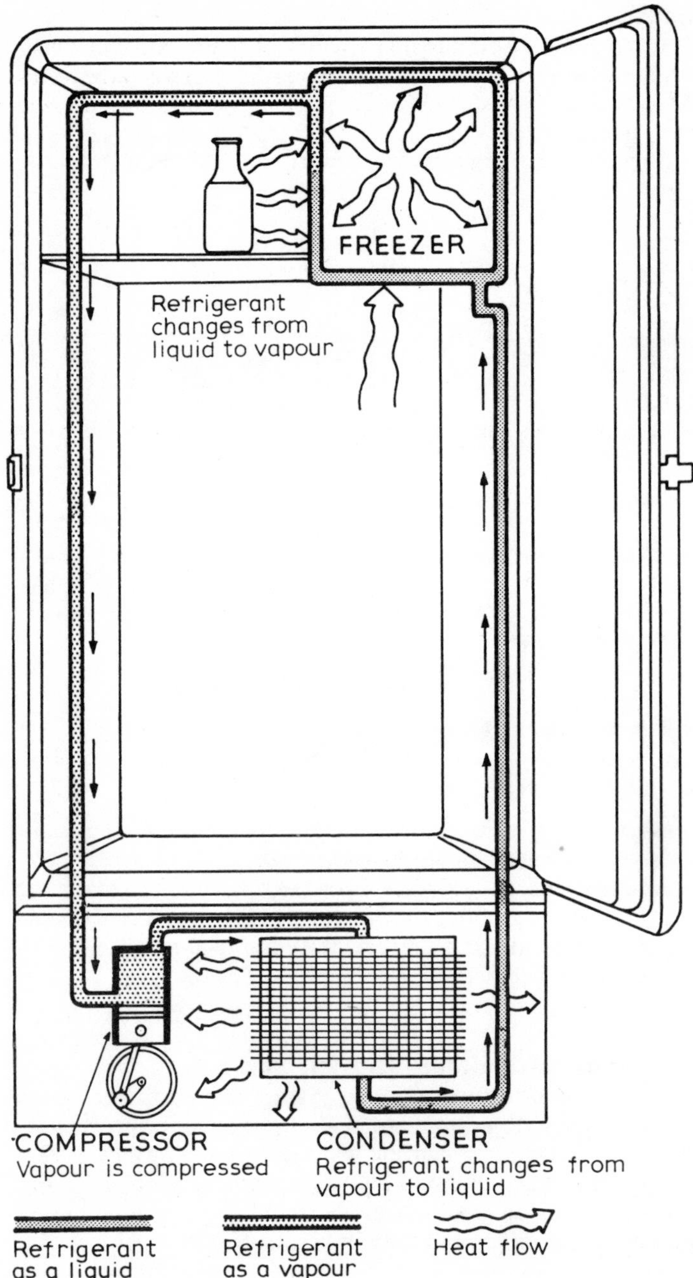

Compression-type refrigeration

Refrigerants

A refrigerant must have a boiling-point below the temperature at which ice forms and it should be non-corrosive and non-explosive. Substances such as sulphur dioxide, methyl chloride and dichloro-difluoromethane are used as refrigerants.

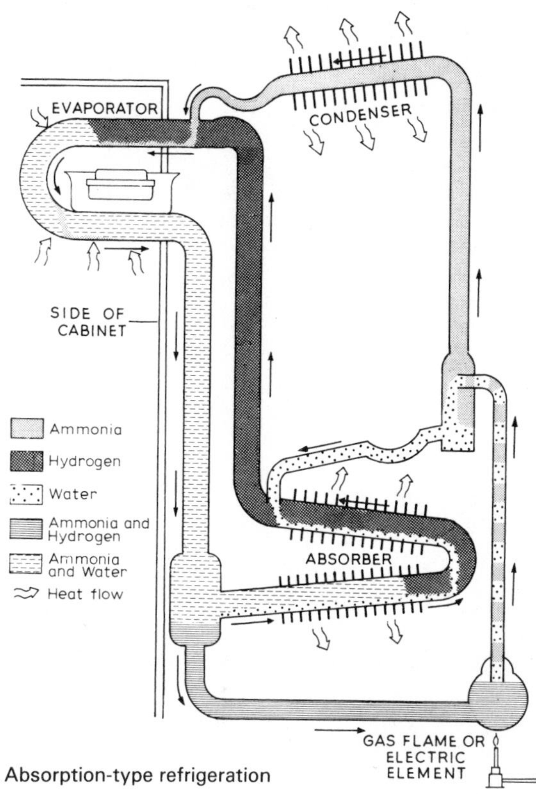

Absorption-type refrigeration

The cabinet

This must be well insulated to prevent entry of heat from outside. In modern refrigerators, insulations such as expanded polystyrene and foamed polyurethane are used extensively.

Cold rooms, chill rooms, deep-freeze cabinets and compartments

In large establishments it is necessary to have refrigerated space at different temperatures. The cold rooms may be divided into separate rooms: one for chill temperature for storing salads, fruits, certain cheeses; one for meats, poultry, game and tinned food which have to be refrigerated; one for deep-frozen foods. Frequently, the cold-room storage is designed so that after proceeding from the chill room into the cold room the deep-freeze compartment is reached. Refrigerated cabinets, thermostatically controlled to various desired temperatures, are also used in large larders. Deep-freeze cabinets are used where a walk-in deep-freeze section is not required and they maintain a temperature of – 18°C. Deep-freeze cabinets require defrosting twice a year.

Further information:
Electricity Council, 30 Millbank, London, SW1 P 4RD.

Charcoal

This is the black porous residue of burnt wood. Traditionally charcoal was the most often used fuel for grilling and it was considered the best because of the flavour it gave to the meat being grilled. The most popular alternative is a gas-fired grill and thinking in terms of economy, what have to be considered are:

a) comparative costs of charcoal and gas,
b) lack of flexibility in use of charcoal,
c) ease of flexibility in use of gas,
d) labour requirements for charcoal grill (lighting, refuelling, cleaning),
e) no labour requirement for gas grill,
f) comparison of flavour of meats grilled by charcoal and gas,
g) customer likes and dislikes.

Comparison of fuels

Many factors need to be taken into consideration in the comparison of fuels. For example, the cheapest fuel to buy may not be the cheapest to burn. When comparing electricity with coal, the initial cost of electricity will be much higher, but when considering coal, one has to think of the storage space required, the cost of labour to move it, the cost of labour to clean the dirt it causes, etc.

A list of advantages and disadvantages of each fuel helps to make comparisons.

Gas cooking

Advantages

1 Convenient, labour saving.
2 Free from smoke and dirt.
3 Easily controlled with immediate full heat and the flames are visible.
4 Special utensils are not required.
5 No fuel storage required.

Disadvantages

1 Some heat is lost into the kitchen.
2 Regular cleaning is necessary for efficiency.

Electric cooking

Advantages

1 Clean to use and maintain.
2 Easily controlled and labour saving.
3 A good working atmosphere for kitchen staff as no oxygen is required to burn electricity.
4 Little heat is lost.
5 No fuel storage is required.

Disadvantages

1 Time is taken to heat up.
2 Initial cost of equipment and maintenance costs.
3 Special utensils are required.

Steam cooking

Advantages

1 Good heat for boiling liquids.
2 Low maintenance costs.

Disadvantages

1 Methods of cooking are limited.
2 High initial cost of installation.

Solid fuel cooking

Advantage
Low maintenance costs.

Disadvantages

1 Cannot meet all cooking requirements.
2 Storage of fuel.
3 Dirt and dust from fuel.
4 Labour costs to move fuel.
5 Difficulty of control of heat.

Oil cooking

Advantages

1 Clean and convenient.

2 Labour saving.

Disadvantages

1 Need for large storage tanks.
2 Sources of supply may be affected.

The storage space required by various fuels (metric)

Oil	per tonne 11 m^3 storage space
Anthracite	per tonne 13 m^3 storage space
Coal	per tonne 14 m^3 storage space
Coke	per tonne 26 m^3 storage space

Gas and electricity do not require storage space; this advantage is paid for by higher initial cost.

Comparative fuel costs

Comparative fuel costs can be calculated by using the following formulae, inserting current prices of all four fuels and arriving at the cost of a 'useful therm' in each case.

1 therm = 100 000 British thermal units (Btu's).

A Btu is a measure of heat, the amount of heat required to raise 1 lb of water by 1°F.

Coke

If coke cost £*x* per ton (insert the current price of coke for *x*)
Divide £ . . . by 20 = cost per cwt.
Divide cost per cwt by 112 = cost per lb.
Coke produces 12 000 Btu's per lb.

$$\text{Therefore 1 therm costs } \frac{\text{cost per lb} \times 100\,000}{12\,000} = \ldots \text{p.}$$

A useful therm is the amount of heat put to good use. Coke is calculated to be 60% efficient.

$$\text{Therefore a useful therm costs } \frac{\text{cost per therm} \times 100}{60} = \ldots \text{p.}$$

Gas

An average cost could be *x*p per therm (substitute current price for *x*). Gas is caculated to be 80% efficient.

$$\text{Therefore a useful therm costs } \frac{\text{cost per therm} \times 100}{80} = \ldots \text{p.}$$

Electricity

1 unit of electricity produces 3412 Btu's.

1 unit of electricity costs *x*p (substitute current price for *x*)
Electricity is calculated to be 100% efficient.

$$\text{Therefore 1 therm costs } \frac{\text{cost per unit} \times 100\,000}{3412} = \ldots \text{p.}$$

Oil

If oil costs *x*p per gallon (substitute current price for *x*). (There are approximately 250 gallons to 1 ton)
1 gallon produces 165 000 Btu's.
Oil is calculated to be 75% efficient.

$$\text{Therefore 1 therm costs } \frac{\text{cost per gallon} \times 100\,000}{165\,000} = \ldots \text{p.}$$

$$\text{1 useful therm } \frac{\text{cost per therm} \times 100}{75} = \ldots \text{p.}$$

Charcoal-flavour grill
Gas heated with refractory stones, fitted with flame failure device and pilot assembly, produces same taste effect as charcoal and is suitable for all types of gas

4
Water

Having read the chapter these learning objectives, both general and specific, should be achieved.

General objectives Be aware of the sources of water and understand how it is stored and circulated in an establishment.

Specific objectives Distinguish between hot and cold water supply systems. Explain how taps and cisterns work and the function of U bends and grease traps.

By law, water authorities are required to provide a supply of clean, wholesome water – that is, water free from: suspended matter; odour and taste; all bacteria which are likely to cause disease; mineral matter injurious to health.

Water is obtained from rainfall which is collected in the following ways: natural lakes, rivers, artifical reservoirs, underground lakes, wells, springs.

Water is collected in storage reservoirs and is then given several cleansing processes before being piped to the consumer.

Primary filter

1 The water passes through a filtration system designed to remove larger solid matter such as twigs and leaves.
2 The water is stored in reservoirs which are deep and much suspect matter settles to the bottom.
3 The water is drawn from the reservoir as required and passed through a sand or micromesh filter to remove particles still in suspension.
4 Water is treated with chloride to kill any disease-bearing bacteria.
5 Samples are taken to ensure the water is fit for human consumption.
6 Water is then pumped into the street mains.

Hard water

In certain districts water is hard, and as this may be detrimental to water pipes, the use of a water softener is sometimes recommended.

Causes of hardness

1 Temporary hardness is due to the natural presence of calcium or magnesium bicarbonates dissolved in the water. This can be removed by boiling, but the vessel in which it boils becomes coated with 'fur'. Fur can be removed from kettles by commercial products or by boiling vinegar in the kettle and then boiling with clean water. It can be removed from pipes and boilers by a plumber, who will pass a strong acid through them.

2 Sulphates or chlorides of calcium and magnesium, when dissolved into the water as it trickles through the earth, is called permanent hardness. This is much more difficult to remove than temporary hardness.

Hardness may be removed from water by the following methods.

a) Soap added to water – the first amount softens the water and any more added will form a lather.
Disadvantages: uneconomical; scum forms, therefore unsuitable for washing.

b) Soda – cheap, quite effective. Roughly the correct amount should be added, eg London water with 16° of hardness requires 28.39g soda to 45.4 litres of water. When using soda and soap together the soda should be added first and allowed to dissolve so that it softens the water.

c) Ammonia solution – works in the same way as soda. These methods will remove temporary hardness as well as permanent hardness.

d) Use of water softener – called the base exchange method. A water softener is an apparatus which by a chemical action can take some of the hardness out of water. After a period of time, dependent on the quality of water passed through the softener, the chemical action slows, the filter becomes clogged and the agent, (usually zeolite) needs cleaning with a brine solution. When the whole of the water supply is softened, the drinking tap is usually placed on the pipe before it passes through the softener, so that the drinking water is not treated.

Proprietor's cold water system (supply)

The tap supply to the sink brings water directly from the main, therefore it is suitable for drinking.

The Water Board's stopcock is situated beyond the boundary of the proprietor's premises. Another stopcock should be fitted just inside the boundary to enable the consumer to turn off the water in an emergency.

The proprietor's stopcock is usually found inside the premises. There are usually other stopcocks throughout the premises enabling water supplies at different points to be cut off. If the ball-valves fail to function, the water would flow out through the overflow pipe. There is not always a cold water cistern. The entire supply sometimes comes direct from the main, and may be considered drinkable.

The cold water cistern

Purpose

To provide an emergency water supply should the main supply be turned off, or lowered.

In the average house there is usually a day's supply in the cistern, provided the water is carefully used. (Where the main supply has to be pumped by the water authority (not relying on gravity) then fluctuations in pressure at peak times - breakfast, lunch, dinner - will occur. Therefore the water authority require cold water storage cisterns to be fitted.)

Disadvantages

Stored water should always be considered suspect as it may become contaminated even though covered.

Position

It must be above the level of all taps; the higher the cistern is placed, the greater the pressure of water then supplied to the taps. The pressure depends upon the distance between the cistern and the taps, and the length of run and changes in direction.

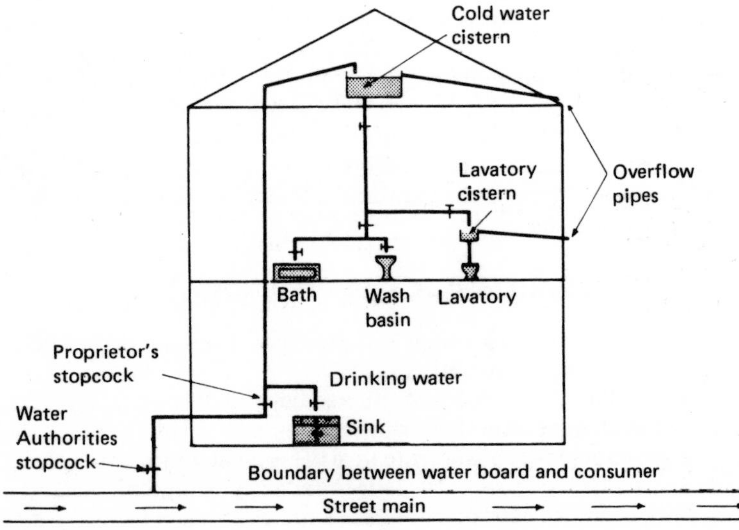

Cold water system

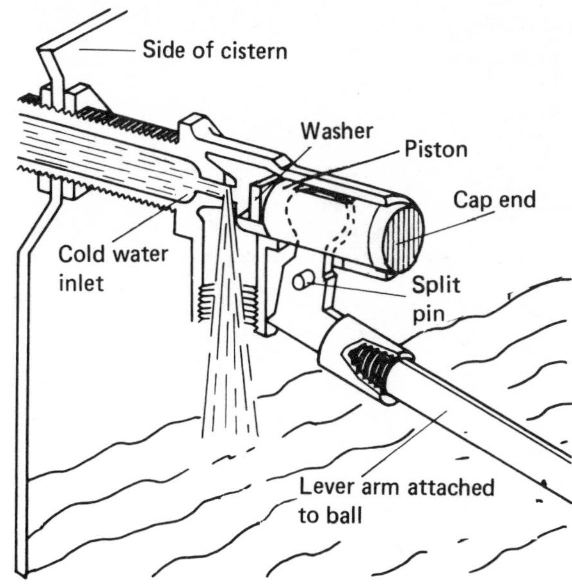

Operation of the ball-valve

Construction

A modern cistern is usually made of galvanised steel or plastic (it also could be made of wood or slate, lined with copper, lead or zinc, or of plastic). It should have a lid to keep the water clean, and this should be easily removed for repair purposes. It has to have an overflow pipe at the top in case anything should go wrong with the ball-valve. The cistern should be insulated with fire resisting material like fibreglass and the thickness will need to be increased if in an exposed position.

Ball-valve

The purpose of this is to keep the cistern filled with a supply of water. When the cistern is full the washer is pressed firmly against the water entrance. Sufficient pressure is maintained by the rod (which connects the lever and the ball) being of adequate length with a correct sized float. As the water level in the cistern drops the ball lowers and opens the valve.

Faults likely to occur

1 The ball may become punctured and therefore will drop to the bottom of the cistern and the valve will remain open.
2 Piston may be seized up in either position. To repair, turn off water at main, take valve to pieces and clean the piston, then replace.
3 Washer may wear out.
4 Sealing may be pitted or worn.

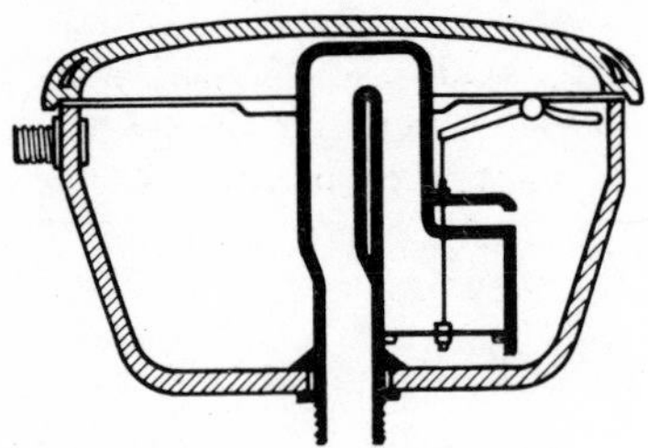

4.6 Litre Flush – Pull and let go
9 Litre Flush – Pull and hold down until the flush stops

A

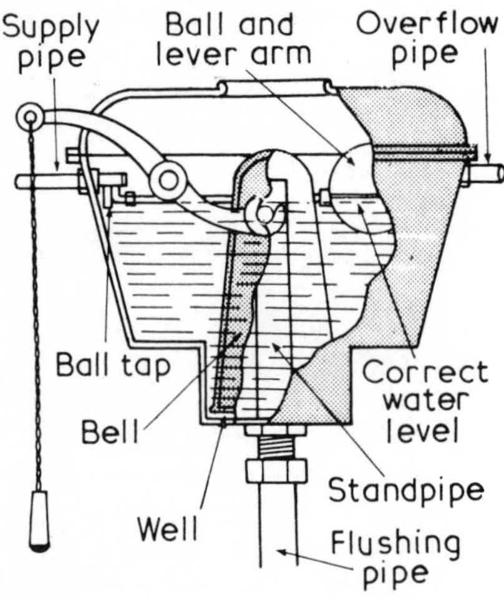

B

Flushing cisterns (water waste preventer)
Flushing cisterns should be so designed that they will discharge 2 gallons of water in 5 seconds. (See **A** dual flush system.)

In diagram **B** the cistern is full and the level of the water is just below the level of the flush-pipe. The bell is at its lowest level. When the lever is pulled, the bell is raised and the water inside the bell is drawn up so that it is now above the level of the flush-pipe. The water then flows down the flush-pipe, pushing down the air which is there. This continues until the cistern is emptied. At the same time the ball is dropping with the level of the water and the ball-valve opens and the cistern begins to fill again. When the cistern is emptied the ball will drop to its original position. In most water authority areas the new flushing cisterns should be dual flush with alternative 4:6 and 9 litre flushes designed to save water.

The water tap

It will be found that after studying the appearance and function of various components, most repairs to a tap will be comparatively simple.

The washers for modern taps are of two sizes and replacements are normally a composite material similar to rubber.

The dome is merely a decorative cover to the tap.

Rewashering of a tap

1 If on a hot water supply turn off the heat source.
2 Turn off the water supply to the tap.
3 Turn the tap on fully a) to make sure the water is turned off, b) to allow space for the dome to be lifted.
4 Unscrew the dome, if any, and lift as high as possible.
5 Unscrew the nut by gripping it at point A and remove from tap. If the washer plate is fixed to the spindle it will lift out with it, if not, remove it separately.
6 Remove old washer by nut B and replace it with a new washer.
7 Reassemble the tap and turn on water supply to test that the new washer is working efficiently.

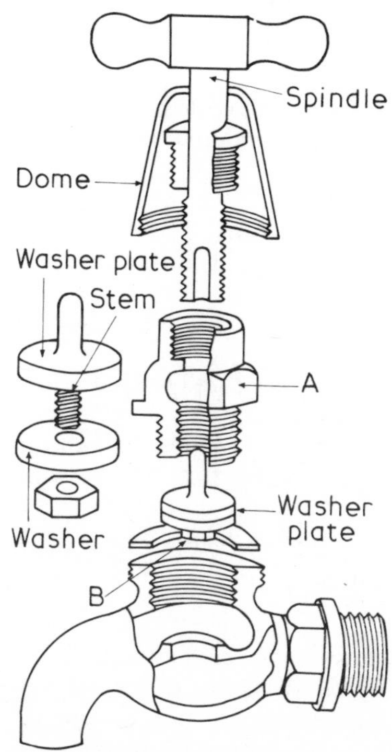

Tap components

Stopcock (stop-valve)
This works and is constructed in exactly the same way as the tap but just stops the water supply.

The Supatap
This is a more modern type of tap which has the big advantage that a new washer may be fitted in a minute or so without tools and without shutting off the water supply or draining the system.

Hot water systems

Methods of heating water

A number of methods and fuels are used and these will vary according to the type of premises and fuel used.

1 Centralized
Where all water is heated by one heat source and distributed to all taps from this source.

2 Decentralized
Where water is heated by a number of different heat sources, often adjacent to the fitting to be served or near to a group of fittings.

Each of the above groups may be sub-divided into either:

Storage heaters
Water is heated by the fuel and stored in a container designed for the purpose and suitably insulated. Water in larger quantities may then be drawn off as required.
Instantaneous heaters
Often small heaters, fitted adjacent to a fitting (single point) or group of fittings (multipoint). Water is heated and supplied direct to the fitting on demand, therefore cutting down heat loss.

Fuels

The fuel used for hot water supply may vary and generally includes gas, electricity, oil and solid fuel. Heat sources which use alternative energy sources may be used often as a pre-heat with the above fuels topping up to the required temperature of 60°C in hard water areas and 70°C in soft water areas for normal domestic use.

These alternative energy sources include solar energy, heat pumps and waste heat from other sources.

Choice of system and fuel
The variations in fuel costs fluctuate as does the way in which fuel is used and therefore each individual system may vary in design and fuel according to the requirements of the consumer. Advice on type of system may be obtained from the fuel suppliers, gas, electricity board etc or a qualified plumber (see Appendix).

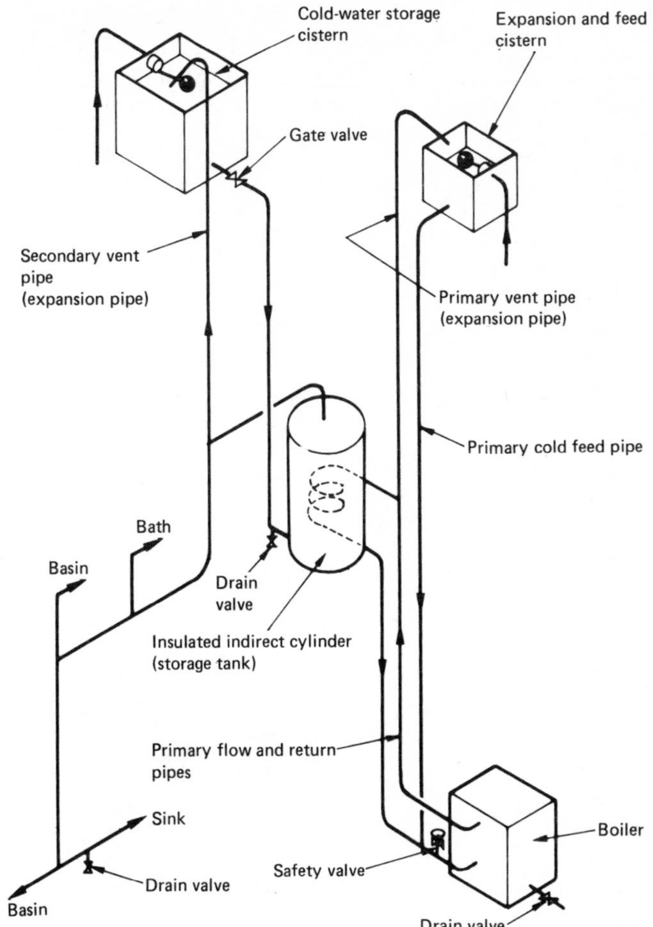

Indirect hot-water supply system

Drainage of water

Drain pipes and waste pipes

Any fitting which discharges water from it should have a drain or waste pipe which has a diameter of not less than the size of the outlet.

Each fitting, or range of fittings, must be fitted with a water seal (trap), when connected to a drainage system, to prevent unpleasant smells rising from the pipe into the room.

The two types of trap used commonly are the 'P' with a horizontal outlet and the 'S' with a vertical outlet (see page 94).

Blockage occurs most frequently because of misuse, such as depositing liquid grease, tea leaves and other waste products in the sink. If a blockage occurs then a rubber water plunger should be used. If this is not satisfactory then the cleaning eye at the bottom, if fitted, may be removed or with modern two piece traps, the connecting nuts undone, draining the water into a receptacle placed underneath and the blockage removed and cleared with flexible wire.

Modern plastic traps (refer *Clean Catering*, HMSO) are very smooth and reduce danger of blockage.

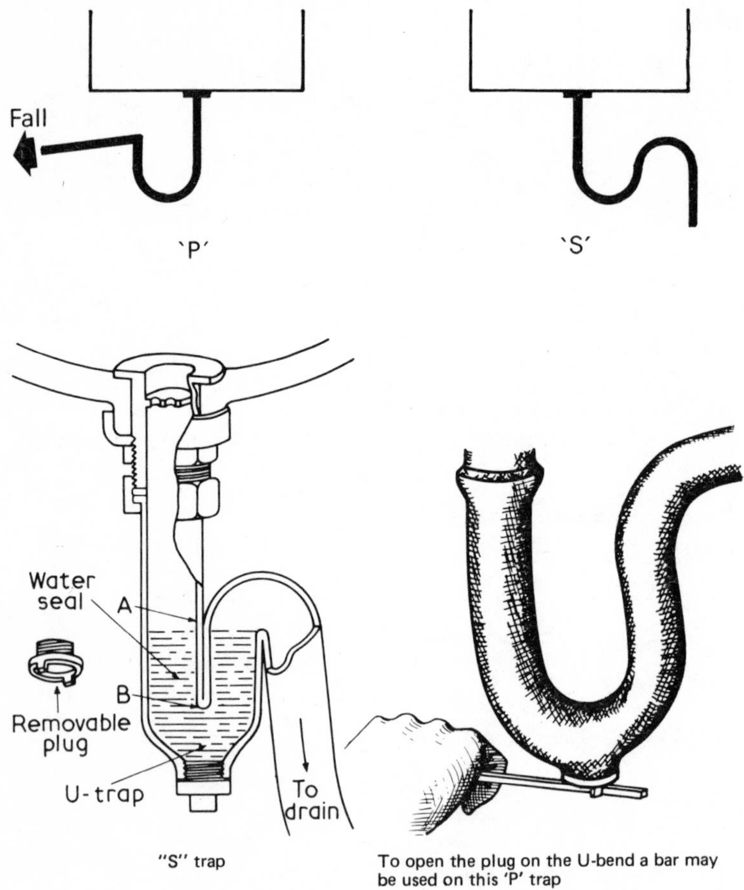

"S" trap

To open the plug on the U-bend a bar may be used on this 'P' trap

Freezing of pipes

The reason for pipes bursting when they freeze is that water expands when frozen. If expansion takes place along the length of the pipe, there will probably be no harm done, but if ice plugs form on either side of A the pipe will bulge.

Precautions

1 Lag all pipes.
2 Keep temperatures above 0°C.

Methods of thawing pipes

1 Wrap a cloth around the frozen part, and pour hot water over it.

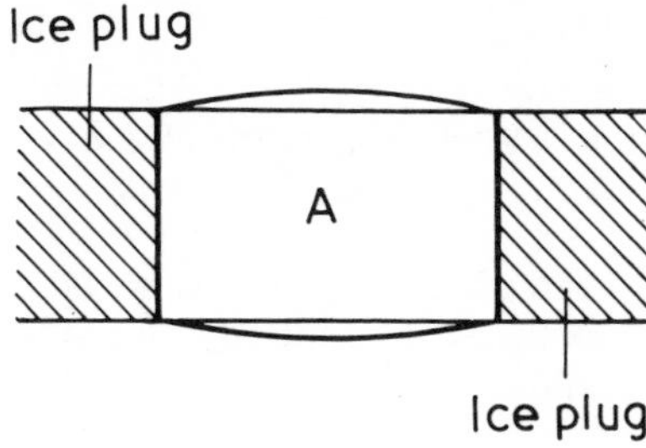

Freezing of water in pipes

2 A blow lamp should only be used by a plumber or experienced person.
3 A heater placed near the frozen part to raise the temperature of the air.
4 If a waste pipe is frozen, pour hot water down it.

For burst pipes

1 Turn off the main tap.
2 Turn off the stopcock between cistern and main pipe.
3 Open all taps, allow the water to escape.
4 Send for a plumber.

Further information: Local Water Board

5
Kitchen equipment

Having read the chapter these learning objectives, both general and specific, should be achieved.

General objectives Know the various pieces of equipment and understand their purpose in the establishment so that they can be used in the practical situation.

Specific objectives Identify each piece of equipment and specify the materials it may be made from. Explain the function and state the cleaning and maintenance of each piece of equipment.

Kitchen equipment is expensive. In order to justify the expense it is essential that maximum use is made of it, which can only be achieved if all the equipment works efficiently and this depends on care and maintenance. It follows that the routine use, care and cleaning of all items of equipment is important, and this should be appreciated and understood. (*Refer to page 32.*)

The kitchen of Simpson's restaurant as it was in 1907

Kitchen equipment at the Royal Lancaster Hotel, London

Kitchen equipment may be divided into three categories:

1 Large equipment – ranges, steamers, boiling pans, fish-fryers, sinks, tables, etc.
2 Mechanical equipment – peelers, mincers, mixers, refrigerators, dishwashers, etc.
3 Utensils and small equipment – pots, pans, whisks, bowls, spoons, etc.

Manufacturers of large and mechanical kitchen equipment issue instructions on how to keep their apparatus in efficient working order, and it is the responsibility of everyone using the equipment to follow these instructions, which should be displayed in a prominent place near the machines.

Arrangements should be made with the local gas board for regular checks and servicing of gas-operated equipment. Similar arrangements should be made with the electricity board in respect of electrical equipment. It is a good plan to keep a log-book of all equipment, showing where each item is located, when servicing takes place, noting any defects that arise, and instructing the fitter who may have carried out any work on the equipment to sign the log-book and to indicate exactly what has been done.

Large equipment

Stoves

A large variety of stoves is available operated by gas, electricity, solid fuel, oil, microwave or microwave plus convection.

Solid tops should be washed clean, or wiped clean with a pad of sacking. When cool the stove tops can be more thoroughly cleaned by washing and using an abrasive. Emery paper can also be used if necessary. After any kind of cleaning a solid top should always be lightly greased.

On the open type of stove all the bars and racks should be removed, immersed in hot water with a detergent, scrubbed clean, dried and put back in place on the stove. All gas jets should then be lit to check that none are blocked. All enamel parts of stoves should be cleaned while warm with hot detergent water, rinsed and dried.

The insides of ovens and oven racks should be cleaned while slightly warm, using detergent water and a mild abrasive if necessary. In cases of extreme dirt or grease being baked on to the stove or oven a caustic jelly may be used, but thorough rinsing must take place afterwards.

Oven doors should not be slammed as this is liable to cause damage.

Convection oven

The unnecessary lighting or the lighting of ovens too early can cause wastage of fuel, which is a waste of money. This is a bad habit common in many kitchens.

When a solid-top gas range is lit, the centre ring should be removed, but it should be replaced after approximately 5 minutes, otherwise unnecessary heat is lost (see page 6 – explosions).

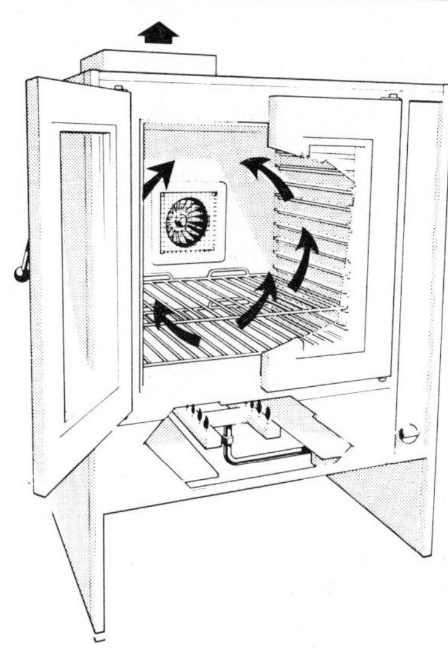

Forced air convection oven

Convection ovens

These are ovens in which a circulating current of hot air is rapidly forced around the inside of the oven by a motorized fan or blower. As a result a more even and constant temperature is created throughout the oven which allows food to be cooked successfully in any part of the oven. This means that the heat is used more efficiently, cooking temperatures can be lower, cooking times shortened and overall fuel economy achieved.

Forced air convection can be described as fast conventional cooking; conventional in that heat is applied to the surface of the food, but fast since moving air transfers its heat more rapidly than does static air. In a sealed oven fast hot air circulation reduces evaporation losses, keeping shrinkage to a minimum, and gives the rapid change of surface texture and colour which are traditionally associated with certain cooking processes.

There are four types of convection oven:

1 Where forced air circulation within the oven is accomplished by means of a motor-driven fan, the rapid air circulation ensures even temperature distribution to all parts of the oven.

2 Where low velocity, high volume air movement is provided by a power blower and duct system.
3 A combination of a standard oven and a forced convection oven designed to operate as either by the flick of a switch.
4 A single roll-in rack convection oven with heating element and fan housed outside the cooking area. An 18-shelf mobile oven rack makes it possible to roll the filled rack directly from the preparation area into the oven.

Further information : Forced Air Convection Ovens, Cornwell, Greene, Belfield, Smith & Co., 20 Kingsway, London, WCl.

Microwave cookers (See also *Practical Cookery*, page 11)
Microwave is a method of cooking and heating food by using high frequency power. The energy used is the same as that which carries television from the transmitter to the receiver but is at a higher frequency.

The waves disturb the molecules or particles of food and agitate them, thus causing friction which has the effect of cooking the whole of the food, whereas in the conventional method of cooking, heat

A microwave convection oven

penetrates the food only by conduction from the outside. Food being cooked by microwave needs no fat or water and is placed in a glass, earthenware, plastic or paper container before being put in the oven. Metal is not used as the microwaves are reflected by it.

All microwave ovens consist of a basic unit of various sizes with varying levels of power. Some feature additions to a standard model such as automatic defrosting systems, browning elements, Stay-Hot controls and revolving turntables.

The oven cavity has metallic walls, ceiling, and floor which reflect the microwaves. The oven door is fitted with special seals to ensure that there is the minimum of microwave leakage. A cut-out device automatically switches off the microwave energy when the door is opened.

When cleaning do not allow the cleaning agent to soil or accumulate around the door seal as this could prevent a tight seal when the door is closed. Never use an abrasive cleaner to clean the interior of the oven as it can scratch the metallic walls. Do not use aerosols either as these may penetrate the internal parts of the oven. Follow the manufacturer's instructions carefully for cleaning. **Further information**: The Microwave Oven Association, 16a The Broadway, London SW19.

Combination convection and microwave cooker

This cooker combines forced air convection and microwave, either of which can be used separately, but which are normally used simultaneously thereby giving the advantages of both systems: speed, colouration and texture of food. Traditional metal cooking pans may also be used without fear of damage to the cooker.

Induction cookers

These are solid top plates made of vitroceramic material which provide heat only when pans are put on them and which stop the heat immediately the pans are removed.

A generator creates a two-way magnetic field at the level of the top. When a utensil with a magnetic base is placed on the top a current passes directly to the pan. This means that a far more efficient use is made of the energy than with conventional cooking equipment. Since the ceramic top is not magnetic but merely a tray to stand the pots and pans on, it never heats up. Tests indicate more than 50 % energy saving. If a pan of water is to be brought to the boil there is no delay waiting for the top to heat up; the transmission of energy through the pan is immediate. When shallow frying, cold oil and the food can be put into the pan together without affecting the quality of the food as the speed of heating is so rapid.

Induction tops have a number of advantages over ranges using conventional sources:

energy saving,
flexible,

faster cooking time,
easy maintenance,
hygienic,
safe,
improved working environment (less heat in the kitchen).

However, induction tops are expensive and special cooking utensils are required. Any non-magnetic material does not work and aluminium and copper are unsuitable. Stainless steel, enamelled ware, iron and specially adapted copper pans are suitable.

Further information : Stangard Metal Workers Ltd, Herringham Road, Charlton, London SE7 8NW.

Steamers

There are four types of steaming ovens:

a) atmospheric steamer,
b) pressure steamer,
c) high compression steamer,
d) pressureless convection steamer.

The atmospheric steamer is pressureless. It has a boiling water bath in the bottom of the steaming compartment and a vent so that the steam does not rise above atmospheric pressure. For this reason the door can be opened safely at any time, although some steam is lost. Heat source can be gas or electricity.

The pressure steamer is constructed with a pressure safety valve which only allows steam to escape on reaching a certain pressure. Foods cooked in this type of steamer cook quicker than in the atmospheric steamer. Care must be taken when opening the door, it should be opened slowly so as to allow pressure to go down and no one should be close to the escaping steam. When opening the door stand on the hinge side.

If the atmospheric and pressure-type steamers are operated by gas or electricity then an automatic water supply by ball valve is provided to ensure a constant level of water in the steam generating tank. It is important to see that the tap controlling the supply of water to the ball valve is working correctly and that the ball valve arm and washer are both in efficient working order. If these precautions are not taken there is danger of the generating tank burning dry and becoming damaged as a result.

Some pressure steaming ovens heat source is from the main steam supply. This type of equipment is fitted with a gauge which registers the steam pressure being supplied and an overflow valve which gives a warning whistle if pressure is allowed to rise to a dangerous level. It is essential that both the gauge and valve are checked by a qualified engineer to ensure that they are working correctly.

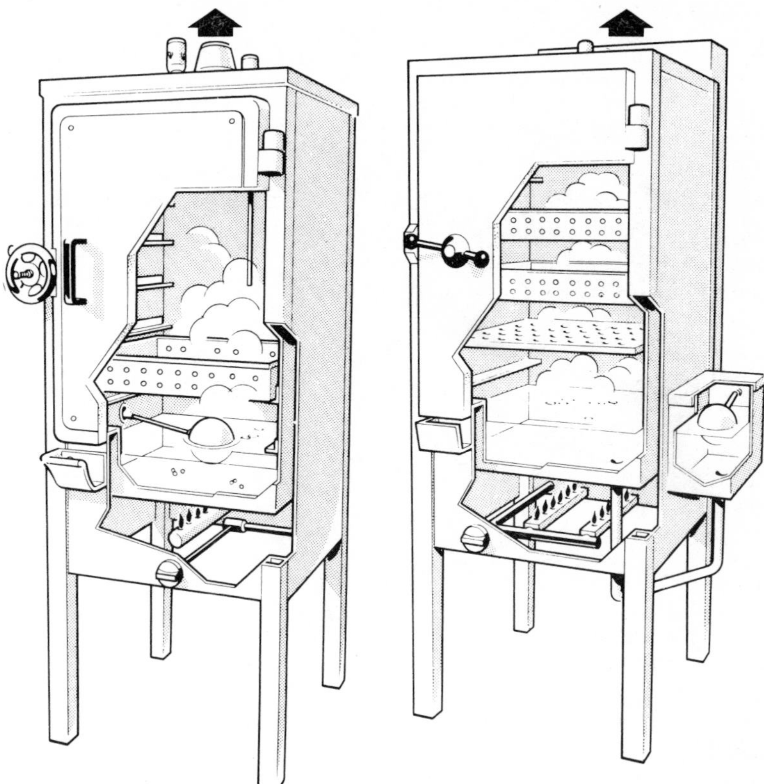

Steamers

The high compression steamer is similar in principle to the pressure steamer but works at a higher pressure, therefore the food cooks more quickly. This equipment is usually fitted with a timer and is designed to batch cook fresh or frozen vegetables in 1-5 minutes, however it does not have a large capacity. (See page 104)

Pressureless convection steamers cook at a low temperature with a convection fan in a pressureless air-free compartment. The steam generator is fitted under the steamer in a separate compartment and it generates purified steam under pressure which is introduced into the cooking compartment.

Cleanliness of steamers is essential; trays and runners should be washed in hot detergent water and then rinsed. The water generating chamber should be drained, cleaned and refitted and the inside of the steamer cleaned with detergent water and rinsed. Steamer door controls should be lightly greased occasionally and the door left open slightly to allow air to circulate when the steamer is not in use.

High compressor steam cooker

Bratt pan

The bratt pan is the most versatile piece of cooking equipment in the kitchen because it is possible to use it for shallow frying, deep frying, stewing, braising and boiling. Because of the large surface area a bratt pan can cook many items of food at one time. A further advantage is that it can be tilted so that the contents can be quickly and efficiently poured out on completion of the cooking process. Bratt pans are heated by gas or electricity and several models are available incorporating various features to meet differing catering requirements.

Boiling pans (See page 114)

Many types are available in different metals – aluminium, stainless steel, etc, in various sizes, 10, 15, 20, 30 and 40 litre capacity, and they may be heated by gas or electricity or from the main steam supply. As they are used for boiling or stewing large quantities of food it is important that they do not allow the food to burn; for this reason the steam-jacket type boiler is the most suitable. Many of these are fitted with a tilting device to facilitate the emptying of the contents.

After use, the boiling pan and lid should be thoroughly washed with mild detergent solution and then well rinsed. The tilting apparatus should be greased occasionally and checked to see that it tilts easily. If gas fired the gas jets and pilot should be inspected to ensure correct working. If a pressure gauge and safety valve are fitted these should be checked to see that they are working correctly.

Electrolux tilting frying table (Bratt pan)

Deep fat-fryers

These are among the items of equipment which are used extensively in many catering establishments. The unskilled or careless worker can cause money to be lost by food or fat being spilt through misuse of a deep fat-fryer.

Fryers are heated by gas or electricity and incorporate a thermostatic control in order to save fuel and prevent overheating. There is a cool zone below the source of heat into which food particles can sink without burning and thus spoiling other foods being cooked. This form of heating also saves fat. (See page 106)

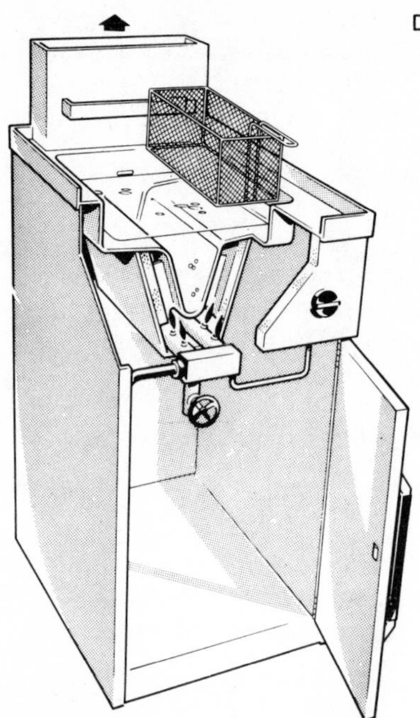

Deep fat fryer

Pressure fryers
Food is cooked in an air-tight frying vat thus enabling food to be fried a lot faster and at a lower oil temperature.

Hot air rotary fryers
These are designed to cook batches of frozen blanced chips or battered foods without any oil in 4-6 minutes.

Computerized fryers are available which may be programmed to control automatically cooking temperatures and times, on and off switches, basket lifting and product holding times. Operational information is fed from a super-sensitive probe which is immersed in the frying medium and passes information about temperature and rates of temperature change which may be caused by: the initial fat temperature, amount of food being fried, fryer efficiency and capacity, fryer recovery rate, quantity and condition of fat, product temperature and water content.

With all the above information the fryer computes exact cooking times and an automatic signalling device indicates the end of a cooking period.

Deep fat-fryers should be cleaned daily after use by:

1 Turning off the heat and allowing the fat to cool.
2 Draining off and straining the fat.
3 Closing the stopcock, filling the fryer with hot water containing

detergent and boiling for 10-15 minutes.

4 Draining off the detergent water, refilling with clean water plus $\frac{1}{8}$ litre of vinegar per 5 litres of water and reboiling for 10-15 minutes.

5 Draining off the water, drying the fryer, closing the stopcock and refilling with clean fat.

Hot-cupboards and bains-marie

Hot-cupboards (commonly referred to in the trade as the hotplate) are used for heating plates and serving dishes and for keeping food hot. Care should be taken to see that the amount of heat fed into the hot-cupboard is controlled at a reasonable temperature. This is important, otherwise the plates and food will either be too hot or too cold and this could obviously affect the efficiency of the service. A temperature of 60°C–76°C is suitable for hot-cupboards and a thermostat is a help in maintaining this.

Hot-cupboards may be heated by steam, gas or electricity. The doors should slide easily, and occasional greasing may be necessary. The tops of most hot-cupboards are used as serving counters and should be heated to a higher temperature than the inside. These tops are usually made of stainless steel and should be cleaned thoroughly after each service.

Bains-marie are open wells of water used for keeping foods hot, and are available in many designs, some of which are incorporated into hot-cupboards, some in serving counters, and there is a type which is fitted at the end of a cooking range. They may be heated by steam, gas or electricity and sufficient heat to boil the water in the bain-marie should be available. Care should be taken to see that a bain-marie is never allowed

Hot-plate area of a central kitchen

to burn dry when the heat is turned on. After use the heat should be turned off, the water drained off and the bain-marie cleaned inside and outside with hot detergent water, rinsed and dried. Any drain-off tap should then be closed.

Grills and salamanders

The salamander or grill heated from above by gas or electricity probably causes more wastage of fuel than any other item of kitchen equipment through being allowed to burn unnecessarily for long unused periods. Most salamanders have more than one set of heating elements or jets and it is not always necessary to have them all turned on full.

Salamander bars and draining trays should be cleaned regularly with hot water containing a grease solvent such as soda. After rinsing they should be replaced and the salamander lit for a few minutes to dry the bars.

For under-fired grills to work efficiently they must be capable of cooking food quickly and should reach a high temperature 15-20 minutes after lighting, and the heat should be turned off immediately after use. When the bars are cool they should be removed and washed in hot water containing a grease solvent, rinsed, dried and replaced on the grill. Care should be taken with the fire bricks if they are used for lining the grill as they are easily broken.

Contact grills

These, sometimes referred to as double-sided or infra-grills, have two heating surfaces arranged facing each other. The food to be cooked is placed on one surface and is then covered by the second. These grills are electrically heated and are capable of cooking certain foods very quickly. Because of this, extra care is needed, particularly when cooks are using this type of grill for the first time.

Under-fired grill

Large-scale systems catering unit

Large-scale frozen-food reheat unit

Fry plates, griddle plates

These are solid metal plates heated from below. They are used for cooking individual portions of meat, hamburgers, eggs, bacon, etc. They can be heated quickly to a high temperature and are suitable for rapid and continuous cooking. When cooking is first commenced on the griddle plates, a light film of oil should be applied to the food and the griddle plate to prevent sticking. To clean griddle plates, warm the plate and scrape off loose food particles. Rub the metal with pumice stone or griddle stone, following the grain of the metal. Clean with hot detergent water, rinse with clean hot water and wipe dry. Finally reseason (prove) the surface by lightly oiling with vegetable oil.

Sinks

Different materials are used for sinks according to the purpose for which they are intended:

a) heavy galvanised iron for heavy pot wash,
b) stainless steel for general purposes.

Tables

Wooden tables should be scrubbed clean with hot soda water, rinsed and wiped as dry as possible to avoid warping.

Formica or stainless steel topped tables should be washed with hot detergent water, rinsed with hot water and dried.

Marble slabs should be scrubbed with hot water and rinsed. All excess moisture should be removed with a clean dry cloth.

No cutting or chopping should be allowed on table tops; chopping boards should be used.

Hot pans should not be put on tables; triangles must be used to protect the table surface.

The legs and racks or shelves of tables are cleaned with hot detergent water and then dried. Wooden table legs require scrubbing.

Butcher's or chopping block

A scraper should be used to keep the block clean.

After scraping, the block should be sprinkled with a few handfuls of common salt in order to absorb any moisture which may have penetrated during the day.

Do not use water or liquids for cleaning unless absolutely necessary as water will be absorbed into the wood and cause swelling.

Storage racks

All types of racks should be emptied and scrubbed or washed periodically.

Mechanical equipment

If a piece of mechanical equipment can save time and physical effort and still produce a good end result then it should be considered for purchase or hire. The performance of most machines can be closely controlled and is not subject to human variations so that it should be easier to obtain uniformity of production over a period of time.

The caterer is faced with two considerations:

1 The cost of the machine, installation, maintenance, depreciation and running cost.
2 The possibility of increased production and a saving of labour cost.

The mechanical performance must be carefully assessed and all the manufacturer's claims as to the machine's efficiency thoroughly checked. The design should be fool-proof, easy to clean and operated with the minimum effort.

When a new item of equipment is installed it should be tested by a qualified fitter before being used by catering staff. The manufacturer's instructions must be displayed in a prominent place near the machine. The manufacturer's advice regarding servicing should be followed and a record book kept showing when and what maintenance the machine is receiving. The following list includes machines typically found in catering premises which are classified as dangerous under the Prescribed Dangerous Machines Order, 1964.

I Power-driven machines of the following types:

1 Worm-type mincing machines.
2 Rotary knife bowl-type chopping machines.
3 Dough mixers.
4 Food mixing machines when used with attachments for mincing, slicing, chipping and any other cutting operation, or for crumbling.
5 Pie and tart making machines.
6 Vegetable slicing machines.

II The following machines whether power-driven or not:

7 Circular knife slicing machines used for cutting bacon and other foods (whether similar to bacon or not).
8 Potato chipping machines.

Before cleaning, the machine should be switched off and the plug removed from the socket.

Potato-peelers

1 Potatoes should be free of earth and stones before loading into the machine.
2 Before any potatoes are loaded the water spray should be turned on and the abrasive plate set in motion.
3 The interior should be cleaned out daily and the abrasive plate removed to ensure that small particles are not lodged below.

4 The peel trap should be emptied as frequently as required.
5 The waste outlet should be kept free from obstruction.

Refrigerators (see also pages 68, 80-81)

In order to maintain a refrigerator at peak efficiency the following points should be observed:

1 Defrost weekly. If the refrigerator is not of the automatic defrosting type the control should be turned to defrost; the racks should be emptied and racks and interior surfaces washed, rinsed and dried.
   If the refrigerator is not defrosted regularly excess frost accumulates on the cooling system, acts as an insulator and causes the refrigerator motor to work longer than is necessary, thus shortening the life of the components.
2 The door or doors should be kept closed as much as possible, otherwise if too much warm air is allowed to enter, the refrigerator plant overworks and excess frost can accumulate on the cooling system.
3 Food should be stored sensibly and in such a way that the cold air can circulate all round. Excessive packing of food into a refrigerator should be avoided.
4 A qualified service engineer should be called in at the first sign of any defect in the machinery operating a refrigerator.

Food-mixer

This is an important labour-saving, electrically operated piece of equipment used for many purposes, for example mixing pastry, cakes, mashing potatoes, beating egg whites, mayonnaise, cream, mincing or chopping meat and vegetables.

1 It should be lubricated frequently in accordance with manufacturer's instructions.
2 The motor should not be overloaded. Overloading can be caused by obstruction to the rotary components. For example, if dried bread is being passed through the mincer attachment without sufficient care the rotary cog can become so clogged with bread that it is unable to move. If the motor is allowed to run damage can be caused to the machine.
3 All components as well as the main machine should be thoroughly washed and dried. Care should be taken to see that no rust occurs on any part. The mincer attachment knife and plates will rust if not given sufficient care.

Vertical high speed cuttermixer

This is an extremely fast, versatile labour-saving machine which can deal with a great amount of the repetitive, time consuming work required in some kitchen operations.

Vertical high speed cutter mixer

Liquidizer
This is a versatile, labour saving piece of kitchen machinery which uses a high speed motor to drive specially designed stainless steel blades to chop, purée or blend foods efficiently and very quickly. As a safety precaution food must be cooled before being liquidised.

Food-slicers and choppers
Food-slicers are obtainable both manually and electrically operated. They are labour-saving devices which can be dangerous if not operated with care. Because of this the working instructions should be placed in a prominent position near the machine.

1 Care should be taken that no material likely to damage the blades is included in the food to be sliced or chopped. It is easy for a careless worker to overlook a piece of bone which, if allowed to come into contact with the cutting blade, could cause severe damage.
2 Each section in contact with food should be cleaned and carefully dried after use.
3 The blade or blades should be sharpened regularly.
4 Moving parts should be lubricated, but oil must not come into contact with the food.
5 Extra care must be taken when blades are exposed.

Chipper (hand or electric)

The manual type should be washed and dried after use. Care should be taken with the interior of the blades; they should be cleaned with a folded cloth. When chipping potatoes, pressure should be applied gradually to prevent damage to the cutting blades which can be caused by violent jerking.

The electric chipper should be thoroughly cleaned and dried after use, particular attention being paid to those parts which come into contact with food. Care should be taken that no obstruction prevents the motor from operating at its normal speed. Moving parts should be lubricated according to the maker's instructions.

Masher (hand or electric)

The hand type should be washed immediately after use, then rinsed and dried.

The electric masher should have the removal sections and the main machine washed and dried after use, extra care being taken over those parts which come into contact with food. The same care should be taken as with electric chippers regarding obstruction and lubrication.

Tilting boiling pans

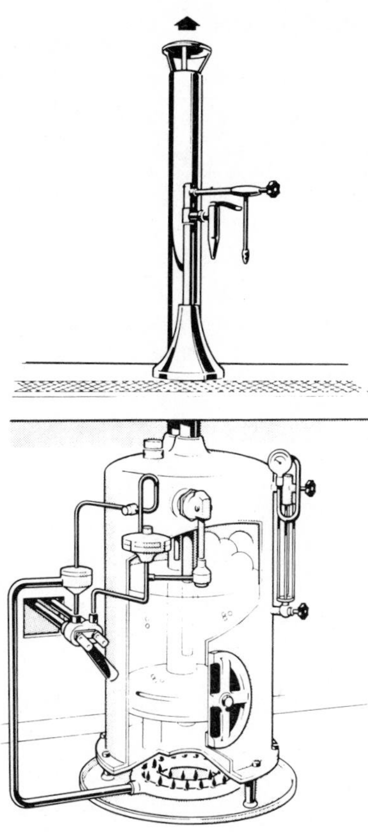

Pressure boiler

Water boiling appliances for tea- and coffee-making

There are two main groups of water boilers: bulk boilers from which boiling water can only be drawn when all the contents have boiled and automatic boilers which provide a continuous flow of boiling water.

Bulk boilers

These are generally used when large quantities of boiling water are required at a given time.They should be kept scrupulously clean, covered with the correct lid to prevent anything falling in and when not used for some time they should be left filled with clean cold water.

Automatic boilers

These boilers have automatic waterfeeds and can give freshly boiled water at intervals. It is important to see that the water supply is efficiently maintained, otherwise there is a danger of the boiler burning dry and being damaged.

Pressure boilers

This is the type that operates many still sets, consisting of steam heating milk boilers and pressure boiler providing boiling water. Care should be taken with the pilot light to see that it is working efficiently. As with all gas-fired equipment it is essential that regular inspection and maintenance is carried out by gas company fitters. (See page 115)

Coffee and milk heaters

Water-jacket boilers are made for the storage of hot coffee and hot milk with draw-off taps from the storage chamber. Inner linings may be of glazed earthenware, stainless steel or heat-resistant glass. It is very important that the storage chambers are thoroughly cleaned with hot water after each use and then left full of clean cold water. The draw-off taps should be cleaned through regularly with a special brush.

Dishwashing machines

For hygienic washing up the generally recognised requirements are a good supply of hot water at a temperature of 60°C for general cleansing followed by a sterilising rinse at a temperature of 82°C for at least one minute. Alternatively low-temperature equipment is available which sterilises by means of a chemical, sodium hypochlorite (bleach). **Further information**, Lever Industrial.

Dishwashing machines take over an arduous job and save a lot of time and labour, ensuring that a good supply of clean, sterilised crockery is available.

There are three main types:

1 *Spray types* in which the dishes are placed in racks which slide into the machines where they are subjected to a spray of hot detergent water at 48°C–60°C from above and below.

   The racks move on to the next section where they are rinsed by a fresh hot shower at 82°C. At this temperature they are sterilised, and on passing out into the air they dry off quickly.

2 *Brush-type machines* use revolving brushes for the scrubbing of each article in hot detergent water; the articles are then rinsed and sterilised in another compartment.

3 *Agitator water machines* in which baskets of dishes are immersed in deep tanks and the cleaning is performed by the mechanical agitation of the hot detergent water. The loaded baskets are then given a sterilising rinse in another compartment.

Dishwashing machines are costly and it is essential that the manufacturer's instructions with regard to use and maintenance are followed at all times.

Food waste disposers

Food waste disposers are operated by electricity and take all manner of rubbish, including bones, fat, scraps and vegetable refuse. Almost every type of rubbish and swill with the exception of rags and tins is finely ground, then rinsed down the drain. It is the most modern and hygienic method of waste disposal. Care should be taken by handlers not to push waste into the machine with a metal object as this can cause damage.

Small equipment and utensils

Small equipment and utensils are made from a variety of materials such as non-stick coated metal, iron, steel, copper, aluminium, wood, etc.

Iron

Items of equipment used for frying, such as fritures and frying-pans of all types, are usually made of heavy, black wrought iron.

Fritures should be washed in a strong grease-solvent solution, then thoroughly rinsed and dried, or they can be thoroughly cleaned with clean cloth or sacking.

Frying-pans (poêles) should not be washed. When new they should be 'proved' or 'seasoned'; they are spread with a thick layer of salt and placed on a hot stove or hot oven for 15–20 minutes, then wiped firmly with a wad of clean sacking. The salt is removed and a little fat or oil is added and they are wiped with a clean cloth. If an abrasive is necessary to clean the pan, salt may be used; if not, a good firm rub with dry sacking and a final light greasing are sufficient.

Frying-pans are available in several shapes and many sizes; for example:

Omelet-pans
Oval fish frying-pans
Frying-pans
Pancake-pans

Baking sheets are made in various sizes of black wrought steel. The less they are washed the less they are likely to cause food to stick. New baking sheets should be well heated in a hot oven, thoroughly wiped with a piece of clean sacking and then lightly oiled. Before being used baking trays should be lightly greased with a pure fat or oil. Immediately after use and while still warm they should be cleaned by scraping and dry-wiping. If washing is necessary hot soda or detergent water should be used.

Tartlet and barquette moulds and cake tins should be cared for in the same way as for baking sheets.

Tinned steel

A number of items are made from this metal; for example:

Conical strainer *(chinois)*, used for passing sauces and gravies.
Fine conical strainer *(chinois fin)*, used for passing sauces and gravies.
Colander, used for draining vegetables.
Vegetable reheating container *(passoir)*, used for reheating

vegetables.

Soup machine and mouli strainer, used for passing thick soups, sauces and potatoes for mash.

Sieves *(tamis)* .

Examples of equipment which is difficult to clean

1 Jelly bag
2 Mouli
3 & 4 Mouli attachments
5 Muslin
6 Sieve
7 Colander
8 Conical strainer – coarse
9 Conical strainer – fine
10 Potato ricer
11 12″ ruler

All the above items should be thoroughly washed immediately after each use and dried; if this is done, washing is simple and quick. If the food or liquid clogs and dries in the mesh it is difficult to clean; the easiest way to wash a sieve is to hold it upside down under running water and tap vigorously with the bristles of a stiff scrubbing brush. If the sieve is moved up and down quickly in water, clogged food will be loosened.

Care should be taken when using sieves; they should be the right way up when food is passed through, the food should be stroked through with a wooden mushroom, not banged, as this can damage the mesh. Only foodstuffs such as flour should be passed through the sieve upside down.

Copper

Pans of copper, lined with tin, are made in various shapes, sizes and capacities used to cook practically every kind of food.

Shallow saucepan with sloping sides	*sauteuse*
Shallow flat round pan with vertical sides	*plat à sauter*
Saucepan	*russe*
Stockpot	*marmite*
Large round deep pan	*rondeau*
Rectangular braising-pan	*braisière*
Roasting tray	*plaque à rôtir*
Turbot kettle	*turbotière*
Salmon kettle	*saumonière*
Gravy, soup, sauce storage pans	*bain-marie*
Moulds of various sizes and shapes	*dariole, charlotte, savarin, bombe, timbale*

Examples of copper equipment
1 Salmon kettle
2 Saucepan
3 Sauteuse
4 Sauté pan
5 Sugar boiler
6 Pomme Anna mould
7 Bowl
8 Braising pan
9 Dariole mould
10 Savarin mould
11 12″ ruler

Copper equipment is expensive, but it is first-class for cooking. This is because copper is a good conductor of heat; also, food burns less easily in copper pans than in pans of many other metals.

The disadvantages of copper are that it tarnishes easily and looks dirty. The tin lining of copper pans can be damaged by misuse – excessive dry heat can soften the tin and spoil the lining. Putting a pan on a fierce fire without liquid or fat is bad practice that can damage the tin lining. Retinning is expensive.

Copper equipment should be inspected periodically to see if the tin is being worn away; if so, it should be collected by a tinsmith and retinned.

Certain items of copper equipment, for example large vegetable boilers, sugar boilers, mixing bowls and egg white bowls, are not lined with tin but made wholly of copper.

The cleaning of copper equipment To keep large quantities of copper equipment clean the following points should be observed:

1 Two large sinks into which the pots may be completely immersed should be available. The water in one sink should be capable of being raised to boiling-point.
2 All dirty pans should be well soaked for a few minutes in boiling water to which a little soda has been added.
3 They should be well scoured, using either a brush or wire wool or similar agent with a scouring powder.
4 The pans are then rinsed in clean hot water and placed upside down to dry.
5 The copper surfaces, if tarnished, may be cleaned with a paste made from $\frac{1}{3}$ silver sand, $\frac{1}{3}$ salt and $\frac{1}{3}$ flour mixed with vinegar; the pans are then thoroughly rinsed and dried.

Aluminium

Saucepans, stockpots, sauteuses, sauté pans, braising pans, fish kettles and large round deep pans and dishes of all sizes are made in cast aluminium. They are expensive, but one advantage is that the pans do not tarnish; also, because of their strong heavy construction, they are suitable for many cooking processes.

A disadvantage is that in the manufacture of aluminium, which is a soft metal, other metals are added to make pans stronger. As a result certain foods can become discoloured; for example, care should be taken when making white sauces and white soups. A wooden spoon should be used for mixing; then there should be no discoloration. The use of metal whisks or spoons must be avoided.

Water boiled in aluminium pans is unsuitable for tea-making as it gives the tea an unpleasant colour. Red cabbage and artichokes should not be cooked in aluminium pans as they will take on a dark colour.

Cleaning of cast aluminium pans

1 All pans should be well soaked in hot detergent water, soda should not be used.
2 After a good soaking, pans should be scoured with a hard bristle brush or rough cloth with an abrasive powder if necessary. Harsh abrasives should be avoided if possible.
3 After scouring, the pans are rinsed in clean hot water and thoroughly dried.

Examples of stainless steel equipment
1 Saucepan
2 Mixing bowl
3 Tray
4 Stockpot
5 Bowl
6 Basin
7 Mandolin
8 12″ ruler

Stainless steel

Stainless steel is a relatively poor conductor of heat, which together with the high price makes it a questionable metal for kitchen pans. To overcome the first point manufacturers add a thick layer of copper to the base of pans. This overcomes the problem of poor heat conduction, but also makes the pan still more expensive. Stockpots and pans of various shapes and sizes are available with or without copper bases.

Stainless steel is used for many small items of equipment.

Non-stick metal

An ever-increasing variety of kitchen utensils eg saucepans, frying pans, baking and roasting tins are available which may be suitable for certain types of kitchen operation such as small scale or à la carte. Particular attention should be paid to the following otherwise the non-stick properties of the equipment will be affected:

a) excessive heat should be avoided;
b) use plastic or wooden spatulas or spoons when using non-stick pans so that contact is not made to the surface with metal;
c) extra care is needed when cleaning non-stick surfaces, the use of cloth or paper is most suitable.

Examples of metal equipment
1 Frying basket
2 Friture and draining wire
3 Baking tray
4 Double grill wires
5 Cooling wire
6 Raised pie moulds
7 Frying pan
8 Omelet pan
9 Grater
10 Pancake pan
11 Flan ring
12 Deep tartlet mould
13 Shallow tartlet mould
14 Boat-shaped mould
15 12″ ruler

Wood, rubber and compound materials

Wooden cutting boards are an important item of kitchen equipment which should be kept in use on all table surfaces to protect the table and the edges of cutting knives.

Wooden cutting boards will warp or splinter if the following points are not observed:

1 A strong, well-constructed board should be used.
2 After use, boards should be scrubbed with a bristle brush, using hot detergent water, rinsed with clean water and dried as much as possible.
3 The boards should not be put over a stove or in a hot-cupboard. Excess heat and water cause wood to warp.
4 Heavy chopping should not occur on boards as this causes splintering. The place for heavy chopping is on the chopping block.

Wooden cutting boards do have the following disadvantages:

1 They are porous and therefore retain taste, smell, bacteria, grease and dirt.

2 They expand and contract when washed and allow small particles of food to become trapped.
3 The cut and scored surface also allows food particles and bacteria to become embedded.

Rubber: Cutting boards are also made of hard rubber and rubber compounds, eg rubber, styrene and clay. These are hygienic because they are solid, in one piece and should not warp, crack or absorb flavours. They are cleaned by scrubbing with hot water and then drying.

Rolling-pins, wooden spoons and spatulas: These items should be scrubbed in hot detergent water, rinsed in clean water and dried.

Rolling-pins should not be scraped with a knife; this can cause the wood to splinter. Adhering paste can be removed with a cloth.

Wooden sieves and mandolins: When cleaning, care of the wooden frame should be considered in the light of the previous remarks. The blades of the mandolin should be kept lightly greased to prevent rust (stainless steel mandolins are available).

Examples of wooden equipment
1 Chopping board
2 Sieve
3 Triangle
4 Salt box
5 Rolling pin
6 Spoon
7 Spatula
8 Mushroom
9 12″ ruler

Examples of small equipment

1 Spider
2 Skimmer
3 Tenon saw
4 Iron spatule
5 Butcher's saw
6 Chopper
7 Fish slice
8 Ladle
9 Metal spoon
10 Perforated spoon
11 Fish scissors
12 Oyster knife
13 Balloon whisk
14 Small whisk
15 Ravioli wheel
16 Trussing needle
17 Larding needle
18 Vegetable peeler
19 Solferino cutter
20 Parisienne cutter
21 Olivette cutter
22 Egg slicer
23 Four bladed chopper
24 Meat bat
25 12" ruler
26 Carving knife
27 Steel

Materials

Muslin (Mousseline).

The tammy cloth (étamine), which is made from calico. Both muslin and tammy cloth are used for straining soups and sauces.

The jelly bag made from thick flannel or nylon for straining jellies.

Piping bags (poche) are made from linen, nylon or plastic and are used for piping preparations of all kinds.

All materials should be washed immediately after use in hot detergent water, rinsed in hot clean water and then dried. Tammy cloths, muslins and linen piping bags must be boiled periodically in detergent water.

Kitchen cloths, papers and foils etc

Kitchen cloths

a) General purpose – for washing up and cleaning surfaces.

b) Tea towel (teacloth) – for drying up and general purpose hand cloths.
c) Bactericide wiping cloths – are impregnated with bactericide to disinfect work surfaces. The cloths have a coloured pattern which fades and disappears when the bactericide is no longer effective; the cloth should then be discarded.
d) Oven cloths – thick cloths designed to protect the hands when removing hot items from the oven. Oven cloths must only be used dry, never damp or wet otherwise the user is likely to be burned.

It is essential that all kitchen cloths are washed or changed frequently, otherwise accumulating dirt and food stains may cause cross contamination of harmful bacteria/germs on to clean food.

Papers

a) Greaseproof – for lining cake-tins, making piping bags and wrapping greasy items of food.
b) Kitchen – white absorbent paper for absorbing grease from deep-fried foods and for lining trays on which cold foods are kept.
c) General purpose - thick absorbent paper for wiping and drying equipment, surfaces, food etc.
d) Towels – disposable - for drying of hands.

Examples of china and earthenware
1 Casserole
2 Oval dish
3 Ravier
4 Bowl
5 Soufflé dish
6 Egg cocotte
7 12″ ruler
8 Pie dish
9 Sole dish
10 & 11 Egg dishes
12 Basin

Foils etc

a) Clingwrap - thin transparent material for wrapping sandwiches, snacks, hot and cold foods. Clingwrap has the advantage of being very flexible and easy to handle and seal.

b) Metal - thin pliable silver coloured material for wrapping and covering foods and for protecting oven roasted joints during cooking.

Further information: Modern Cooking Equipment, Young Northwood Books. Napleton, *Microwave Cookery* (Northwood Publications).

Convection and Microwave Cookery (Mealstream (UK) Ltd, 38 Woodham Lane, New Haw, Weybridge, Surrey).

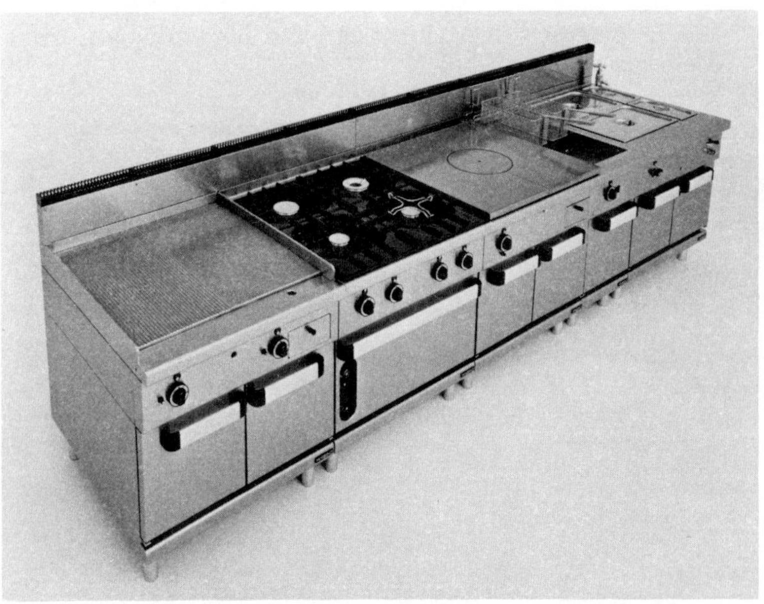

Compact griddle, open top, solid top, deep fryer and oven range

6

Elementary nutrition and food science

Having read the chapter, these learning objectives, both general and specific, should be achieved.

General objectives Understand the basic principles of nutrition and know the reasons why it is necessary to have this understanding so as to be able to apply this knowledge.

Specific objectives Explain the digestive system and state the function of nutrients. List the foods containing the various nutrients and state the effect of heat upon those nutrients. Explain basal metabolism and specify the value of foods in the diet.

Food and nutrients

A food is any substance, liquid or solid, which provides the body with materials –

a) for heat and energy,
b) for growth and repair,
c) to regulate the body processes.

These materials are known as *nutrients*. They are:

proteins	vitamins
fats	minerals
carbohydrates	water

The study of these nutrients is termed nutrition. Only those substances containing nutrients are foods. Most foods contain several nutrients, a few foods contain only one nutrient, for example sugar.

For the body to obtain the maximum benefit from food it is essential that everyone concerned with the buying, storage, cooking and serving of food and the compiling of menus should have some knowledge of nutrition.

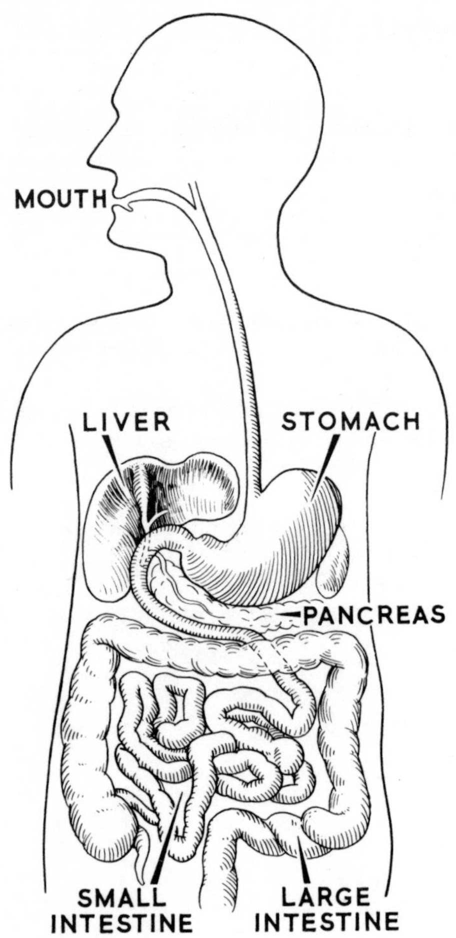

The digestive tract

Digestion

This is the breaking down of the food and takes place:

a) in the mouth, where food is mixed with saliva, and starch is broken down by the action of an enzyme in saliva;

b) in the stomach, where the food is mixed and gastric juices are added, and proteins are broken down;

c) in the small intestine, where proteins, fats and carbohydrates are broken down further and additional juices are added.

Absorption

To enable the body to benefit from food it must be absorbed into the blood-stream; this absorption occurs after the food has been broken down; the product then passes through the walls of the digestive tract into the blood-stream.

This occurs in:

a) the stomach where simple substances are formed as a result of digestion and pass through the stomach lining into the blood stream;
b) in the small intestine where more of the absorption of nutrients takes place due to a further breakdown of the food;
c) in the large intestine, where water is re-absorbed from the waste.

For the body to obtain the full benefit from foods it should be remembered that to stimulate the flow of saliva and gastric juices food must smell, look and taste attractive.

When the full benefit is not obtained or there is a lack of one or more nutrients then this leads to a state of malnutrition.

The main function of nutrients

Energy	*Growth and repair*	*Regulation of body processes*
Carbohydrates	Proteins	Vitamins
Fats	Minerals	Minerals
Proteins	Water	Water

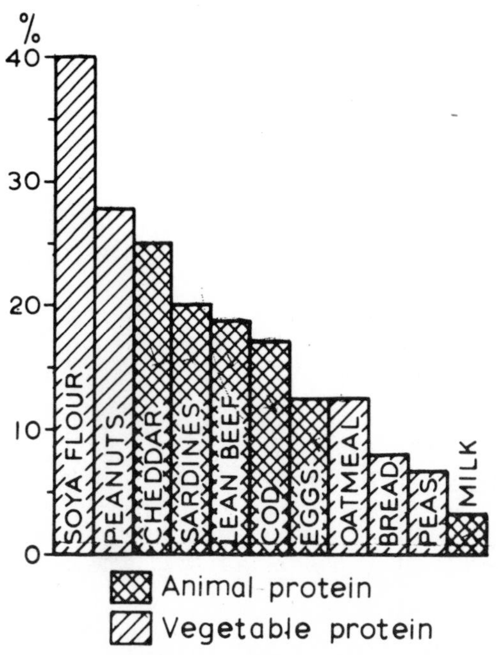

Proportion of protein in some foods

Proteins

Protein is an essential part of all living matter; it is therefore needed for the growth of the body and for the repair of body tissues.

There are two kinds of protein:

a) Animal protein found in meat, game, poultry, fish, eggs, milk, cheese, eg myasin, collagen (meat, poultry and fish), albumin, ovovitellin (eggs), casein (milk and cheese).

b) Vegetable protein found mainly in the seeds of vegetables. The proportion of protein in green and root vegetables is small. Peas, beans and nuts contain most protein and the grain of cereals, such as wheat, have a useful amount because of the large quantity eaten, eg gliadin and glutenin forming gluten with water (wheat and rye).

Main supply of protein in the average diet

It follows that as protein is needed for growth, growing children and expectant and nursing mothers will need more protein than other adults whose requirements are mainly for repair. Any spare protein is used for producing heat and energy.

What protein is

Protein is composed of amino-acids, and the protein of cheese is different from the protein of meat because the number and arrangement of the acids are not the same. A certain number of these amino-acids are essential to the body and have to be provided by food. Proteins containing all the essential amino-acids in the correct proportion are said to be of high biological value. The human body is capable of changing the other kinds of amino-acids to suit its needs.

It is preferable that the body has both animal and vegetable protein, so that a complete variety of the necessary amino-acids is available. During digestion protein is split into amino-acids; these are absorbed into the blood-stream and used for building body tissues and to provide some heat and energy.

Cooking effects on protein

On being heated, the different proteins in foods set or coagulate at different temperatures; above these temperatures shrinkage occurs, this is particularly noticeable in grilling or roasting meat. Moderately cooked protein is the most easy to digest – for example, a lightly cooked egg is more easily digested than a raw egg or a hard-boiled egg.

Fats

There are two main groups of fats, animal and vegetable. The function of fat is to protect vital organs of the body, to provide heat and energy, and certain fats also provide vitamins.

Fats can be divided into:

a) solid fat;
b) oils (fat which is liquid at room temperature).

Fats are obtained from the following foods:

aminal origin: dripping, butter, suet, lard, cheese, cream, bacon, meat fat, oily fish;
vegetable origin: margarine, cooking fat, nuts, soya-beans.

Oils are obtained from the following foods:

animal origin: halibut and cod-liver oil;
vegetable origin: from seeds or nuts.

Fats are composed of glycerol to which is attached three fatty acids (hence the name triglyceride). Fats differ because of the fatty acids from which they are derived. These may be, for example, butyric acid in butter, stearic acid in solid fat, such as beef suet, oleic acid in most oils. These fatty acids affect the texture and flavour of the fat.

To be useful to the body, fats have to be broken down into glycerol and fatty acids so that they can be absorbed; they can then provide heat and energy.

The food value of the various kinds of fat is similar, although some animal fats contain vitamins A and D.

Main supply of fat in the average diet

Fats should be eaten with other foods such as bread, potatoes, etc, as they can then be more easily digested and utilised in the body.

Certain fish, such as herrings, mackerel, salmon, sardines, contain oil (fat) in the flesh. Other fish, such as cod and halibut, contain the oil in the liver.

Vegetables and fruit contain very little fat, but nuts have a considerable amount.

Cooking effects on fat
Cooking has little effect on fat except to make it more digestible.

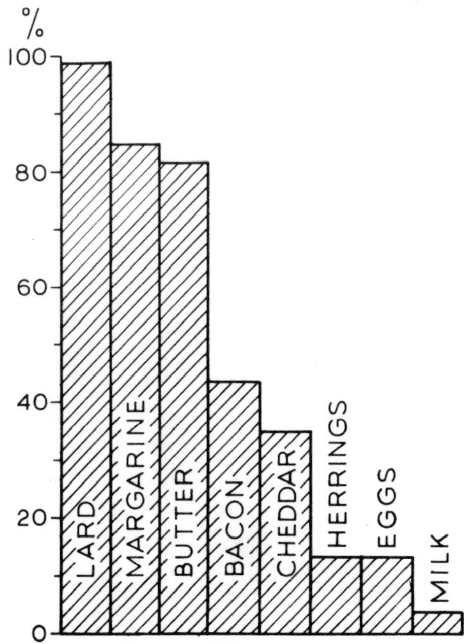

Proportion of fat in some foods

Carbohydrates

There are three main groups of carbohydrates:

a) sugar,
b) starch,
c) cellulose.

The function of carbohydrates is to provide the body with most of its energy. Starch is composed of a number of glucose molecules (particles), and during digestion starch is broken down into glucose.

a) Sugar

There are several kinds of sugar:

Glucose: found in the blood of animals and in fruit and honey.
Fructose: found in fruit, honey and cane sugar.
Sucrose: found in beet and cane sugar.

Lactose: found in milk.
Maltose: produced naturally during the germination of grain.

Sugars are the simplest form of carbohydrate and the end-products of the digestion of carbohydrates. They are absorbed in the form of glucose and simple sugars and used to provide heat and energy.

Main supply of carbohydrate in the average diet

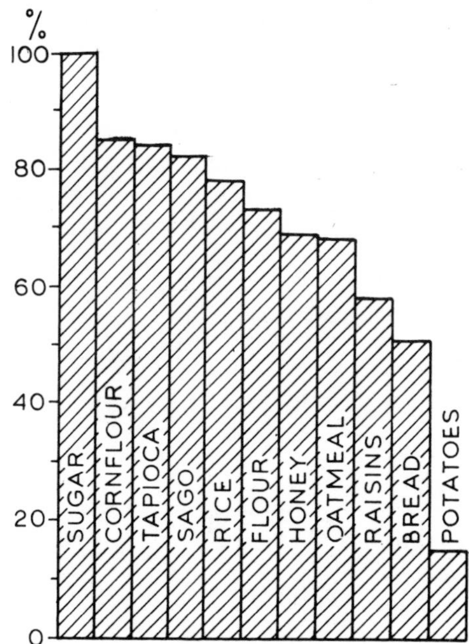

Proportion of carbohydrate in some foods

b) Starch

Starch is contributed to the diet by the following foods:

Whole grains: rice, barley, tapioca.
Powdered grains: flour, cornflour, ground rice, arrowroot.
Vegetables: potatoes, parsnips, peas, beans.
Unripe fruit: bananas, apples, cooking pears.
Cereals: cornflakes, shredded wheat, etc.
Cooked starch: cakes, biscuits.
Pastes: macaroni, spaghetti, vermicelli.

Cooking effects on starch

Uncooked starch is not digestible.

Foods containing starch have cells with starch granules, covered with a cellulose wall which breaks down when heated or made moist. When browned, as with the crust of bread, toast, roast potatoes, skin on rice pudding, etc., the starch forms dextrins and these taste sweeter. On heating with water or milk, starch granules swell and absorb liquid thus thickening the product eg thickened gravy or cornflour sauce. This thickening process is known as gelatinisation of starch.

c) Cellulose

Cellulose is the coarser structure of vegetables and cereals which is not digested but is used as roughage in the intestine. It is often now referred to as dietary fibre.

Vitamins

Vitamins are chemical substances which are vital for life, and if the diet is deficient in any vitamin, ill-health results. As they are chemical substances they can be produced synthetically.

General function of vitamins

To assist the regulation of the body processes, eg:

1 To help the growth of children.
2 To protect against disease.

Vitamin A

Function

1 Assists in the growth of children.
2 Helps to resist infection.
3 Enables people to see better in the dark.

Vitamin A is fat soluble, therefore it is to be found in fatty foods. It can be made in the body from carotene, the yellow substance found in many fruits and vegetables.

Dark green vegetables are a good source of vitamin A, the green colour masking the yellow of the carotene. Carotene is gradually destroyed by light (hence the fading of orange coloured spices and vegetables on prolonged storage).

Foods in which vitamin A is found

Halibut-liver oil	Margarine (to which vitamin A is added)	Carrots
Cod-liver oil		Spinach
Kidney	Cheese	Watercress
Liver	Eggs	Tomatoes
Butter	Milks	Apricots
	Herrings	

Fish-liver oils have the most vitamin A. The amount of vitamin A in dairy produce varies. Because cattle eat fresh grass in summer and stored feeding-stuffs in winter, the dairy produce contains the highest amount of vitamin A in the summer.

Kidney and liver are also useful sources of vitamin A.

Vitamin D

Function

Vitamin D controls the use the body makes of calcium. It is therefore

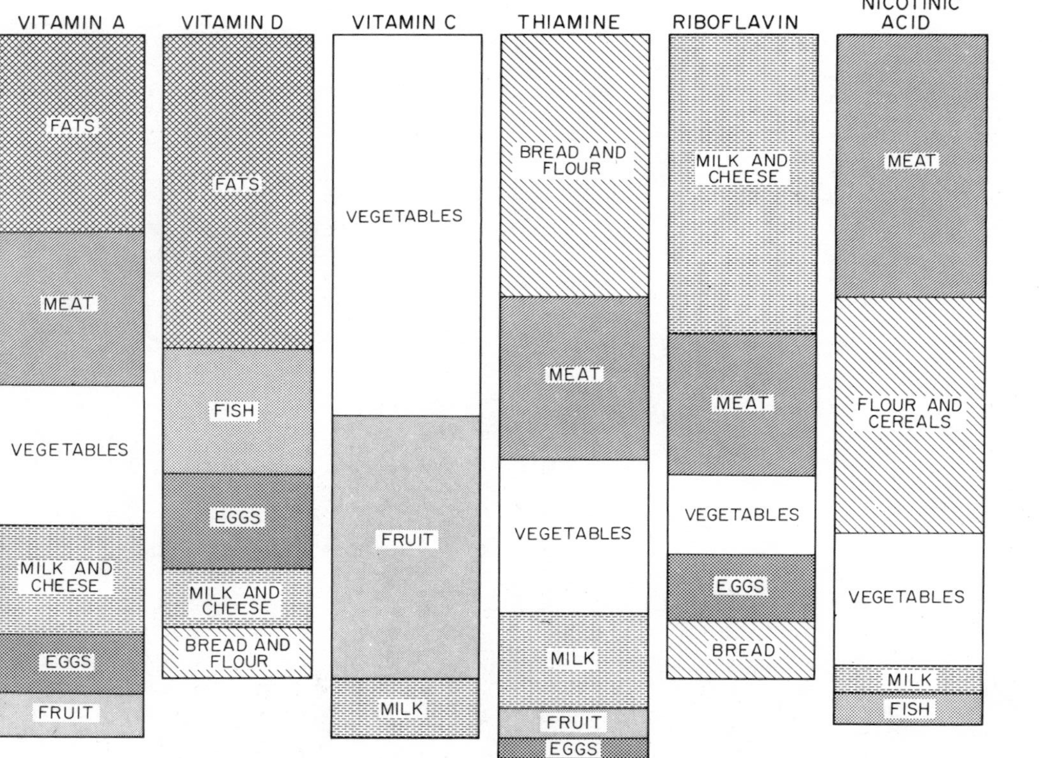

Main supply of vitamins in the average diet

necessary for healthy bones and teeth.

Like vitamin A it is fat soluble.

Sources of vitamin D

An important source of vitamin D is from the action of sunlight on the deeper layers of the skin.

Fish-liver oils	Margarine (to which vitamin D is added)
Oily fish	Dairy produce
Egg yolk	

Compared with vitamin A there are fewer sources of vitamin D, the fish-liver oils being the most important.

Vitamin B

When first discovered vitamin B was thought to be one substance only; it is now known to consist of at least eleven substances, the three main ones are:

Thiamine (B_1)
Riboflavin (B_2)
Nicotinic acid, or niacin
Others include Cyanocobalamin (B_{12}), folic acid and Pyridoxine (B_6).

Function

Vitamin B is required to:

1 Keep the nervous system in good condition.
2 Enable the body to obtain energy from the carbohydrates.
3 Encourage the growth of the body.

Vitamin B is water soluble and can be lost in cooking water.

Some foods in which vitamin B is found

Thiamine (B_1)	*Riboflavin (B_2)*	*Nicotinic acid*
Yeast	Yeast	Meat extract
Bacon	Liver	Brewers' Yeast
Oatmeal	Meat extract	Liver
Peas	Cheese	Kidney
Wholemeal bread	Egg	Beef
Bacon		

Vitamin C (ascorbic acid)

Function

1 Vitamin C is necessary for the growth of children.
2 Assists in the healing of cuts and the uniting of bones.
3 Prevents gum and mouth infection.

Vitamin C is water soluble and can be lost during cooking or soaking in water. It is also lost by bad storage (keeping foods for too long, bruising, or storing in a badly ventilated place) and by cutting vegetables into small pieces.

Some foods in which vitamin C is found

Blackcurrants	Potatoes	Brussels sprouts and other greens
Strawberries	Lemons	Oranges
Grapefruit	Tomatoes	

The major sources in the British diet are potatoes and green vegetables.

Mineral elements

There are nineteen mineral elements, most of which are required by the body in very small quantities. The body has at certain times a greater demand for certain mineral elements and there is a danger then of a deficiency in the diet. Calcium, iron and iodine are those most likely to be deficient.

Calcium – is required for:

1 Building bone and teeth.
2 Clotting of the blood.
3 The working of the muscles.

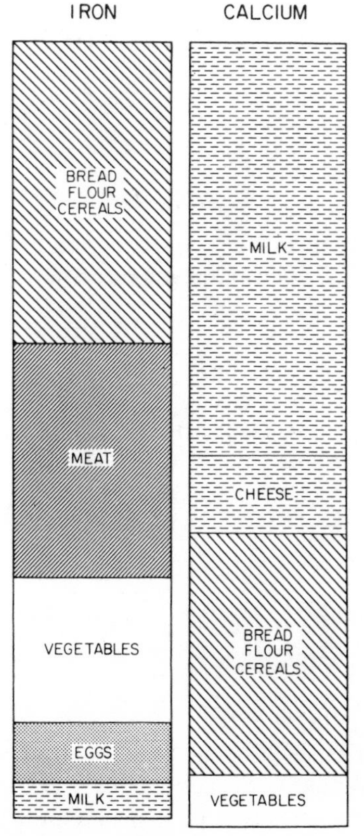

Main supply of iron and calcium in the average diet

The use the body makes of calcium is dependent on the presence of vitamin D.

Sources of calcium

Milk and milk products.

The bones of tinned oily fish.

Wholemeal bread and white bread (to which calcium is added). Note that the practice of adding calcium, iron, thiamin or nicotinic acid to flour may be discontinued as a DHSS report number 23 states there are no longer valid reasons for adding them.

Vegetables (greens).

It may also be present in drinking water.

Although calcium is present in certain foods (spinach, cereals) the body is unable to make use of it as it is not in a soluble form and therefore cannot be absorbed.

Owing to extra growth of bones and teeth, infants, adolescents, expectant and nursing mothers have a greater demand for calcium.

Phosphorus – is required for:

1 Building the bones and teeth (in conjunction with calcium and vitamin D).
2 The control of the structure of the brain cells.

Sources of phosphorus

Liver, kidney
Eggs, cheese
Bread, fish

Iron – is required for building the haemoglobin in blood and is therefore necessary for transporting oxygen and carbon dioxide round the body.

Sources of iron

Lean meat, offal
Egg yolk
Wholemeal flour
Green vegetables
Fish

Iron may also be present in drinking water and obtained from iron utensils in which food is prepared.

As the haemoglobin in the blood should be maintained at a constant level, the body requires more iron at certain times than others, after loss of blood, etc.

Sodium – is required in all body fluids, and is found a salt (sodium chloride). Excess salt is continually lost from the body in urine. The kidneys control this loss. We also lose sodium in sweating, over which loss we have no control.

Sources of sodium
Many foods are cooked with salt or have salt added (bacon and cheese) or contain salt (meat, eggs, fish).

Iodine – is required for the functioning of the thyroid gland which regulates basal metabolism (see page 143).

Sources of iodine
Sea foods
Vegetables grown near the sea
Drinking water obtained near the sea

Potassium, magnesium, sulphur and copper are some of the other minerals required by the body.

Water

Water is required for:

1 All body fluids
2 Digestion.
3 Absorption.
4 Metabolism.
5 Excretion.
6 Secretion.
7 Playing a part in regulation of body temperatures by evaporation of perspiration.

Sources of water

a) Drinks – beverages
b) Foods – lettuce; cabbage; apples; potatoes; eggs; beef; bread; cheese; margarine.
c) Combustion or oxidation – when fats, carbohydrates and protein are used for energy a certain amount of water (metabolic water) is produced within the body.

The cooking of nutrients

Protein
When protein is heated it coagulates and shrinks. Too much cooking can spoil the appearance of the food, eg scrambled eggs, as well as causing destruction of certain vitamins. See chapter 8 page 131 for cooking effects on protein.

Carbohydrate
Unless starch is thoroughly cooked it cannot be digested properly. (For example, insufficiently cooked pastry or bread.) When cooked, the starch granules swell, burst and then the starch can be digested. (This is called gelatinisation of starch). See also page 135.

When sugar is heated it melts and with further heating loses water, gradually turning brown, dark brown and then black. This is known as caramelisation of sugar.

Fat

The nutritive value of fat is not affected by cooking. During cooking processes a certain amount of fat may be lost from food when the fat melts as with the grilling of meat, for example.

Mineral elements

There is a possibility of some minerals being lost in the cooking liquor, so diminishing the amount available in the food. This applies to soluble minerals such as salt, but not to calcium or iron compounds which do not dissolve in the cooking liquor.

Iron

Iron may be acquired from foods cooked in iron utensils. The iron in foods is not affected by the cooking.

Calcium

Cooking foods in hard water may very slightly increase the amount of calcium in food.

Vitamins

Vitamins A and D withstand cooking temperatures, and they are not lost in the cooking.

Vitamin B_1 (thiamine) can be destroyed by high temperatures and by the use of bicarbonate of soda. It is soluble in water and can be lost in the cooking. Vitamin B_2 (riboflavin) is not destroyed easily by heat but bright sunlight can break it down.

Vitamin C is lost by cooking and by keeping food warm in a hot place. It is also soluble in water (the soaking of foods for a long time and bruising are the causes of losing vitamin C). It is unstable and therefore easily destroyed in alkaline conditions (bicarbonate of soda must not be used when cooking green vegetables.).

Food requirements

Energy is required to enable the heart to beat, for the blood to circulate, the lungs and other organs of the body to function, for every activity such as talking, eating, standing, sitting and for strenuous exercise and muscular activity.

Young and active people require a different amount of food from elderly, inactive people because they expend more energy, and this energy is obtained from food during chemical changes taking place in the body.

The energy value of a food is measured by a term called a kilocalorie or Calorie (this term should be written with a capital C although popularly is often written with a small c). This is the amount of heat required to raise the temperature of 1000 grammes of water from 15° to 16°.

A new unit is gradually replacing the Calorie. This is the joule. Since the joule is too small for practical nutrition, the kilojoule (kJ) is used.

1 Calorie = 4.18 kJ

(Both units will be given here and for ease of conversion 1 Calorie will be taken to equal 4.0 kJ.

Food contain certain amounts of the various nutrients which are measured in grams.

The energy value of nutrients is:

1 gram carbohydrate produces 4 Calories	(16 kJ);
1 gram protein produces 4 Calories	(16 kJ);
1 gram fat produces 9 Calories	(36 kJ).

The energy value of a food, diet or menu is calculated from the nutrients it contains; for example, 28 grams of food containing:

10 grams carbohydrate	will produce	10 × 4	=	40 Calories (160 kJ)
2 grams protein	will produce	2 × 4	=	8 Calories (32 kJ)
5 grams fat	will produce	5 × 9	=	45 Calories (180 kJ)
		Total		93 Calories (372 kJ)

Foods having a high fat content will have a high energy value; those containing a lot of water, a low energy value. All fats, cheese, bacon and sugar have a high energy value.

Men require more Calories (kJ) than women, big men and women require more than small men and women, and people engaged on energetic work require more Calories (kJ) than those with sedentary occupations.

Basal metabolism

Basal metabolism is the term given to the amount of energy required to maintain the functions of the body and to keep the body warm when it is still and without food. The number of Calories (kJ) required for basal metabolism is affected by the size, sex and general condition of the body. The number of Calories (kJ) required for basic metabolism is approximately 1700 per day.

In addition to the energy required for basal metabolism, energy is also required for everyday activities, such as getting up, dressing, walking, etc, and the amount required will be closely related to a person's occupation.

The approximate energy requirements per day for the following examples are:

Clerk	2000 Calories (8 000 kJ)
Carpenter	3000 Calories (12 000 kJ)
Blacksmith	4000 Calories (16 000 kJ)

The table below indicates the recommended daily allowance of Calories to provide a healthy diet for the categories of people shown.

Age and sex	*Calories*	*Kilojoules*
Boys and girls:		
0 – 1 year	1000	4000
2 – 6 years	1500	6000
7 – 10 years	2000	8000
Boys:		
11 – 14 years	2750	11 000
15 – 19 years	3500	14 000
Men:		
20 + years	3000	12 000
(for average activity)		
Girls:		
11 – 14 years	2750	11 000
15 – 19 years	2500	10 000
Women:		
20 + years	2500	10 000
(for average activity)		

Value of foods in the diet

Milk

Cows' milk is almost the perfect food for human beings; it contains protein, carbohydrate, fat, minerals, vitamins and water.

When milk is taken into the body it coagulates in the same way as in the making of junket. This occurs in the stomach when digestive juices (containing the enzyme rennin) are added. Souring of milk is due to the bacteria feeding on the milk sugar (lactose) and producing lactic acid from it, which brings about curdling.

Composition of milk

Approximately	87%	water
	3 – 4%	protein (casein)
	3 – 4%	fat
	4 – 5%	sugar
	0.7%	minerals (particularly calcium)
	Vitamins A, B and D	

(There are legal minimum requirements for the various grades of milk. In the Channel Islands the percentage of fat is 4%, all others is 3%.).

Milk, therefore, is a body-building food because of its protein, an energy food because of the fat and sugar, and a protective food as it contains vitamins and minerals. Owing to the large amount of water, while it is a suitable food for babies, it is too bulky to be the main source of protein and other nutrients after the first few months of life. It is also deficient in iron and vitamin C.

However, it should be included in everyone's diet as a drink and it may be used in a variety of ways.

Cream

Cream is the fat of milk and the minimum fat content of single cream is approximately 18%, for double cream 48% and clotted cream 60%.

Cream is therefore an energy-producing food which also supplies vitamins A and D. It is easily digested because of the small globules of fat.

Butter

Butter is made from the fat of milk and contains vitamins A and D, the amount depending on the season. Like cream it is easily digested.

Composition of butter

Approximately	85% fat
	13% water
	1% salt
	Vitamins A and D

Butter is also an energy-producing food and a protective food in so far as it provides vitamins A and D.

Margarine

Margarine, which is made from vegetable and animal oils, and skimmed milk has vitamins A and D added to it. The composition and food value of margarine are similar to butter.

Cheese

Cheese is made from milk, the composition varying according to whether the cheese has been made from whole milk, skimmed milk or milk to which extra cream has been added.

The food value of cheese is exceptional because of the concentration of the various nutrients it contains. The minerals in cheese are useful, particularly the calcium and phosphorus. Cheese is also a source of vitamins A and D.

It is a body-building, energy-producing and protective food because of its protein, fat and mineral elements and vitamin content.

Cheese is easily digested, provided it is eaten with starchy foods and eaten in small pieces as when grated. The composition of Cheddar cheese is: 25% protein; 35% fat; 30% water; also calcium and vitamins A and D.

Meat, poultry and game

Meat consists of fibres which may be short, as in a fillet of beef, or long, as in the silverside of beef. The shorter the fibre the more tender and easily digested the meat. Meat is carved across the grain to assist mastication and digestion of the fibres.

Hanging of the meat helps to make the flesh of meat more tender; this is because acids develop and soften the muscle fibres. Marinading in wine or vinegar prior to cooking also helps to tenderise meat so that it is more digestible. Expensive cuts of meat are not necessarily more nourishing than the cheaper cuts.

Meat contains proteins, variable amounts of fat, water, also iron and thiamine. It is therefore an important body-building food. (Bacon is particularly valuable because of its thiamine.)

Liver and kidney are also protein foods. They contain less fat than most meats, but are a good source of iron and vitamin A.

Sweetbreads are protein foods, particularly valuable to invalids because they are easily digested.

Tripe in addition to its protein is a good source of calcium as it is treated with lime during its preparation. It is also easily digested.

Fish

Fish is as useful a source of animal protein as meat.

The amount of fat in different fish varies: oily fish contain 5 – 18%, white fish less than 2%.

When the bones are eaten calcium is obtained from fish (tinned sardines or salmon).

Oily fish is not so easily digested as white fish because of the fat; shellfish is not easily digested because of the coarseness of the fibres.

Fish is important for body building, and certain types (oily fish) are energy producing and protective because of the fat and vitamins A and D.

Eggs

Egg white contains protein known as egg albumin and the amount of white is approximately twice the amount of the yolk.

The yolk is more complex; it contains more protein than the white, also fat, vitamins A and D, thiamine, riboflavin, calcium, iron, sulphur and phosphorus. Lecithin (an emulsifying agent) and Cholesterol are also present.

Composition of eggs (approximate)

	Whole egg	*White*	*Yolk*
Water	73%	87%	47%
Protein	12%	10%	15%
Fat	11%		33%
Minerals	1%	.5%	2%
Vitamins			

Because of the protein, vitamins, mineral elements and fat, eggs are body building, protective and energy-producing food.

Fruit

The composition of different fruit varies considerably; for example, avocado pears contain a large amount of fat, whereas most other fruits contain none. In unripe fruit the carbohydrate is in the form of starch which changes to sugar as the fruit ripens.

The cellulose in fruits acts as roughage.

Fruit is valuable because of the vitamins and minerals it contains. Vitamin C is present in certain fruits, particularly citrus varieties (oranges, grapefruit) and blackcurrants and other summer fruits. Dried fruits such as raisins and sultanas are a useful source of energy because of their sugar content, but they contain no vitamin C.

Composition of fruit

Approximately:	Water	85%
	Carbohydrate	5 - 10%
	Cellulose	2 - 5%
	Minerals	.5%
	Vitamin C	

Very small amounts of fat and protein are found in most fruits.

Fruit is a protective food because of its minerals and vitamins.

Nuts

Nuts are highly nutritious because of their protein, fat and minerals. Vegetarians rely on nuts to provide the protein in their diet.

Nuts are not easily digested because of their fat content and cellulose.

Vegetables

Green vegetables

Green vegetables are particularly valuable because of their vitamins and minerals; they are therefore protective foods. The most important minerals they contain are iron and calcium. Green vegetables are rich in carotene, which is made into vitamin A in the body.

The greener the vegetable the greater its nutritional value. Vegetables which are stored for long periods or are damaged or bruised quickly lose their vitamin C value, therefore they should be used as quickly as possible.

Green vegetables also act as roughage in the intestines.

Root vegetables

Compared with green vegetables most root vegetables contain starch and sugar; they are therefore a source of energy. Swedes and turnips contain a little vitamin C and carrots and other yellow-coloured vegetables contain carotene, which is changed into vitamin A in the body.

Potatoes

Potatoes contain a large amount of starch (approximately 20%) and a small amount of protein just under the skin. Because of the large quantities eaten, the small amount of vitamin C they contain is of value in the diet.

Onions

The onion is used extensively and contains some sugar, but its main value is to provide flavour.

Peas and broad beans

These vegetables contain carbohydrate, protein and carotene.

Cereals

Cereals contain from 60% to 80% carbohydrate in the form of starch and are therefore energy foods. They also contain 7 – 13% protein, depending on the type of cereal, and 1 – 8% fat.

The vitamin B content is considerable in stone ground and whole meal four, and B vitamins are added to other wheat flours. Fortification of flour may be discontinued (see page 150).

Oats contain good quantities of fat and protein, but are not used in large quantities nowadays.

Sugar

There are several kinds of sugar, such as those found in fruit (glucose), milk (lactose), cane and beet sugar (sucrose).

Sugar, with fat, provides the most important part of the body's energy requirements.

Saccharine, although sweet, is chemically produced and has no food value.

Liquids

Water

Certain waters contain mineral salts; for example, hard waters contain soluble salts of calcium. Some spas are known for the mineral salts contained in the local water. Fluoride may be present naturally in some waters, and makes children's teeth more resistant to decay.

Artificial mineral waters

These consist of sweetened water flavoured with acid.

Tea and coffee

These drinks have no food value, but they do act on the nervous system as a stimulant.

Cocoa

Cocoa contains some fat, starch and protein, also some vitamin B and mineral elements.

Because tea, coffee and cocoa are usually served with milk and sugar they do have some food value.

Balanced diet

A balanced diet provides adequate amounts of the various nutrients for energy, growth, repair and regulation of body processes.

As a guide to a nutritionally satisfactory diet these points should be observed:

Milk – $\frac{1}{2}$ litre (1 pint) daily (more for children).
Meat, fish, poultry – once a day.
Eggs, cheese, pulses – daily.
Fruit – once a day (preferably citrus fruit).
Vegetables – daily, as well as potatoes.
Fats (butter or margarine) – daily.
Cereals (wholemeal bread or oatmeal) – daily.
Sugar – daily.
Water – 1 $\frac{1}{2}$ litre liquid daily, some in the form of water.

In the compiling of a balanced diet the *protective foods* are considered first:

1 *Dairy produce* (including eggs) to provide calcium, vitamins and iron (from eggs), among other things.
2 *Fresh fruit and vegetables* for calcium, iron, vitamin C and some vitamin A.
3 *Whole grain cereals* for vitamin B.
4 *Oily fish and liver* for vitamins A and D and iodine.

Next, the *body-building foods* are considered:

Meat (including bacon, fish, poultry, game and pulses).

Finally, the *energy-producing foods* are added:

Fats, sugar and starchy foods.

Provided the protective and bodybuilding foods are well represented in the diet the appetite can normally be left to determine the energy-producing food requirements.

Food additives

These can be divided into 12 categories, and except for purely 'natural' substances, their use is subject to certain legislation.

1 Preservatives – for example natural ones include salt, sugar, alcohol and vinegar; synthetic ones are also widely used.
2 Colouring agents – natural, including cochineal, caramel and

saffron, and many synthetic ones.

3 Flavouring agents – synthetic chemicals to mimic natural flavours, for example monosodium glutamate to give a meaty flavour to foods.
4 Sweetening – for example saccharin and sorbitol.
5 Emulsifying agents (to stop separation of salad creams, ice cream etc); examples are lecithin and glyceryl mono stearate (GMS).
6 Antioxidants – to delay the onset of rancidity in fats due to exposure to air, for example Vitamin E and BHT.
7 Flour improvers – to strengthen the gluten in flour, for example chlorine dioxide and Vitamin C.
8 Thickeners.

a) animal – gelatine,
b) marine – agar-agar,
c) vegetable – gum tragacanth (used for pastillage), pectin,
d) synthetic products.

9 Humectants – to prevent food drying out, for example glycerine (used in some icings).
10 Polyphosphate – injected into poultry before rigor mortis develops. It binds water to the muscle and thus prevents 'drip', giving a firmer structure to the meat.
11 Nutrients – for example vitamins and minerals added to breakfast cereals, Vitamins A and D added to margarine.
12 Miscellaneous – anti-caking agents added to icing sugar and salt; firming agents (for example calcium choride) added to tinned fruit and vegetables to prevent too much softening in the processing; mineral oils added to dried fruit to prevent stickiness.

Further information:

Mottram, *Human Nutrition* (Arnold),
Manual of Nutrition (HM Stationery Office),
Wells, *Focus on Food* (Forbes),
Matthrews & Wells, *2nd Book of Food and Nutrition* (Forbes),
Hildreth, *Elementary Science of Food* (Allman),
Kilgour, *Science for Catering Students* (Heinemann),
Gaman and Sherrington, *Science of Food* (Pergamon).

Foods containing the various nutrients and their use in the body

Name	Food in which it is found	Use in body
Protein	Meat, fish, poultry, game, milk, cheese, eggs, pulses, cereals	For-building and repairing body tissues. Some heat and energy
Fat	Butter, margarine, cooking-fat, oils, cheese, fat meat, oily fish	Provides heat and energy
Carbohydrate	Flour, flour products and cereals, sugar, syrup, jam, honey, fruit, vegetables	Provides heat and energy
Vitamin A	Oily fish, fish-liver oil, dairy foods, carrots, tomatoes, greens	Helps growth. Resistance to disease
Vitamin B_1 – thiamine	Yeast, pulses, liver, whole grain cereals, meat and yeast extracts	Helps growth. Strengthens nervous system
Vitamin B_2 – riboflavin	Yeast, liver, meat, meat extracts, whole grain cereals	Helps growth, and helps in the production of energy
– nicotinic acid (Niacin)	Yeast, meat, liver, meat extracts, whole grain cereals	Helps growth
Vitamin C – ascorbic acid	Fruits such as strawberries, citrus fruits, green vegetables, root vegetables, salad vegetables, potatoes	Helps growth, promotes health
Vitamin D – sunshine vitamin	Fish-liver oils, oily fish, dairy foods	Helps growth. Builds bones and teeth
Iron	Lean meat, offal, egg yolk wholemeal flour, green vegetables, fish	Building up the blood
Calcium (lime)	Milk and milk products, bones of fish, wholemeal bread	Building bones and teeth, clotting the blood, the working of the muscles
Phosphorus	Liver and kidney, eggs, cheese bread	Building bones and teeth, regulating body processes
Sodium (salt)	Meat, eggs, fish, bacon, cheese	Prevention of muscular cramp

7
Commodities

Having read the chapter, these learning objectives, both general and specific, should be achieved.

General objectives Know how to purchase the various commodities and understand their storage and use. Be aware of the variety available and appreciate their versatility in the practical situation. This applies to all fresh and processed items.

Specific objectives Where appropriate identify the commodity and recognise the points of quality. State sources of origin.

List suitable uses in cooking. Specify when in season and explain how the items should be stored. State if and how they may be preserved. Where suitable translate into French and give examples of how expressed on the menu. Specify purchasing unit and costs.

1 Meat
2 Offal
3 Poultry
4 Game
5 Fish
6 Vegetables
7 Fruits
8 Nuts
9 Eggs
10 Milk
11 Fats and oils
12 Cheese
13 Cereals
14 Raising agents
15 Sugar
16 Cocoa and chocolate
17 Coffee
18 Tea
19 Pulses
20 Herbs
21 Spices
22 Condiments
23 Colourings, flavourings, essences
24 Grocery, delicatessen
25 Confectionery goods

When studying commodities students are recommended to explore the markets not only for detailed knowledge of fresh foods but also for any and every possible substitute in the form of convenience or ready prepared foods. Comparison should be made of the merits of various brands of foods as should comparison of convenience foods with fresh unprepared foods. Factors to be considered in comparison should include quality, price, hygiene, labour, cost, time, space required and disposal of waste.

Students are advised to be cost conscious at the outset in all their studies and to form the habit of keeping up to date with current prices of all commodities, equipment, labour and overheads. A trained inbuilt awareness of costs is an important asset to any successful caterer.

The student of catering should begin to form opinions as to when and in what circumstances fresh, convenience or possibly a combination of both foods should be used. Convenience or ready foods are not new and have been in use for many years; however, there are many more products on the market today and the wise caterer is the one who makes a thorough study of all types available and, if and when they are suitable, incorporates them into his organisation.

It is possible that the most important factor when considering the use of convenience (ready) foods is the same as for traditional foods. Who is the food intended for? What price are they able to pay? Having considered these points it may then be decided whether the customer will accept, reject or possibly even prefer convenience foods.

Owing to price fluctuations, and lack of information on metric purchasing units, we have left vacant spaces for commodity prices and sizes. It is hoped that the completion of these spaces will be a useful exercise for students.

1 Meat *La viande*

Meat is probably the most important food that we use, accounting as it does for a major share of our total expenditure on food.

Cattle, sheep and pigs are reared for fresh meat and certain pigs are specifically produced for bacon. The animals are humanely killed and prepared in hygienic conditions, the skins or hides are removed, the innards are taken out of the carcass and the offal is put aside.

			Wholesale cost		Season
	French	Purchasing unit	Home-killed	Imported	home-killed
lamb	l'agneau	carcass or joint			April-May
mutton	le mouton	carcass or joint			all year round
veal	le veau	carcass, side or joint			all year round
beef	le bœuf	side, quarter or joint			all year round
pork	le porc	carcass, side or joint			September-April

The carcasses of beef are split into two sides and those of lambs, sheep, pigs and calves are left whole; they are then chilled in a cold room before being sent to market.

To cook meat properly it is necessary to know and understand the structure of meat. Lean flesh is composed of muscles, which are numerous bundles of fibres held together by connective tissue. The size of these fibres is extremely small, especially in tender cuts or cuts from young animals, and only the coarsest fibres may be distinguished by the naked eye. The size of the fibres varies in length, depth and thickness and this variation will affect the grain and the texture of the meat.

Wholesale purchasing of meat

The quantity of connective tissue binding the fibres together will have much to do with the tenderness and eating quality. There are two kinds of connective tissue, the yellow (*elastin*) and the white (*collagen*). The thick yellow strip that runs along the neck and back of animals is an example of elastin. Elastin is found in the muscles, especially in older animals or those muscles receiving considerable exercise. Elastin will not cook, but it must be broken up mechanically by pounding or mincing. The white connective tissue collagen can be cooked, as it decomposes in moist heat to form gelatine. The amount of connective tissue in meat is determined by the age, breed, care and feed given to the animal.

The quantity of fat and its condition are important factors in determining eating quality. Fat is found on the exterior and interior of the carcass and in the flesh itself. Fat deposited between muscles or between the bundles of fibres is called marbling. If marbling is present, the meat is likely to be tender, of better flavour and moist. Much of the flavour of meat is given by fats found in lean or fatty tissues of the meat. Animals absorb flavour from the food they are given, therefore the type of feed is important in the end eating quality of the meat.

Extractives in meats are also responsible for flavour. Muscles that receive a good deal of exercise have a higher proportion of flavour extractives than those receiving less exercise. Shin, shank, neck and other parts receiving exercise will give richer stock and gravies and the meat with more flavour than the tender cuts.

Tenderness, flavour and moistness are increased if beef is hung after slaughter (see page 156), or pre-tenderised (see page 156). Pork and veal are hung for 3–7 days according to the temperature. Meat is generally hung at a temperature of 1 °C.

Textured vegetable proteins (TVP)

With meat and fish prices continually increasing many caterers are finding it difficult to keep within cost budgets, particularly the welfare caterer who has to provide high minimum levels of nutrition. By partially replacing the meat used in traditional dishes with TVP, it is possible to reduce costs, provide nutrition and minimise cooking losses, yet still serve a meal that is acceptable in appearance.

TVP can be made of protein derived from wheat, oats, cotton-seed, soya bean and other vegetable sources. The major source for TVP, however, is the soya bean, due to its high protein content, yield per acre of land and world-wide availability.

There are two methods for texturing soya proteins, extrusion and spinning, and TVP is available in several forms such as: fine mince, granules, flakes, nuggets, small chunks, large chunks and strips. TVP can also be obtained in light brown, brown and pink colours, and flavours varying from bland (enabling the product to absorb flavour from accompanying stock and cooking juices) to beef, ham and pork.

As TVP is still comparatively speaking a relatively new product, research and development are continuously being carried out by a number of food manufacturers. The chief uses for TVP at present are as a meat extender varying from approximately 10–60% replacement of fresh meat, and TVP is being used in such dishes as casseroles, stews, pies, pasties, sausage rolls, pâté, hamburgers, meat loaf and curries.

A recommendation by the Committee on Medical Aspects of Food Policy states: ‘Any substance promoted as a replacement or an alternative to a natural food should be the nutritional equivalent in all but the unimportant aspects of the natural food it would simulate.’

This means that the manufacturer of novel protein foods such as TVP are required to add nutrients to their products in sufficient quantity and variety so that no nutritional value is lost by consumers of these products.

The manufacturers of these products are able to supply full details of the composition, nutritive values and costed recipes. Students are advised to ensure that they obtain latest up-to-date information before studying or preparing to use textured vegetable protein.

Butchers' meat

Main sources of supply
Lamb and mutton: England, Scotland, New Zealand, and Australia.
Beef: England and Scotland.
Veal: England, Scotland and Holland.
Bacon: England and Denmark.

Storage
Fresh meat

1 Fresh meat must be hung to allow it to become tender.
2 The time for hanging depends on the temperature of the cold store. The lower the temperature, the longer it can be hung.
3 The time for hanging at 1°C would be up to 14 days.
4 Meat should be suspended on hooks.

Bacon

1 Bacon is kept in a well-ventilated cold room.
2 Joints of bacon should be wrapped in muslin and hung, preferably in a cold room.
3 Sides of bacon are also hung on hooks.
4 Cut bacon is kept on trays in the refrigerator or cold room.

Beef *Le boeuf*
Approximately 80% of beef used in Britain is home produced.

The hanging or maturing of beef at a chill temperature of 1°C for up to 14 days has the effect of increasing tenderness and flavour. This hanging process is essential as animals are generally slaughtered around the age of 18–21 months, which is a sufficient period of time for the beef to be tough. Also a short time after death an animal's muscles stiffen, a condition known as *rigor mortis*. After a time chemical action caused by enzymes and increasing acidity relax the muscles and the meat becomes soft and pliable. As meat continues to hang in storage *rigor mortis* is lost and tenderness, flavour and moistness increase. (Pork, lamb and veal are obtained from young animals so that toughness is not a significant factor.)

Tenderness can also be achieved by injecting an enzyme into the live animal before slaughter. When the cooking temperature of the meat reaches a certain degree, the enzyme is activated and when the temperature has reached a high level the enzyme is destroyed. This pre-tenderising process has two advantages: it can reduce the amount of hanging time thus saving both storage space and time, and it can double the amount of meat from the animal that can be grilled or roasted.

Large quantities of beef are prepared as chilled boneless prime cuts, vacuum packed in film. This process has the following advantages. It extends the storage life of the cuts, and allows the tenderising process to continue within the vacuum pack because the action of the enzyme does not require air. The cuts are boned and fully trimmed thus reducing labour cost and storage space.

It is essential to store and handle vacuum packed meat correctly. Storage temperature should be 0°C with the cartons the correct way up so that the drips cannot stain the fatty surface. A good circulation of air should be allowed between cartons.

When required for use the vacuum film should be punctured in order to drain any blood present before the film is removed. On opening the film a slight odour usually occurs, but this should quickly disappear on exposure to the air. The vacuum packed beef has a deep red colour, but when the film is broken the colour should change to its normal characteristic red within 20–30 minutes. Once the film is punctured the meat should be used as soon as possible.

Quality

1 Lean meat should be bright red, with small flecks of white fat (marbled).
2 The fat should be firm, brittle in texture, creamy white in colour and odourless.
3 Home-killed beef is best.

Veal *Le veau*

Originally most top quality veal came from Holland, but as the Dutch methods of production are now used extensively in Britain, supplies of home-produced veal are available all the year round. Good quality carcasses weighing around 100 kg can be produced from calves slaughtered at 12 to 24 weeks old. This is the quality of veal necessary for first-class cookery. Calves which are not considered by the producer to be suitable for the production of quality veal or beef are however, slaughtered within 10 days after birth and are known as 'Bobby' calves. The meat obtained is suitable for stewing, pie, casseroles, etc.

1 The flesh of veal should be pale pink, firm, not soft or flabby.
2 Cut surfaces must not be dry, but moist.
3 Bones in young animals should be pinkish white, porous and with a small amount of blood in their structure.
4 The fat should be firm and pinkish white.
5 The kidney ought to be firm and well covered with fat.

Lamb and mutton *L'agneau et Le mouton*

In Britain, five times as much lamb and mutton is eaten than in any other European country. Approximately 40% of the lamb and mutton consumed is home produced and the balance comes from Australia and New Zealand. As the seasons in Australia and New Zealand are opposite to those in Britain these supplies can be integrated with our own. Most lamb carcasses imported are from animals aged between 4–6 months.

1 Lamb is under one year old - after one year it is termed mutton.
2 The carcass should be compact and evenly fleshed.
3 The lean flesh of lamb and mutton ought to be firm and of a pleasing dull red colour and of a fine texture or grain.
4 The fat should be evenly distributed, hard, brittle, flaky and clear white in colour.
5 The bones should be porous in young animals.

Pork *Le porc*

Approximately 95% of pork used in Britain is home produced. The keeping quality of pork is less than that of any other meat, therefore the greatest care is needed in the handling, preparation and cooking of pork. Pork must always be *well* cooked. This is necessary because TRICHINAE (parasitic worms) may be present and must be destroyed by heat. If they are present in the meat and are not destroyed in cooking they will find their way into the voluntary muscles of those who eat pork and they will continue to live in the human body.

1 Lean flesh of pork should be pale pink.
2 The fat should be white, firm, smooth and not excessive.
3 Bones must be small, fine and pinkish.
4 The skin, or rind, ought to be smooth.

Sucking pigs weigh 10 to 20 lbs dressed and are usually roasted whole.

Bacon *Le lard*

Bacon is the cured flesh of a baconer pig.

A baconer pig is the type that is specifically reared for bacon because its shape and size yield economic bacon joints.

The curing process consists of salting either by a dry method and smoking or by soaking in brine followed by smoking.

Green bacon is brine cured but not smoked, it has a milder flavour and does not keep as long as smoked bacon.

1 There should be no sign of stickiness.
2 There must be no unpleasant smell.
3 The rind should be thin, smooth and free from wrinkles.
4 The fat ought to be white, smooth and not excessive in proportion to the lean.
5 The lean meat of bacon should be deep pink in colour and firm.

Ham *Le jambon*

Ham is the hind leg of a pig cut round from the side of pork with the aitch bone; it is preserved by curing or pickling in brine and then dried and smoked. York, Bradenham (Wiltshire) and Suffolk are three of the most popular English hams.

Bradenham is easily distinguished by its black skin. The Bradenham and Suffolk hams are sweet and mild cured. Imported hams include the

Parma, Bayonne and Westphalian, all of which are carved paper thin and eaten raw as hors d'œuvre.

Food value
Meat, having a high protein content, is valuable for the growth and repair of the body and as a source of energy.

Preservation
Salting: Meat can be pickled in brine; this method of preserving meat may be applied to silverside, brisket and ox-tongues. Salting is also used in the production of bacon, before the sides of pork are smoked. This also applies to hams.

Chilling: This means that meat is kept at a temperature just above freezing-point in a controlled atmosphere. Chilled meat cannot be kept in the usual type of cold room for more than a few days, and this is sufficient time for the meat to hang, enabling it to become tender.

Freezing: Small carcasses, such as lamb and mutton, can be frozen and the quality is not affected by freezing. They can be kept frozen until required and then thawed out before being used.

Some beef is frozen, but it is inferior in quality to chilled beef.

Canning: Large quantities of meat are canned and corned beef is of importance since it has a very high protein content. Pork is used for tinned luncheon meat.

See following pages for diagrams of cuts of meat, French and English terms, and menu examples.

Further information: Institute of Meat, 91/93 Charterhouse Street, London ECIM 6HR.

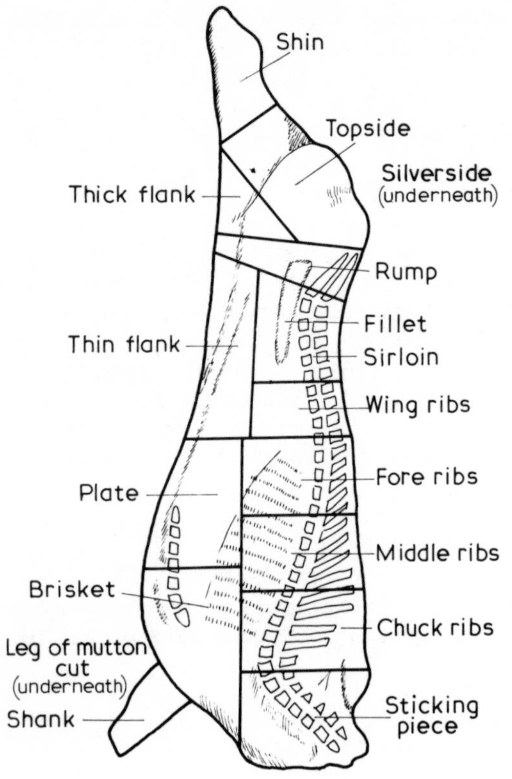

Side of beef Forequarter of beef Hindquarter of beef

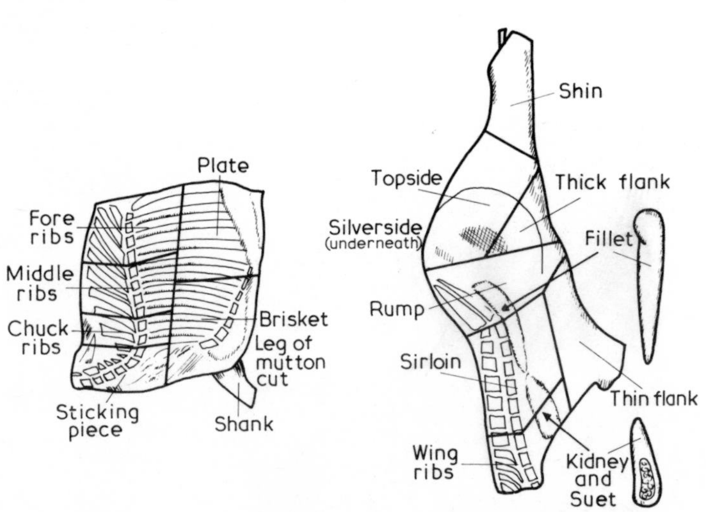

Beef — *Le bœuf*

Joint	French	Use	Menu example
Hindquarter:			
shin	la jambe	consommé, beef tea, stewing	Consommé royale
topside	la tranche tendre	braising, stewing, second-class roasting	Bœuf braisé aux nouilles
silverside	la gite à la noix	pickled and boiled	Boiled silverside, carrots and dumplings
thick flank	la tranche grasse	braising and stewing	Ragoût de bœuf aux légumes
rump	la culotte de bœuf	grilling and frying as steaks	Grilled rump steak
sirloin	l'aloyau de bœuf	roasting, grilling and frying as steaks	Contrefilet de bœuf rôti: entrecôte grillé
wing ribs	la côte de bœuf	roasting, grilling and frying as steaks	Côte de bœuf rôti à l'anglaise
thin flank	la bavette	stewing, boiling, sausages	Bœuf bouilli à la française
fillet	le filet de bœuf	roasting, grilling, frying	Tournedos chasseur, filet de bœuf bouquetière
Forequarter:			
fore-ribs middle ribs	les côtes	roasting, braising	Côte de bœuf roti
chuck ribs	les côtes du collier	stewing, braising	Beef steak pie
sticking piece	la collier	stewing, sausages	Sausage toad in the hole
plate and brisket	la poitrine	pickled and boiled	Pressed beef
leg of mutton cut	la tallon du collier	braising and stewing	Hachis de bœuf duchesse
shank	la jambe	consommé, beef tea	Consommé aux profiterolles

Veal **Le veau**

Joint	French	Use	Menu example
leg	le cuissot de veau	See below*	Cuissot de veau rôti
loin	la longe de veau	roasting, frying, grilling	Côte de veau napolitaine
best end	la carré de veau	roasting, frying, grilling	Côte de veau milanaise
shoulder	l'épaule de veau	braising, stewing	Goulash de veau à l'hongroise
neck end	le cou de veau	stewing	Fricassée de veau à l'ancienne
scrag	le cou de veau	stock, stewing	Stock
breast	la poitrine de veau	stewing, roasting	Blanquette de veau aux nouilles
Leg of Veal:			
knuckle	le jarret de veau	stewing	Osso buco
cushion	la noix de veau	escalopes, roasting, sauté, braising	Noix de veau braisé Belle Hélène
under cushion	la sous noix	escalopes, roasting, sauté, braising	Escalope de veau viennoise
thick flank	le quasi	escalopes, roasting, sauté, braising	Sauté de veau Marengo

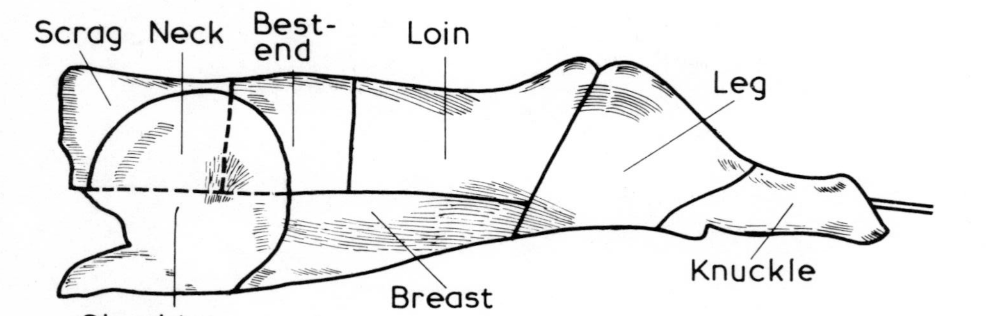

Side of veal

Lamb and mutton — L'agneau et mouton

Joint	French	Use	Menu example
shoulder	l'épaule d'agneau ou de mouton	roasting, stewing	Epaule d'agneau boulangère
leg	le gigot d'agneau	roasting. (mutton – boiled)	Gigot d'agneau rôti; sauce menthe; Gigot de mouton bouilli; sauce aux câpres
breast	la poitrine	stewing, roasting	Irish stew
middle neck	le cou	stewing	Navarin d'agneau printanier
scrag end	le cou	broth	Mutton broth
best end	le carré d'agneau	roasting, grilling, frying	Carré d'agneau persillé
saddle	la selle d'agneau	roasting, grilling, frying	Selle d'agneau niçoise
loin	la longe d'agneau	roasting, grilling, frying	Longe d'agneau farci
chop	la côte d'agneau	grilling, frying	Grilled loin chop
cutlet	la côtelette d'agneau	grilling, frying	Côtelette d'agneau Réforme
fillet	le filet mignon	grilling, frying	Filet mignon fleuriste

1 shoulders (2)
2 legs (2)
3 breasts (2)
4 middle neck
5 scrag end
6 best end
7 saddle

Side of lamb

Bacon *Le lard*

Joint	Use	Menu example
collar	boiling, grilling	boiled bacon, pease pudding and parsley sauce
hock	boiling, grilling	
back	grilling, frying	Œuf au lard
streaky	grilling, frying	Canapé Diane
gammon	boiling, grilling, frying	Jambon braisé au madère Grilled gammon and pineapple

Further information can be obtained from British Bacon Curers Federation, Upper Ickfield Way, Tring, Herts.

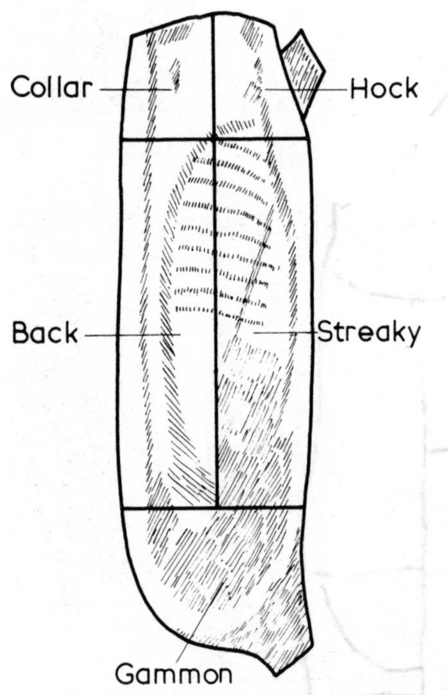

Side of bacon

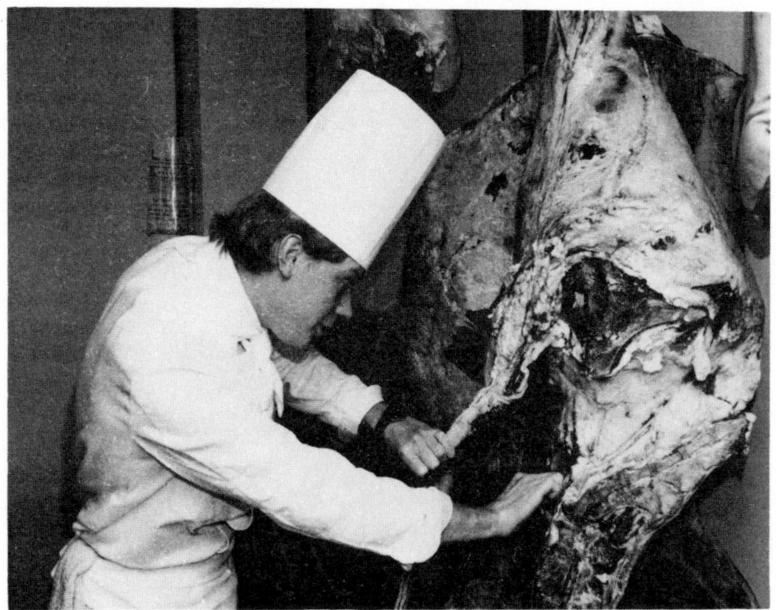

Inspecting a freshly delivered hindquarter of beef

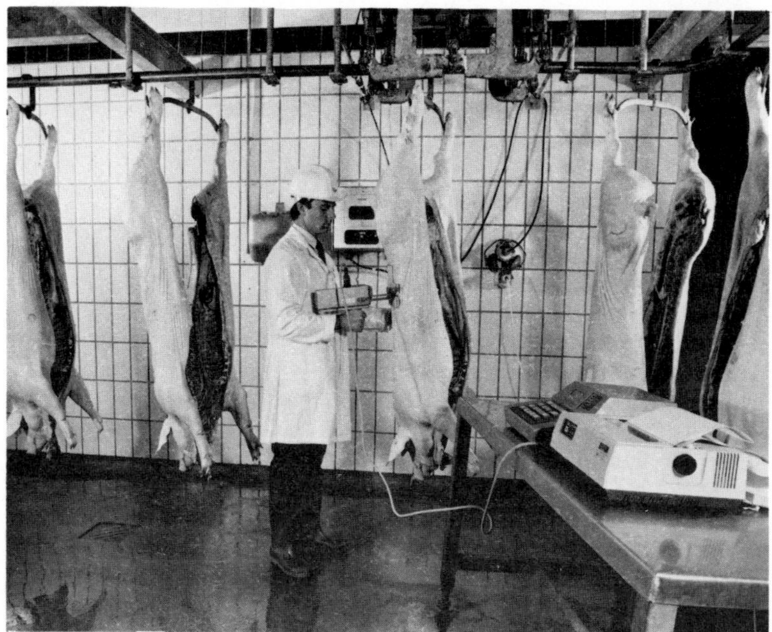

Meat technician using probe meter (connected to computer print-out) to measure depth of fat in pig carcass

Pork — *Le porc*

Joint	French	Use	Menu example
leg	le cuissot de porc	roasting, boiling	Cuissot de porc rôti; sauce pommes
loin	la longe de porc	roasting, frying, grilling	Longe de porc rôti; Côte de porc à la flamande
spare rib	la basse côte	roasting, pies	Pork pie
belly,	la poitrine	pickling, boiling	Boiled belly of pork and pease pudding
shoulder	l'épaule de porc	roasting, sausages, pies	Saucisse de porc grillé, sauce charcutière

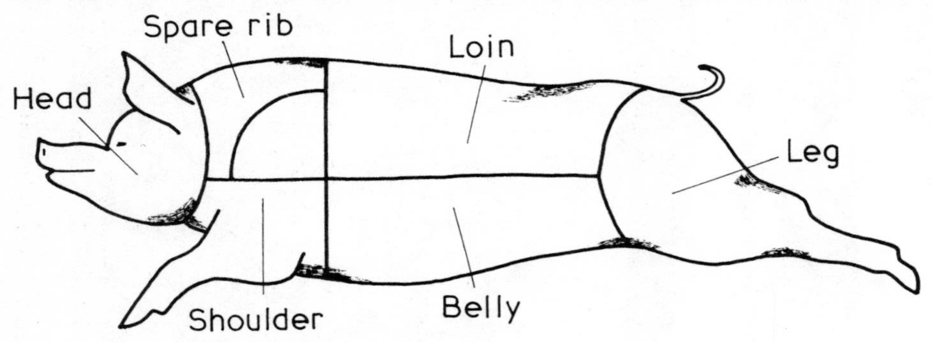

Side of pork

2 Offal and other edible parts of the carcass

Offal is the name given to the edible parts taken from the inside of the carcass: liver, kidney, heart, sweetbread. Tripe, brains, oxtail, tongue and head are sometimes included under this term. Fresh offal (unfrozen) should be purchased as required and must be kept under refrigeration since it does not keep as well as muscle meat. Frozen offal should be kept frozen until required.

Tripe — *La tripe*

Tripe is the stomach lining or white muscle of the ox. Honeycomb tripe is the second compartment of the stomach and considered the best. Smooth tripe is the first compartment of the stomach and is considered of lesser eating quality than honeycomb tripe. Sheep tripe, darker in colour, is obtainable in some areas. It may be boiled or braised.

Menu example: Tripe and onions

Oxtail — *3–5lbs* — *La queue de bœuf*

Oxtails should be of good size and lean with no signs of stickiness. They are usually braised or used for soup.

Menu examples: Queue de bœuf braisée au primeurs
Queue de bœuf liée

Head — *La tête*

Sheep's heads can be used for stock, and pigs' heads for brawn. The calf's head is used for soup and certain dishes such as Tête de veau vinaigrette.

Sheep's and calves' heads should not be sticky; they should be well fleshed and odourless.

Suet — *La graisse de rognon*

Beef suet should be creamy white, brittle and dry. It is used for suet paste. Other fat should be fresh and not sticky. Suet and fat may be rendered down for dripping.

Marrow — *La moelle*

Beef marrow is obtained from the bones of the leg of beef. It should be of good size, firm, creamy white and odourless. It is used as a garnish as with Entrecôte bordelaise and for savouries.

Menu example: Croûte baron

Brains — *La cervelle*

Calf's brains are those usually used; they must be fresh.

Menu example: Cervelle au beurre noir

Bones — *Les os*

Bones must be fresh, not sticky; they are used for stock.

Liver — *Le foie*

Calf's liver is the most expensive and is considered the best in terms of tenderness and delicacy of flavour and colour.
Lamb's liver is mild in flavour, tender and light in colour.
Sheep's liver, being from an older animal, is firmer in substance, darker in colour and has a stronger flavour than lamb or calf's liver.
Pig's liver has a strong flavour and if used for ordinary frying purposes it is not generally as acceptable as lamb or calf's liver.
Ox or beef liver is the cheapest and if taken from an older animal can be coarse in texture and strong in flavour.

English	French	Purchasing unit	Cost		Use	Menu example
			Wholesale	Retail		
calf's liver	le foie de veau				frying	Foie de veau au lard
lamb's liver	le foie d'agneau				frying	Brochette de foie d'agneau
pig's liver	le foie de porc				frying, pâté	Pâté de foie
ox liver	le foie de bœuf				braising, frying	Foie de bœuf lyonnaise

Quality

1 Liver should appear fresh and have an attractive colour.
2 It must not be dry or contain tubers in the flesh.
3 It should be smooth in texture.
4 The smell must be pleasant.

Food value: Liver is valuable as a protective food; it consists chiefly of protein and contains useful amounts of vitamin A and iron.

Kidney — *Le rognon*

Lamb's kidney is light in colour, delicate in flavour and is ideal for grilling and frying.
Sheep's kidney is darker in colour and has a stronger flavour than lamb.
Calf's kidney is light in colour, delicate in flavour and can be used in a wide variety of dishes.
Ox kidney is dark in colour, strong in flavour and is generally used mixed with beef, steak and kidney pie or pudding.
Pig's kidney is smooth, longer and flat by comparison with sheep's kidney. It has a strong flavour which makes it unpopular with a large number of people.

English	French	Purchasing unit	Cost		Use	Menu example
			Wholesale	Retail		
calf's kidney	le rognon de veau				stewing (Pies, puddings)	Steak and kidney pudding
lamb's kidney	le rognon d'agneau				grilling, sauté	Rognons sautés Turbigo
pig's kidney	le rognon de porc				grilling, sauté	Rognon de porc
ox kidney	le rognon de bœuf				stewing, soup	Soupe aux rognons

Quality

Ox kidney

1 Ox kidney should be fresh.
2 The suet is left on the kidney until it is to be prepared for cooking.
3 If the suet is removed the kidney does not remain moist.
4 The colour should be deep red.

Lamb's kidney

1 Lamb's kidney should be covered in fat, which is removed just before use.
2 The fat should be crisp and the kidney moist.
3 Both the fat and kidney must have a pleasant smell.

Food Value: The food value of kidney is similar to liver: it is a protective food containing vitamin A and iron.

Hearts — *Le cœur*

The *ox or beef* heart is the largest of all hearts, used for cooking. It is dark coloured, solid in texture and tends to be dry and tough.
Calf's heart, coming from a younger animal, is lighter in colour and more tender.
Sheep's heart is dark and dense in texture and can be dry and tough unless cooked carefully.
Lamb's heart is smaller and lighter than the sheep heart and is normally served whole, whereas the larger heart would be sliced for serving.

English	French	Purchasing unit	Cost: Wholesale	Cost: Retail	Use	Menu example
sheep's heart	le cœur d'agneau				braising	Cœur d'agneau farci
ox heart	le cœur de bœuf					Cœur de bœuf braisé

Quality: Hearts should not have an excessive amount of fat or tubes. When cut they should be moist.
Food value: They have a high protein content and are therefore valuable for growth and repair of the body.

Tongues — *La langue*

English	French	Purchasing unit	Cost: Wholesale	Cost: Retail	Use	Menu example
lamb's tongue	la langue d'agneau				boiling, braising	Langue d'agneau poulette
ox tongue	la langue de bœuf				boiling, braising	Langue de bœuf braisée au madère

Quality

1 Tongues must be fresh.
2 They should not have an excessive amount of waste at the root-end.
3 They must smell pleasant.
4 Ox tongues may be used fresh or salted.
5 Sheep's tongues are used unsalted.

Sweetbreads

Le ris

There are two kinds, unequal in shape as in quality. The heart bread is round and plump and of far superior quality than the throat or neckbread which is longer and uneven in shape. Calf's heart breads can weight up to $1\frac{1}{2}$ lbs ($\frac{3}{4}$ kg), lamb's heart breads up to 4 oz (100 g).

English	French	Purchasing unit	Cost		Use	Menu example
			Wholesale	Retail		
lamb's sweetbreads	le ris d'agneau				Braising, frying	Ris d'agneau braisé Clamart
calf's sweetbreads	le ris de veau				Braising, frying	Ris de veau bonne-maman

Quality

1. The heart bread and the neck breads should be fleshy and large.
2. They ought to be creamy white in colour.
3. They must have a pleasant smell.
4. The heart breads are superior in quality to the neck breads.

Food value: Both neck and heart sweetbreads are valuable as foods, particularly for hospital diets. They are very easily digested and useful for building and repairing body tissues.

3 Poultry — *La volaille*

Poultry is the name given to domestic birds specially bred to be eaten and for their eggs.

Season

Owing to present-day methods of poultry production by the battery method, and to deep freezing, poultry is available all the year round.

Food value

The flesh of poultry is more easily digested than that of butchers' meat. It contains protein and is therefore useful for building and repairing body tissues and providing heat and energy.

Storage

Fresh poultry must be hung by the legs in a well-ventilated room for at least 24 hours, otherwise it will not be tender.

The innards are not removed until the bird is required.

Frozen birds must be kept in a deep-freeze cabinet until required; they are then allowed to thaw out completely before cooking.

Quality

1 The breast of the bird should be plump.
2 The vent-end of the breast-bone must be pliable.
3 The flesh should be firm.
4 The skin to be white, unbroken and with a faint bluish tinge.
5 The legs should be smooth, with small scales and small spurs.
6 Old birds have large scales and large spurs on the legs.

Use

Baby chicken (spring chicken) — *Le poussin*

Spring chickens are from 4 to 6 weeks old; they may be roasted, pot-roasted or grilled.

Menu examples: Poussin polonaise
Poussin grillé au lard

Small roasting chicken — *Le poulet de grain*

These 3–4-month-birds are roasted, pot-roasted or grilled.

Menu examples: Poulet rôti à l'anglaise
Poulet de grain grillé à l'américaine

Medium roasting chicken (broiler) — *Le poulèt reine*

These are prime tender chickens; they are used for sautés, pies, suprêmes, roasting, grilling and pot-roasting. They are also referred to as broilers, which is an American term. To broil means to grill.

Menu examples: Chicken pie
Poulet sauté Parmentier
Suprême de volaille princesse
Poulet en casserole bonne-femme

Large roasting or boiling bird — *La poularde*

These are the fully grown birds which are roasted, pot-roasted and boiled. They are also used for sandwiches, salads and bouchées.

Menu examples: Galantine de volaille
Vol-au-vent de volaille
Poularde pochée au riz, sauce suprême
Poularde poêlé aux champignons
Poularde en chaud-froid

Capon — *Le chapon*

A capon is a bird which has been specially treated (caponised) so as to produce a large tender roasting chicken.

Menu example: Roast Surrey capon

Boiling fowl — *La poule*

This is an old hen which has finished laying eggs; it is suitable for stocks and soups because of the excellent flavour that an older bird gives.

Menu examples: Crème reine
Sauce suprême
Consommé de volaille
Cock-a-leekie

Further information: British Poultry Federation, 52 High Holborn, London, WC2.

Duckling — *Le caneton*

Duck — *Le canard*

Gosling — *l'oison*

Goose — *l'oie*

Quality

1 The feet and bills should be bright yellow.
2 The upper bill should break easily.
3 The web feet must be easy to tear.

Use: Ducks and geese may be roasted or braised.

Menu examples: Roast Aylesbury duckling
Caneton à l'orange
Canard braisé aux petits pois
Roast goose, sage and onion stuffing
Oison braisé aux navets

English	French	Undrawn weight (approx)	Number of portions (approx)	Cost		Season	At best
				Wholesale	Retail		
single baby chicken (spring chicken)	le poussin	360 g– $\frac{1}{2}$ kg	1			spring	
double baby chicken	le poussin double	$\frac{1}{2}$ kg– $\frac{3}{4}$ kg	2			spring	
small roasting chicken	le poulet de grain	$\frac{3}{4}$ kg– 1 kg	3–4			spring and early summer	
medium roasting chicken	le poulet reine	1 kg– 2 kg	4–6			spring, summer	
large roasting or boiling chicken	la poularde	2 kg– 3 kg	6–8			all year round	autumn
capon	le chapon	3 kg– $4\frac{1}{2}$ kg	8–12			all year round	autumn
boiling fowl	la poule	$2\frac{1}{2}$ kg– 4 kg	8–12			all year round	
young turkey	le dindonneau	$3\frac{1}{2}$ kg	**			September	
turkey	la dinde	$3\frac{1}{2}$ kg–20 kg	**			Sept.–Mar.	winter
duckling	le caneton	1 kg– $1\frac{1}{2}$ kg	3–4			Mar.–Sept.	summer
duck	le canard	$1\frac{1}{2}$ kg– $2\frac{1}{2}$ kg	4–6			Aug.–Feb.	autumn
gosling	l'oison	2 kg– $3\frac{1}{2}$ kg	**			Aug.–Oct.	summer
goose	l'oie	$3\frac{1}{2}$ kg– 7 kg	**			Sept.–Feb.	autumn
guinea fowl	le pintarde	1 kg– $1\frac{1}{2}$ kg	3–4			all year round	summer
pigeon	le pigeon	$\frac{3}{4}$ kg– 1 kg	1–2			all year round	spring

Drawn poultry loses approximately 25% of its original weight.
All poultry is bought by number and weight.
** For turkey and goose allow $\frac{1}{2}$ kg undrawn weight per portion.

Turkey

Le dindonneau
La dinde

Quality: The breast should be large, the skin undamaged and with no signs of stickiness. The legs of young birds are black and smooth, the feet supple with a short spur. As the bird ages the legs turn reddish grey and become scaly. The feet become hard.
Use: Turkeys are usually roasted.

Menu examples: Roast Norfolk turkey and chestnut stuffing
Dindonneau rôti, Sauce airelles
Emincé de dinde à la king
Turkey sandwich

Further information: The British Turkey Federation, The Bury, Church Street, Chesham, Bucks.

Guinea fowl *Le pintarde*

When plucked these grey-and-white feathered birds resemble a chicken with darker flesh. The young birds are known as squabs.

The points relating to chicken apply to guinea fowl.

Menu examples: Pintarde en cocotte grand-mère
Pintarde rôti

Pigeon *Le pigeon*

Pigeon should be plump, the flesh mauve-red in colour and the claws pinkish. Tame pigeons are smaller than wood pigeons.

Menu examples: Pigeon pie
Pigeon braisé aux olives

4 Game *Le gibier*

Game is the name given to certain wild birds and animals which are eaten; there are two kinds of game:

a) feathered,
b) furred.

Food value

As game is less fat than poultry or meat it is easily digested. Water fowl are not so easily digested, owing to their oily flesh. Game is useful for building and repairing body tissues and for energy.

Storage

1 Hanging is to some degree essential for all game. It drains the flesh of blood and begins the process of disintegration which is essential to make the flesh soft, edible and also develop flavour.
2 The hanging time is determined by the type, condition and age of the

game and the storage temperature.

3 Old birds need to hang for a longer time than young birds.
4 Game birds are not plucked or drawn before hanging.
5 Venison and hare are hung with their skins on.
6 Game must be hung in a well-ventilated, dry, cold storeroom; this need not be refrigerated.
7 Game birds should be hung by the neck with the feet down.

Game availability

Game is available fresh in season between the following dates and frozen for the remainder of the year:

Grouse	August 12 – December 10
Snipe	August 12 – January 31
Partridge	September 1 – February 1
Wild duck	September 1 – January 31
Pheasant	October 1 – February 1
Woodcock	October 31 – February 1

Venison, hares, rabbits and pigeons are available throughout the year.

Quality

Venison

Joints of venison should be well fleshed and a dark brownish-red colour.

Hares and rabbits

The ears of hares and rabbits should tear easily. With old hares the hare lip is more pronounced than in young animals. The rabbit is distinguished from the hare by shorter ears, feet and body.

Birds

1 The beak should break easily.
2 The breast plumage is soft.
3 The breast should be plump.
4 Quill feathers should be pointed, not rounded.
5 The legs should be smooth.

Venison *La venaison*

Venison is the flesh of the deer and the lean meat is a dark blood red colour. The surface of the carcass is usually dusted with a mixture of flour, salt and black pepper before being hung in a cold room for 2–3 weeks. The roebuck (Fr *chevreuil*) is the deer which is frequently used.

Venison is usually roasted or braised in joints. Small cuts may be fried. Before cooking it is always marinaded to counteract the toughness and dryness of the meat.

Menu examples: Roast haunch of venison: Cumberland sauce
Selle de chevreuil: sauce poivrade

Game — Le gibier

English	French	Undrawn weight (approx)	Number of portions (approx)	Cost		Season	At best
				Wholesale	Retail		
Furred:							
venison	la venaison	15 kg	15–20			July–Feb.	autumn
hare	le lièvre	$2\frac{1}{2}$–$3\frac{1}{2}$ kg	6–8			Aug.–Feb.	winter
rabbit	le lapin	1 kg	4			Sept.–April	winter
Feathered:							
pheasant	le faisan	$1\frac{1}{2}$–2 kg	4			Oct 1–Feb. 11	winter
partridge	le perdreau	$\frac{1}{4}$–$\frac{1}{2}$ kg	1–2			Sept. 1–Feb. 11	autumn
grouse	la grouse	360 g	1–2			Aug. 12–Dec. 20	Sept.
woodcock	la bécasse	$\frac{1}{4}$ kg–360 g	1			Oct.–Nov.	autumn
snipe	le bécassine	120 g	1			Oct.–Nov.	autumn
quail	le caille	150 g	1			All year round	autumn
wild duck	le canard-sauvage	1–$1\frac{1}{2}$ kg	2–4			Sept.–Mar.	winter
teal	la sarcelle	$\frac{1}{2}$–$\frac{3}{4}$ kg	1–2			Oct.–Jan.	winter
wood pigeon	le pigeon	360 g	1			Mar.–Oct.	autumn

Hare *La lièvre*

Hare is cooked as a stew called jugged hare (*civet de lièvre*) and the saddle (*rable de lièvre*) is roasted.

Menu examples: Civet de lièvre bourguignonne
Rable de lièvre, sauce grand veneur

Pheasant *Le faisan*

Is one of the most common game birds. Average weight $1\frac{1}{2}$–2 kg (2–4 lbs). Young birds have a pliable breast bone and soft pliable feet. Hang for 5–8 days. Used for roasting, braising or pot roasting. *Menu examples:* Salmis de faisan, Faisan poêlé au céleri.

Partridge *Le perdreau*

The most common varieties are the grey legged and the red legged. Average weight 200–400 g ($\frac{1}{2}$–1 lb). Hang for 3–5 days. Used for roasting or braising. *Menu example:* Perdreau aux choux.

Grouse *La grouse*

A famous and popular game bird, the red grouse which is shot in Scotland and Yorkshire. Average weight 300 g (12 oz). Young birds have pointed wings and rounded soft spurs. Hang for 5–7 days. Used for roasting. *Menu example:* Roast grouse.

Woodcock *La bécasse*

Small birds with long thin beaks. Average weight 200–300 g (8–12 oz). Prepare as for snipe.

Snipe *La bécassine*

100 g (4 oz). Hang for 3–4 days. The heads and neck are skinned, the eyes removed and they are then trussed with their own beaks. When drawing the birds only the gizzard, gall-bladder & intestines are removed. The birds are then roasted with the liver and heart left inside. *Menu example:* Bécassine rôtie.

Quail *La caille*

Small birds weighing 50–75 g (2–3 oz) produced on farms, usually packed in boxes of 12. Quails are not hung. They are usually roasted or braised. *Menu examples:* Caille rôtie, Caille en aspic.

Wild Duck *Le canard-sauvage*

Such as mallard and the widgeon, average weight 1–$1\frac{1}{2}$ kg (2–3 lb). Hang for 1–2 days. Usually roasted or braised. *Menu example*: Canard sauvage bigarade.

Teal *La sarcelle*

The smallest duck, weighing 400–600 g (1–$1\frac{1}{2}$ lb). Hang for 1–2 days. Usually roasted or braised. Young birds have small pinkish legs and soft down under the wings. Teal and wild duck must be eaten in season otherwise the flesh is coarse and has a fishy flavour. Teal is a small species of wild duck which is roasted or braised. *Menu example:* Sarcelle rôtie à l'orange

Further information: 238–43 *Practical Cookery*, 5 edition.

5 Fish *Les poissons*

Fish have formed a large portion of man's food because of their abundance and relative ease of harvesting. It is interesting to note that fish consume the smaller organisms in the sea and are themselves the food of the larger organisms. In addition to providing fresh and processed food for human consumption, other valuable products such as oil and isinglass as well as fertiliser come from fish.

Unfortunately fish are not in unlimited supply and due to overfishing it is now necessary to have fish farms, eg trout and salmon, to supplement the natural resources. This is not the only problem; due to contamination by man, the seas and rivers are increasingly polluted, thus affecting supplies and the suitability of fish, and particularly shellfish, for human consumption.

Fish are valuable, not only because they are a good source of protein, but because they are suitable for all types of menus and can be cooked and presented in a wide variety of ways. The range of different types of fish of varying textures, taste and appearance is indispensable to the creative chef.

Types or varieties

1 White fish

a) Round (eg cod, whiting, hake).
b) Flat (eg plaice, sole, turbot).

2 Oily fish
These are all round in shape (eg herring, mackerel, salmon).

3 Shellfish

a) Crustacea (eg lobsters, crabs).
b) Mollusca (eg oysters, mussels).

Seasons
Supplies of fish are subject to weather conditions and to the fact that shoals of certain fish vary their positions at different times of the year. Therefore the fish which are in season all the year round may not be available everywhere. To know the fish in season – that is to say, when they are at their best – see table, p 183.

Available all the year round: Bream, cod, plaice, halibut, whiting, brill, haddock, turbot, sole.

Spring: Eel, mackerel, whitebait, conger eel, salmon.

Summer: Eel, herring, salmon, skate, conger eel, mullet, salmon trout, whitebait.

Autumn: Skate.

Winter: Mackerel, skate, smelt.

Purchasing unit

Fresh fish is bought by the kilogram or by the number of fillets or whole fish of the weight that is required. For example, 30 kg of salmon could be ordered as 2 × 15 kg, 3 × 10 kg or 6 × 5 kg. Frozen fish can be purchased in 15-kg blocks.

Source

As we live on an island, fish is plentiful, although overfishing and pollution are having a worrying effect on the supplies of certain fish. Most catches are made off Iceland, Scotland, the North Sea, Irish Sea and the English Channel. Salmon are caught in certain English and Scottish rivers. Frozen fish is imported from Scandinavia, Canada and Japan; the last two countries send frozen salmon to Britain.

Storage

1. Fresh fish are stored in a fish-box containing ice, in a separate refrigerator or part of a refrigerator used only for fish.
2. The temperature must be maintained just above freezing-point.
3. Frozen fish must be stored in a deep-freeze cabinet or compartment.
4. Smoked fish should be kept in a refrigerator.

Quality points for buying

When buying fish the following points should be looked for to ensure freshness:

1. Eyes: bright, full and not sunken.
2. Gills: bright red in colour.
3. Flesh: firm and resilient so that when pressed the impression goes quickly. The fish must not be limp.
4. Scales: these should lie flat, be moist and plentiful.
5. Skin: this should be covered with a fresh sea slime, or be smooth and moist.
6. Smell: the smell must be pleasant.

Buying points

1. Fish should be purchased daily.
2. If possible, it ought to be purchased direct from the market or supplier.
3. The fish should be well iced so that it arrives in good condition.
4. The flesh of the fish should not be damaged.
5. Fish may be bought on the bone or filleted. (The approximate loss from boning and waste is 50% for flat fish, 60% for round fish.)

6 Fillets of plaice and sole can be purchased according to weight. They are graded from 45 g to 180 g per fillet and go up in weight by 15 g.
7 Medium-sized fish are usually better than large fish, which may be coarse; small fish often lack flavour.

Food value

Fish is as useful a source of animal protein as meat. The oily fish, such as sardines, mackerel, herrings and salmon, contains vitamins A and D in its flesh, but in white fish, such as halibut and cod, these vitamins are present in the liver.

The bones, when eaten, of sardines, whitebait and tinned salmon provide the body with calcium and phosphorus.

Since all fish contains protein it is a good body-building food and the oily fish is useful for energy and as a protective food because of its vitamins.

Owing to its fat content oily fish is not so digestible as white fish and is not so suitable for invalid cookery.

Preservation (see also Chapter 9)

Freezing

Fish is either frozen at sea or as soon as possible after reaching port. It should be thawed out before being cooked. Plaice, halibut, turbot, haddock, sole, cod, trout, salmon, herring, whiting, scampi, smoked haddock and kippers are frozen.

Canning

The oily fish are usually used for canning. Sardines, salmon, anchovies, pilchards, herring and herring roe are canned in their own juice, as with salmon, or in oil or tomato sauce.

Salting

In this country salting of fish is usually accompanied by a smoking process.

Kippers: The herring is split, gutted, washed and placed in a brine solution for 15–30 minutes. The fish are then impaled on tenterhooks and smoked in a kiln for up to 6 hours. The smoke is produced by very slowly burning wood-shavings and sawdust.

Bloaters: Bloaters are salted and lightly smoked whole herring.

Cured herrings: Cured herrings are packed in salt and red herrings are salted and then smoked.

Caviar: Caviar is the slightly salted roe of the sturgeon which is sieved, tinned and refrigerated. Imitation caviar is also obtainable – see page 251.

Smoking

Haddock: Haddock are slit open, salted slightly and then smoked.

Salmon: Salmon and mackerel are filleted and smoked – see page 254.

Cod roes, eel, sprats, trout: These are also smoked whole.

Buckling: Buckling is the name given to whole smoked herring.

Pickling

Herrings: Herring pickled in vinegar are filleted, rolled and skewered and known as rollmops.

Use

Fish is cooked by boiling, poaching, grilling, baking, shallow and deep frying. Certain fish are not cooked, apart from the smoking or curing process. This applies for example to smoked salmon, smoked eel, smoked trout, and buckling.

Oily fish

Anchovies *Les anchois*

Anchovies are small round fish used tinned in this country; they are supplied in 60 g and 390 g tins. They are filleted and packed in oil.

They are used for making anchovy butter and anchovy sauce, for garnishing dishes such as Scotch woodcock and Escalope de veau viennoise. They may be used as a dish in a selection of hors d'œuvre, as a savoury, and they can be used in puff pastry and served at cocktail parties.

Conger eel *Le congre*

Conger eel is a dark grey sea-fish with white flesh which grows up to 3 metres in length. It may be used in the same way as eels, or it may be smoked.

Menu example: Anguille fumée; Sauce raifort

Eel *L'anguille*

Eels live in fresh water and grow up to 1 metre in length. They are found in many British rivers and considerable quantities are imported from Holland. Eels must be kept alive until the last minute before cooking and they are generally used in fish stews.

Menu example: Bouillebaisse; jellied eels

Herring *Le hareng*

Fresh herrings are used for breakfast and lunch menus; they may be grilled, fried or soused. Kippers, which are split, salted, dried and smoked herrings, are served mainly for breakfast and also for a savoury. Average weight, 250 g.

Menu example: Hareng grillé; Sauce moutarde

Fish	French	Season	Purchasing unit	Cost per kg
Oily fish:				
anchovy	l'anchois		60 g, 390 g tins	
conger eel	le congre	all year round; best, summer	by the kilogram	
eel	l'anguille	all year round	by the kilogram	
herring	le hareng	Oct.–Dec.* May–Aug.	number and weight	
mackerel	le maquereau	Winter, spring and summer	number and weight	
salmon	le saumon	Feb.–Aug.	number and weight	
salmon trout	la truite saumonée	Mar.–Aug.	number and weight	
sprat	le sprat	Nov.–Mar.	by the kilogram	
trout	la truite	Feb.–Sept.	number and weight	
tunny	le thon		127 g, 240 g tins	
sardine	la sardine		135 g, 345 g, 795 g tins	
whitebait	la blanchaille	Feb.–Aug.	by the kilogram	
White fish:				
Flat fish:				
brill	la barbue	all year round	whole fish by number and weight	
halibut	le flétan	July–March	whole fish by number and weight	
plaice	la plie	all year round; best, May–Jan.	fillets or number and weight	
skate	la raie	Summer, winter and autumn	by the kilogram	
sole	la sole	all year round	fillets, number and weight	
turbot	le turbot	all year round	number and weight	
Round fish:				
bream	la brème	all year round	fillets or whole fish by number and weight	
cod	le cabillaud	all year round; Sept.–Mar.	fillets or whole fish or weight	
haddock	l'aigrefin	all year round	fillets or whole fish, by number or weight	
hake	le colin	all year round; best, June–Jan.	whole fish or by the pound	
red mullet	le rouget	July–Oct.	number and weight	
smelt	l'éperlan	Oct.–May	boxes, or by the kilogram	
whiting	le merlan	all year round; best, May–Jan.	number and weight	

* Season given is for home-fished herring. Norwegian herring available January and February.

Mackerel — *Le maquereau*

Mackerel are grilled, shallow fried or soused, and may be used on breakfast and lunch menus. They must be used fresh because the flesh deteriorates very quickly. Average weight, 360 g.

Menu example: Maquereau grillé; Beurre d'anchois

Salmon — *Le saumon*

Salmon is perhaps the most famous river fish and is caught in such well-known British rivers as the Dee, Tay, Severn, Avon, Wye and Spey. A considerable number are imported from Scandinavia, Canada, Germany and Japan. Apart from using it fresh, salmon is tinned or smoked. When fresh, it is usually boiled or grilled. When boiled it is cooked in a court-bouillon. Frequently, whole salmon are cooked and when cold decorated and served on buffets. Weight varies from $3\frac{1}{2}$–15 kg. Salmon under $3\frac{1}{2}$ kg are known as grilse.

Menu examples: Darne de saumon pochée; Sauce hollandaise
Darne de saumon grillée; Sauce verte
Mayonnaise de saumon

Salmon trout (sea trout) — *La truite saumonée*

Salmon trout are a sea fish similar in appearance to salmon, but smaller, and they are used in a similar way. Average weight, $1\frac{1}{2}$–2 kg.

Menu example: Truite saumonée froide; Sauce mayonnaise

Sardines — *Les sardines*

Sardines are small fish of the pilchard family which are usually tinned and used for hors d'œuvre, sandwiches and as a savoury. Fresh sardines are also available and may be cooked by grilling or frying.

Sprats — *Le sprat*

Sprats are small fish fried whole and are also smoked and served as an hors d'œuvre.

Trout — *La truite*

Trout live in rivers and lakes and in this country they are cultivated on trout farms. When served *au bleu*, they must be alive just before cooking; they are then killed, cleaned, sprinkled with vinegar and cooked in a court-bouillon. Trout are also served grilled or shallow fried, and may also be smoked and served as an hors d'œuvre. Average weight, 180–230 g.

Menu examples: Truite fumée
Truite de rivière grenobloise

Tunny — *Le thon*

Tunny is a very large fish cut into sections, tinned in oil and is used mainly in hors d'œuvre and salads.

Whitebait *La blanchaille*
Whitebait are the fry of young herring, 2–4 cm long, and they are deep fried.

Menu example: Blanchailles diablées

White fish–flat

Brill *La barbue*
Brill is a large flat fish which should not be confused with turbot. Brill is oval in shape; the mottled brown skin is smooth with small scales. It can be distinguished from turbot by lesser breadth in proportion to length, average weight, 3–4 kg.

It is usually served in the same way as turbot.

Menu examples: Tronçon de barbue grillé; Sauce anchois
Suprême de barbue Mornay

Halibut *Le flétan*
Halibut is a long and narrow fish, brown, with some darker mottling on the upper side; it can be 3 metres in length and weigh 150 kg. Halibut is served on good-class menus as it is valued very highly for its flavour. It is poached, boiled, grilled or shallow fried.

Menu example: Suprême de flétan belle meunière

Plaice *La plie*
Plaice are oval in shape, dark brown colouring with orange spots on the upper side, used on all types of menus; they are usually deep fried or grilled. Average weight, 360–450 g.

Menu example: Plie grillé; Beurre maître d'hôtel

Larger plaice known in French as 'carrelet' are also used for poached, fried and grilled dishes.

Menu example: Filet de carrelet Dugléré

Skate *La raie*
Skate, a member of the ray family, is a very large fish and only the wings are used. It is always served on the bone and either shallow or deep fried or cooked in a court-bouillon and served with black butter.

Menu example: Raie au beurre noir
Fried skate

Sole *La sole*
Sole is considered to be the best of the flat fish. The quality of the Dover sole is well known for its excellence. Other varieties such as lemon sole and the witch (Torbay Sole) are inferior to it.

Soles are cooked by poaching, grilling or frying both shallow and deep. They are served whole or filleted and garnished in a great many ways.

The usual size for soles is 180–750 g; fillets are taken from sole of 500 g and over. A 180–250 g sole is referred to as a slip sole. When serving a whole fish, a 250–500 g sole may be used, the size depending on the type of establishment and the meal for which it is required.

Menu examples: Sole colbert
Sole diéppoise
Filet de sole Waleska
Paupiette de sole Newburg

Turbot — *Le turbot*

Turbot has no scales and is roughly diamond in shape; it has knobs known as tubercules on the dark skin. In proportion to its length it is wider than brill; $3\frac{1}{2}$–4 kg average weight.

Turbot may be cooked whole, filleted or cut into portions on the bone. It may be boiled, poached, grilled or shallow fried.

Menu examples: Turbot poché; Sauce hollandaise
Suprême de turbot florentine
Tronçon de turbot grillé; Beurre maître d'hôtel

Early morning scene at a large wholesale fish market

White fish–round

Bream *La brème*
Sea bream is a short plump reddish fish, with large scales. It is used on many menus of less expensive price; it is usually filleted and deep fried, but other methods of cooking are employed. Average weight $\frac{1}{2}$–1 kg.

Menu examples: Filet de brème frit; Sauce tartare
Filet de brème meunière

Cod *Le cabillaud*
Cod varies in colour but is mostly greenish, brownish or olive grey. It can measure up to 5 feet in length. Cod is cut into steaks or filleted and cut into portions, and it can be deep or shallow fried or boiled. Small cod are known as codling. Average weight of cod, $2\frac{1}{2}$–$3\frac{1}{2}$ kg.

Menu examples: Darne de cabillaud pochée; Sauce persil
Grilled cod steak

Gudgeon *Le goujon*
Gudgeon are small fish found in continental lakes and rivers. They may be deep fried whole. On menus in this country the French term '*en goujon*' refers to other fish such as sole or turbot, cut into pieces the size of gudgeon.

Menu example: Filets de sole en goujons; Sauce tartare

Haddock *L'aigrefin*
Haddock is distinguished from cod by the thumb mark on the side and by the lighter colour. Every method of cooking fish may be applied to haddock, and it appears on all kinds of menus. Apart from fresh haddock, smoked haddock is used a great deal for breakfast: it may also be served for lunch and as a savoury. Average weight, $\frac{1}{2}$–2 kg.

Finnan haddock is the most popular smoked haddock, which takes its name from a fishing village, Findon, south of Aberdeen. For the smoking process the haddock is cleaned, split and the head removed. Then salting takes place for 2 hours, after which the fish are dried for 2–3 hours and finally smoked over peat, hardwood, sawdust or fir cones for 12 hours.

Menu examples: Filet d'aigrefin à l'Orly
Haddock Monte Carlo
Canapé Ivanhoe

Hake *Le colin*
Owing to overfishing hake is not plentiful. It is easy to digest. The flesh is very white and of a delicate flavour. It is usually boiled.

Menu example: Boiled hake and egg sauce

Red mullet — *Le rouget*

Red mullet is on occasion cooked with the liver left in, as it is considered that they help to impart a better flavour to the fish. Mullet are usually cooked whole. Average weight, 360 g–$\frac{3}{4}$ kg.

Menu examples: Rouget grenobloise
Rouget en papillote

Smelt — *L'éperlan*

Smelts are small fish found in river estuaries and imported from Holland; they are usually deep fried or grilled. When grilled they are split open. The weight of a smelt is from 60 to 90 g.

Menu example: Éperlan frit à l'anglaise

Whiting — *Le merlan*

Whiting are very easy to digest and they are therefore suitable for invalid cookery. They may be poached, grilled or deep fried and used in the making of fish stuffing (farce de poisson). Average weight, 360 g.

Menu example: Merlan en colère

Rockfish

Rockfish is the fishmonger term applied to cat fish, coal fish, dog fish, and conger eel etc, after cleaning and skinning. It is usually deep fried in batter.

Shellfish — *Les crustacés*

English	French	Season	Purchasing unit	Cost per unit
shrimp	la crevette grise	all year round		
prawn	la crevette rose	all year round		
Dublin Bay prawn or scampi	la langoustine	all year round	number	
crayfish	l'écrevisse (f.)			
lobster	le homard	all year round; best, summer	number and weight	
crawfish	la langouste	Summer	number and weight	
crab	le crabe	all year round; best, summer	number and weight	
oysters	les huîtres	Sept–April	dozen	
mussels	les moules	Sept–March		
scallop	la coquille St Jacques	Nov–March	number	

Seasons

Available all year round: Shrimps, prawns, lobsters, crabs.
Spring: Oysters, mussels, scallops.
Summer: Lobster, crawfish, crab.

Autumn: Crab.
Winter: Crab, oysters, mussels, scallops.

Food value

Shellfish is a good body-building food. As the flesh is coarse and therefore indigestible a little vinegar is used in the cooking to soften the fibres.

Quality, purchasing points and storage

1 With the exception of shrimps and prawns all shellfish, if possible, should be purchased alive, so as to ensure freshness.
2 They should be stored in a cold room.
3 Shellfish are kept in boxes and covered with damp sacks.
4 Shellfish should be cooked as soon as possible after purchasing.

Shrimps and prawns are usually bought cooked and may be obtained in their shell or peeled. They should be freshly boiled, of an even size and not too small. Frozen shrimps and prawns are obtainable in packs ready for use.

Use

Shrimps — *Les crevettes grises*

Shrimps are used for garnishes, decorating fish dishes, cocktails, sauces, salads, hors d'œuvre, potted shrimps, omelets and savouries.

Menu examples: Cocktail de crevettes
Omelette aux crevettes

Prawns — *Les crevettes roses*

Prawns are larger than shrimps, they may be used for garnishing and decorating fish dishes, for cocktails, canapé moscovite, salad, hors d'œuvre and for such hot dishes as curried prawns.

Menu examples: Crevettes roses
Curried prawns

Scampi, Dublin bay prawn — *La langoustine*

Scampi are found in the Mediterranean. The Dublin bay prawn, which is the same family, is caught around the Scottish coast. These shellfish resemble small lobster about 16 cm long and only the tail flesh is used.

Menu examples: Scampi frites
Scampi provençale

Crayfish — *L'écrevisse*

Crayfish are a type of small fresh-water lobster used for garnishing cold buffet dishes and for recipes using lobster.

Lobster — *Le homard*

Quality and purchasing points

1 Live lobsters are bluish black in colour and when cooked they turn bright red.
2 They should be alive when bought.
3 Lobsters should have both claws attached.
4 They ought to be fairly heavy in proportion to their size.
5 Price varies considerably with size. For example, small $\frac{1}{2}$-kg lobsters are more expensive per kilogram than are large lobsters.
6 Lobster prices fluctuate considerably during the season.
7 Hen lobsters are distinguished from the cock lobsters by a broader tail.
8 There is usually more flesh on the hen, but it is considered inferior to that of the cock.
9 The coral of the hen lobster is necessary to give the required colour for certain soups, sauces and lobster dishes. For these, 1 kg hen lobsters should be ordered.
10 When required for cold individual portions, cock lobsters of $\frac{1}{4}$–$\frac{1}{2}$ kg are used to give two portions.

Use: Lobsters are served cold in cocktails, hors d'œuvre, salads, sandwiches and on buffets. When hot they are used for soup, grilled and served in numerous dishes with various sauces. They are also used as a garnish to fish dishes.

Menu examples: Mayonnaise de homard
Homard Mornay
Bisque de homard (lobster soup)

Crawfish — *La langouste*

Crawfish are like large lobsters without claws, but with long antennae. They are brick red in colour when cooked. Owing to their size and appearance they are used mostly on cold buffets but they can be served hot.

Menu example: Langouste parisienne

Crab — *Le crabe*

Quality and purchasing points

1 Crab should be alive when bought and both claws attached to the body.
2 The claws should be large and fairly heavy.
3 The hen crab has a broader tail, which is pink. The tail of cock is narrow and whiter.
4 There is usually more flesh on the hen crab, but it is considered to be of inferior quality to that of the cock.

Use: Crabs are used for hors d'œuvre, cocktails, salads, dressed crab, sandwiches and bouchées.

Oysters — *Les huîtres*

Whitstable and Colchester are the chief English centres for oysters where they occur naturally and are also cultured. Since the majority of oysters are eaten raw it is essential that they are thoroughly cleansed before the hotels and restaurants receive them.

Quality and purchasing points

1 Oysters must be alive; this is indicated by the firmly closed shells.
2 They are graded in sizes and the price varies accordingly.
3 Oysters should smell fresh.
4 They should be purchased daily.
5 English oysters are in season from September to April (when there is an R in the month).
6 During the summer months oysters are imported from France, Holland and Portugal.

Storage: Oysters are stored in barrels or boxes, covered with damp sacks and kept in a cold room to keep them moist and alive. The shells should be tightly closed; if they are open, tap them sharply, and if they do not shut at once, discard them.
Use: The popular way of eating oysters is in the raw state. They may also be served in soups, hot cocktail savouries, fish garnishes, as a fish dish, in meat puddings and savouries.

Menu examples: Whitstable natives
Huîtres frits; Sauce tartare
Anges à cheval
Steak, kidney and oyster pudding

Mussels — *Les moules*

Mussels are extensively cultivated on wooden hurdles in the sea, producing tender, delicately flavoured, plump fish.
Buying and quality points

1 The shells must be tightly closed.
2 The mussels should be large.
3 There should not be an excessive number of barnacles attached to the mussels.
4 Mussels should smell fresh.

Storage: Mussels are kept in boxes, covered with a damp sack and stored in a cold room.
Use: They may be served hot or cold or as a garnish.

Menu examples: Moules marinière
Moules vinaigrette

Scallops *La coquille St Jacques*

Buying and quality points

1 As scallops are found on the sea bed and are therefore dirty it is advisable to purchase them ready cleaned.
2 If scallops are not bought cleaned the shells should be tightly closed.
3 The orange part should be bright in colour and moist.
4 If they have to be kept, they should be stored in an ice-box or refrigerator.

Use: Scallops are usually poached or fried.

Menu examples: Fried scallops
Coquille St Jacques bonne femme

Fish offal

Liver

An oil rich in vitamins A and D is obtained from the liver of cod and halibut. This is used medicinally.

Roe

Those used are the soft and hard roes of herring, cod, sturgeon and the coral from lobster.

Soft herring roes are used to garnish fish dishes and as a savoury. Cod's roe is smoked and served as hors d'œuvre. The roe of the sturgeon is salted and served raw as caviar and the coral of lobster is used for colouring lobster butter and lobster dishes and also as a decoration to fish dishes.

Further information: Fishmongers' Company, Fishmongers Hall, London Bridge, E.C.4; the White Fish Authority, 2/3 Cursitor Street, London, E.C.4.

6 Vegetables *Les legumes*

Fresh vegetables and fruits are important foods both from an economic and nutritional point of view. On average, each person consumes 125–150 kg per year of fruit and vegetables.

The purchasing of these commodities is difficult because of:

a) the highly perishable nature of the products,
b) changes in market practice owing to varying supply and demand,
c) the effect of preserved foods, eg frozen vegetables.

The high perishability of fresh vegetables and fruits causes problems not encountered in other markets. Fresh vegetables and fruits are living organisms and will lose quality quickly if not properly stored and handled. Improved transportation and storage facilities can reduce loss of quality.

Automation in harvesting and packaging speeds the handling process and helps retain quality.

Vacuum cooling, a process whereby fresh produce is moved into huge chambers, where, for about half an hour, a low vacuum is maintained, inducing rapid evaporation which quickly reduces field heat, has been highly successful in improving quality.

Experience and sound judgement are essential for the efficient buying and storage of all commodities, but none probably more so than fresh vegetables and fruit.

The grading of fresh fruits and vegetables within the EEC

Since 1973 there has been a gradual introduction of the EEC quality grading for fresh fruits and vegetables on the British market. There are four main quality classes for produce:
Extra Class - for produce of top quality,
Class I - for produce of good quality,
Class II - for produce of reasonably good quality,
Class III - for produce of low marketable quality.

Types or varieties

1 Roots

a) Roots: carrots, parsnips, beetroots, swedes, turnips, radishes.
b) Tubers: potatoes, Jerusalem artichokes.
c) Bulbs: onions, shallots, leeks, garlic.

2 Green vegetables

a) Leaves: cabbage, lettuce, sprouts, spinach.
b) Flowers: cauliflower, broccoli, globe artichoke.
c) Fruits: tomatoes, marrow, cucumber, avocado, aubergine.
d) Legumes: peas, beans.
e) Blanched stems: asparagus, sea-kale, celery, chicory.

3 Fungi

Truffles, mushrooms, cèpes, morels.

Food value

Root vegetables: Root vegetables are useful in the diet because they contain starch or sugar for energy, a small but valuable amount of protein, some mineral salts and vitamins. They are also useful sources of cellulose and water.

Seasons

It is difficult to memorise when vegetables are in season. The student is advised to think in terms of seasons of the year for home-grown produce. With the rapid advance and development of air cargo transport services, many vegetables are in season all the year round.

Green vegetables: The food value of green vegetables is not the same as for root vegetables because no food is stored in the leaves, it is only produced there; therefore little protein or carbohydrate is found in green vegetables. They are rich in mineral salts and vitamins, particularly vitamin C and carotene. The greener the leaf the larger the quantity of vitamin present.

The chief mineral salts are calcium and iron.

Spring		
asparagus	greens	broccoli – white and purple
artichokes, Jerusalem	cauliflower	new potatoes
new carrots	new turnips	
Summer		
artichokes, globe	turnips	asparagus
cauliflower	aubergine	cos lettuce
beans, broad	peas	radishes
beans, French	carrots	sea-kale
sweetcorn		
Autumn		
artichokes, globe	parsnips	field mushrooms
artichokes, Jerusalem	aubergine	peppers
beans, runner	cauliflower	red cabbage
salsify	celery	shallots
celeriac	swedes	marrow
turnips		
Winter		
Brussels sprouts	chicory	cabbage
kale	celery	parsnips
cauliflower	broccoli	red cabbage
Savoy cabbage	celeriac	swedes
turnips		
All the year round		
Although the following vegetables are available all the year round, nevertheless at certain times, owing to bad weather, a heavy demand or other circumstances, supplies may be temporarily curtailed.		
beetroot	leeks	cabbage
mushrooms	carrots	onions
cucumber	spinach	lettuce
tomatoes	watercress	potatoes

Quality and purchasing points

Root vegetables

1. Must be clean and free from soil (earth increases weight and consequently the price).
2. They must be firm, sound and free from spade marks.

Green vegetables

1 They must be absolutely fresh.
2 The leaves must be bright in colour, crisp and not wilted.
3 Cabbage and brussels sprouts should have tightly growing leaves and be compact.
4 Cauliflowers should have closely grown flower and firm, white head; not too much stalk or outer leaves.
5 Peas and beans should be crisp and of medium size. Pea-pods should be full, beans not stringy.
6 Blanched stems must be firm, white, crisp and free from soil.

Storage

1 Root vegetables should be emptied from sacks and stored in bins or racks.
2 Green vegetables should be stored on well-ventilated racks.
3 Salad vegetables can be left in their containers and stored in a cool place.

Preservation

Canning: Certain vegetables are preserved in tins: artichokes, asparagus, carrots, celery, beans, peas (fins, garden, processed), tomatoes (whole, purée), mushrooms, truffles.

Dehydration: Onions, carrots, potatoes and cabbage are shredded and quickly dried until they contain only 5% water.

Drying: The seeds of legumes (peas and beans) have the moisture content reduced to 10%.

Pickling: Onions and red cabbage are examples of vegetables preserved in spiced vinegar.

Salting: French and runner beans may be sliced and preserved in dry salt.

Freezing: Many vegetables such as peas, beans, sprouts, spinach and cauliflower are deep frozen.

Use

Vegetables	Uses	Menu example
artichoke, globe	hot vegetable	Artichaut en branche; Sauce hollandaise
	cold vegetable	Artichaut en branche; Sauce vinaigrette
	garnish (trimmed into fonds)	Used in garnish clamart
	cold quartered for hors d'œuvre	Artichauts à la grecque
artichoke, Jerusalem	soup	Crème Palestine
	hot vegetable	Topinambours à la crème

Vegetables	Uses	Menu example
asparagus	hot vegetable	Asperges; Beurre fondu
	cold vegetable	Asperges; Sauce mayonnaise
	soup	Creme d'asperges
asparagus points	garnish for egg, fish, meat, poultry and cold dishes	Omelette aux pointes d'asperges Suprême de volaille princesse
aubergine	hot vegetable	Aubergine frite
	hot hors d'œuvre	Aubergine provençale
	hot stuffed vegetable	Aubergine farcie
	garnish for fish and meat dishes	Sole meunière aux aubergines
avocado (see page 206)	hors d'œuvre	Avocat vinaigrette
beans, broad	hot vegetable	Fèves au beurre
beans, runner	hot vegetable	
beans, French	hot vegetable	Haricots verts sautés au beurre
	salad	Salade niçoise
	hors d'œuvre	
beetroot	soup	Bortch
	hors d'œuvre	
	salads	Salade de betterave
broccoli, white	hot vegetable	Brocolis polonaise
broccoli, green	hot vegetable	Brocolis, see Hollandaise
	hors d'œuvre	Brocolis à la grecque
broccoli, purple	hot vegetable	Brocolis au beurre
brussels sprouts	hot vegetable	Choux de Bruxelles nature
cabbage	hot vegetable	Choux verts
cabbage, red	pickled	
	hot vegetable	Choux rouges à la flamande
cabbage, white	hot vegetable	Choux nature
	salad	Cole slaw
carrots	hot vegetable	Carottes Vichy
	soup	Purée Crécy
	hors d'œuvre and garnishes	
cauliflower	hot vegetable	Chou-fleur; Sauce crème
	soup	Crème Dubarry
	hors d'œuvre	Chou-fleur provençale
celeriac	soup	
	hors d'œuvre	
	salad	Salade Waldorf
celery	soup	Crème de céleri
	hors d'œuvre	Céleri à la grecque
	hot vegetable	Céleri braisé
	salad	
	garnish	
chicory	salad	
	hot vegetable	Endive au jus
Chinese cabbage	salad	
	hot vegetable	
cucumber	hors d'œuvre	
	salad	
	garnish	Filet de sole Doria
curly endive	salad	

Vegetables	Uses	Menu example
fennel	hot vegetable	Fenouil braisé
		Fenouil
kale, curly	hot vegetable	Chou-frisé nature
leeks	soup	Cock-a-leekie
	hors d'œuvre	Poireaux; Sauce vinaigrette
	hot vegetable	Poireaux braisés
lettuce, cos	salads	Laitue romaine
lettuce, round or	salads	
cos	hot vegetable	Laitue braisée
	garnish to meat dishes	
mangetout	hot vegetable	Mangetout nature
marrow	hot vegetable	Courge provençale
	hot vegetable	Courge persillées
marrow, small	hot vegetable	Courgette farci
mushrooms	soup	Crème de champignons
	sauces	Sauce chasseur
	hot vegetables	Champignons grillés
	garnish	
	savouries	
	hot hors d'œuvre	Champignons à la crème
onion	soup	Soupe à l'oignon
	sauces	Sauce Soubise
	hors d'œuvre	
	salads	
	hot vegetables	Oignons frits à la française
	egg dishes	Omelette lyonnaise
parsnips	hot vegetable	Panais au beurre
peas	soup	Purée St.-Germain
	hors d'œuvre	
	hot vegetable	Petits pois à la flamande
	salads	
	garnishes	Clamart
peppers	hors d'œuvre	
(pimentos)	salad	
	hot vegetable	Piment farci
potatoes	soup	Purée Parmentier
	hors d'œuvre	Salade de pomme de terre
	hot vegetable	
	garnishes	Parmentier
	salads	
radishes	hors d'œuvre	
	salads	
	decorating aspic work	
salsify	hot vegetable	Salsifis au gratin
		Salsifis sautés
sea-kale	hot vegetable	Chou de mer Mornay
	cold vegetable	Chou de mer; Sauce mayonnaise
shallots	hors d'œuvre	
	sauces	
spinach	soup	
	hot vegetable	Epinards en branche
	garnish	Florentine
	soufflé	Soufflé au épinards
spring greens	hot vegetable	Choux de printemps

Vegetables	Uses	Menu example
swedes	hot vegetable	Purée de rutabaga
sweetcorn	hors d'œuvre	
	vegetable	Maïs, beurre fondu
	garnish	Suprême de volaille Maryland
		Oeuf poché Washington
tomatoes	soups	Crème portugaise
	sauces	Sauce tomate
	hors d'œuvre	
	salads	Salade de tomates
	hot vegetable	Tomates farcies
	garnish	
truffles	hot vegetable	
	garnishes	Périgord
	decorating aspic dishes	
turnips	hors d'œuvre	
	hot vegetable	Navets au beurre
	garnishes	
watercress	soup	Purée cressonnière
	salads	
	garnishes	

Further information: Fresh Food and Vegetable Information Bureau, 9 Walton Street, London SW3 25D.

Vegetable	French name	Wholesale		Retail		Season
		Purchasing unit	Cost	Purchasing unit	Cost	
artichoke, globe	l'artichaut	case-18, 24, 36		single		June–Sept
artichoke, Jerusalem	le topinambour	box				Oct–April
asparagus	l'asperge	bundle or box		bundle		April–June
aubergine	l'aubergine	box		single		June–Oct
beans, broad	les fèves	net or box				June–August
beans, runner		net or box				August–Sept
bean, French	les haricots verts	'chip' box or tray				June–Sept
beetroot	la bétterave	half-bag				all the year
broccoli, white	le brocoli	crate		or head		Jan–March
broccoli, purple	le brocoli	crate				Feb–March
broccoli, green (calabrese)	le brocoli	crate				most of the year
Brussels sprouts	les choux de Bruxelles	net or carton				Oct–March
cabbage	le chou vert	net or box				all the year
cabbage, red	le chou rouge	net or box				Sept–Jan
cabbage, Savoy		net or box				Oct–March
carrots	la carotte	half-bag				*new:* May–June *old:* rest of year
cauliflower	le chou-fleur	crate (10–48)		head		March–Dec
celeriac	le céleri rave	box				Oct–Feb
celery	le céleri			head		Sept–March
chicory	l'endive (f)	box				Nov–March
cucumber	le concombre	tray (12)		single		all the year; at best, May–Sept
kale, curly	le chou frisé	crate				Jan–March
leek	le poireau	crate		single		all the year
lettuce, cos	la laitue romaine	box (12)		single		June-Sept

Vegetable	French name	Wholesale		Retail		Season
		Purchasing unit	Cost	Purchasing unit	Cost	
lettuce, round	la laitue	box (12)		single		all the year *English:* April–Sept *Dutch:* *French:* Oct–March *hot-house:*
marrow	la courge	box (6 pieces)		piece		July–Sept
mushroom, open	le champignon	'chip'				all the year
mushroom, button	le champignon	'chip'				all the year
onion	l'oignon	net				all the year
parsnip	le panais	$\frac{1}{2}$ bag or box				Oct–April
peas	les petits pois	net or box				June–August
peppers	le piment	box		single		Sept–Nov
potatoes, old	le pomme de terre	bag				all the year
potatoes, new	la pomme nouvelle	bag				March–Sept
radishes	le radis	crate (12/18 bunches)		bunch		March–August
salsify	le salsifi	box				Oct–March
sea-kale	le chou de mer	box				*indoor:* Dec–April *outdoor:* April–June
shallot	l'échalote (f.)	bag or net				Sept–Oct
spinach	l'épinard (m.)	crate or box				all the year
spring greens	le chou de printemps	bag or crate				March–May
swedes	le rutabaga	net or box				Oct–March
sweetcorn, or corn on the cob	le maïs	tray or box				July–Sept
tomatoes	la tomate	box				all the year
truffles	la truffe					Oct–March
turnips	le navet	net or $\frac{1}{2}$ bag				*new:* May–July *old:* August–March
watercress	le cresson	'chip'		bunch		all the year

7 Fruits *Les fruits*

For culinary purposes fruit may be divided into the following groups:

Soft fruits: Raspberries, strawberries, loganberries, gooseberries, blackberries, red and black currants.

Hard fruits: Apples and pears.

Stone fruits: Cherries, damsons, plums, apricots, greengages, peaches, nectarines, mangoes.

Citrus fruits: Oranges, lemons, limes, grapefruit, mandarins, clementines, tangerines, satsumas.

Tropical and other fruits: Bananas, pineapples, dates, figs, grapes, melons, rhubarb, cranberries, paw-paws.

Seasons

The chief citrus fruits (oranges, lemons and grapefruit) are available all the year. Mandarins, clementines, satsumas and tangerines are available in the winter.

Rhubarb is in season in the spring and the soft and stone fruit then become available from June in the following order: gooseberries, strawberries, raspberries, cherries, currants, damsons, plums.

Imported apples and pears are available all the year round; home-grown mainly from August to April. Many varieties of fruits are imported from all over the world, frequently using speedy air transport cargo services, and this enables some fruits (eg strawberries) to be in season virtually the whole year round.

Food value

The nutritive value of fruit depends on its vitamin content, especially vitamin C; it is therefore valuable as a protective food.

The cellulose of fruit is useful as roughage.

Storage

Hard fruits, such as apples, are left in boxes and kept in a cool store.

Soft fruits, such as raspberries and strawberries, should be left in their punnets or baskets in a cold room.

Stone fruits are best placed in trays so that any damaged fruit can be seen and discarded.

Peaches are left in their delivery trays or boxes.

Citrus fruits remain in the delivery trays or boxes.

Bananas should not be stored in too cold a place because the skins turn black.

Quality and purchasing points

1 Soft fruits deteriorate quickly, especially if not sound. Care must be taken to see that they are not damaged or too ripe when bought.
2 Soft fruit should appear fresh; there should be no shrinking, wilting or signs of mould.

3 The colour of certain soft fruits is an indication of ripeness (strawberries, gooseberries).
4 Hard fruit should not be bruised. Pears should not be over-ripe.

Preservation

Drying: Apples, pears, apricots, peaches, bananas and figs are dried. Plums when dried are called prunes, and currants, sultanas and raisins are produced by drying grapes.

Canning: Almost all fruits may be canned. Apples are packed in water and known as solid packed apples; other fruits are canned in syrup.

Bottling: Bottling is used domestically, but very little fruit is commercially preserved in this way. Cherries are bottled in maraschino.

Candied: Orange and lemon peel are candied; other fruits with a strong flavour, such as pineapple, are preserved in this way.

The fruit is covered in hot syrup which is increased in sugar content from day to day until the fruit is saturated in a very heavy syrup. It is then allowed to dry slowly until it is no longer sticky.

Glacé: The fruit is first candied and then dipped in fresh syrup to give a clear finish. This method is applied to cherries.

Crystallised: After the fruit has been candied it is left in fresh syrup for 24 hours and then allowed to dry very slowly until crystals form on the surface of the fruit.

Most of the candied, glacé and crystallised fruits are imported from France.

Jam: Fruit which is edible but slightly imperfect is used in the manufacture of jam.

Jelly: Jellies are produced from fruit juice.

Quick freezing: Strawberries, raspberries, loganberries, apples, blackberries, gooseberries, grapefruit and plums are frozen and they must be kept below zero.

Cold storage: Apples are stored at temperatures between 1°C–4°C, depending on the variety of apple.

Gas storage: Fruit can be kept in a sealed store room where the atmosphere is controlled. The amount of air is limited, the oxygen content of the air is decreased and the carbon dioxide increased.

Fruit juices, syrups and drinks

Fruit juices such as orange, lemon, blackcurrant are canned.

Syrups such as rose hip and orange are bottled.

Fruit drinks are also bottled; they include orange, lime and lemon.

Uses

General use: With the exception of certain fruits (lemon, rhubarb, cranberries) fruit can be eaten as a dessert or in its raw state. Some fruits have dessert and cooking varieties – eg apples, pears, cherries and gooseberries.

English	French	Wholesale		Retail		Season
		Unit	Cost	Unit	Cost	
apple	la pomme	box				all year round; cheapest, Oct–Dec
apricot	l'abricot (m.)	box				May–Sept
banana	la banane	box		by number and weight		all year round
blackberry	la mûre de ronce	punnet				Sept–Oct
blackcurrants	le cassis	punnet				July–Sept
red currants	les groseilles rouges	punnet				July–Sept
cherry	la cerise	box				June–August
clementine		box				
cranberries	les airelles	bag or box				Nov–Jan
damson	la prune de damas	box				Sept–Oct
date	la datte	box		box or pkt		
fig	la figue	box or tray		pkt		July–Sept
gooseberry	la groseille à macquereau	box				July–Sept
grapefruit	la pamplemousse	box (approx 40)		number		all year round
grapes	les raisins	box				all year round; best in autumn

English	French	Wholesale		Retail		Season
		Unit	Cost	Unit	Cost	
greengage	la reine-Claude	box				August
lemon	le citron	box		number		all year round
mandarin	la mandarine	box				Nov–June
melon	le melon	box (number varies according to size)		number		
orange	l'orange (f.)	box (approx 210)		number		all year round
peach	la pêche	box or tray		number		Sept
pear	la poire	box or tray				Sept–March
pear, avocado	l'avocat (m.)	tray		number		
pineapple	l'ananas (m.)	box		number		all year round; best in summer
plum	la prune	box		number		July–Oct
raspberry	la framboise	punnet		punnet		June–August
rhubarb	la rhubarbe	box				Dec–June
strawberry	la fraise	punnet		punnet		June–August
		punnet				
tangerine		box or tray				

Stone fruits

Damsons, plums, greengages, cherries, apricots, peaches and nectarines are used as a dessert, stewed (compôte) for jam, pies, puddings and in various sweet dishes. Peaches are also used to garnish certain meat dishes.

Menu examples:	Damson pie	Compôte de reine-Claude
	Condé d'abricot	Pêche melba
	Flan aux prunes	Nectarine au kirsch
	Cerises jubilée	Jambon braisé aux pêches
	Caneton aux cerises	

Hard fruits

Apples: The popular English dessert varieties include Beauty of Bath, Worcester Pearmain, Cox's Orange Pippin, Blenheim Orange, Laxton's Superb and James Grieve. Imported apples include Golden Delicious, Granny Smith and Sturmers. The Bramley is the most popular cooking apple.

Pears: The William Conference and Doyenne du Comice are among the best known pears.

Apples and pears are used in many pastry dishes. Apples are also used for garnishing meat dishes and for sauce which is served with roast pork and duck.

Menu examples:	Apple pie	Apple pudding and custard
	Charlotte aux pommes	Apfelstrudel
	Beignet de pommes	Poire Belle Hélène
	Flan aux poires	

Soft fruit

Raspberries, strawberries, loganberries and gooseberries are used as a dessert. Gooseberries, black and red currants, and blackberries are stewed, used in pies and puddings. They are used for jam and flavourings.

Menu examples: Gooseberry fool
Blackcurrant tart
Stewed red currants and raspberries
Barquette de fraises
Glace aux fraises
Pêches et framboises rafraîchies au kirsch

Citrus fruits

Oranges, lemons and grapefruit are not usually cooked, except for marmalade. Lemons are used for flavouring and garnishing, particularly fish dishes. Oranges are used mainly for flavouring, and in fruit salads; also to garnish certain poultry dishes. Grapefruit are served for breakfasts and as a first course generally for luncheon. Limes, tangerines (mandarines, satsumas, clementines) and tangelos (ugli fruit) are other types of citrus fruit.

Menu examples:	Lime soufflé	Florida cocktail
	Crêpe au citron	Filet de sole frit au citron
	Mandarine glacée	Caneton bigarade
	Bavarois à l'orange	Salade d'orange

Tropical and other fruits

Avocado, bananas, Cape gooseberries, cranberries, dates, figs, grapes, guavas, kiwi fruit, lychees, mangoes, melons, papaya (paw-paw), passion fruit, persimmon, pineapple, rhubarb.

Avocado pear: a green skinned pear - shaped fruit with a bland, mild, nutty flavour. There are two main types, a) smooth skin, soft when ripe, b) rough pebbly skin, green when unripe, purple-black when ripe. Avocado is usually served as a first course, halved, stone removed and filled with a variety of fillings eg crabmeat, prawns or shrimps in a sauce or salad dressing, hot or cold.

Banana: As well as being used as a dessert bananas are grilled for a fish garnish, fried as fritters and served as a garnish to poultry (Maryland). They are used in fruit salad and other sweet dishes.

Menu examples:	Beignets de bananes	Filet de sole caprice
	Flan aux bananes	Poulet Maryland

Cape gooseberries (physalis): a sharp, pleasant flavoured small round fruit dipped in fondant and served as a type of petit four.

Cranberries: These hard red berries are used for cranberry sauce, which is served with roast turkey.

Dates: Whole dates are served as a dessert; stoned dates are used in various sweet dishes and petits fours.

Menu examples:	Date pudding; almond sauce
	Date and apple slice

Figs: Fresh figs may be served as a first course or dessert. Dried figs are used for fig puddings.

Grapes: Black and white grapes are used as a dessert, in fruit salad, as a petits fours and also as a fish garnish (véronique).

Guavas: size varies between that of a walnut to an apple. Ripe guavas have sweet pink flesh. They can be eaten with cream or mixed with other fruits.

Kiwi fruit: have a brown furry skin. The flesh is green with edible black seeds which when thinly sliced gives a pleasant decorative appearance.

Lychees: a Chinese fruit with a delicate flavour. Obtainable tinned in syrup and also fresh.

Mangoes: can be as large as a melon or as small as an apple. Ripe mangoes have smooth pinky-golden flesh with a pleasing flavour. They may be served in halves sprinkled with lemon juice, sugar, rum, or ginger. Mangoes can also be used in fruit salad and for sorbets.

Melon: There are several types of melon. The most popular are:

Honeydew: These are long, oval-shaped melons with dark green skins. The flesh is white with a greenish tinge. Imported from North Africa and Spain. Season – late summer, autumn, winter.

Charentais: Charentais melons are small and round with a mottled green and yellow skin. The flesh is orange coloured. They are imported from France. Season – late summer.

Cantaloup: Cantaloup are large round melons with regular indentations. The rough skin is mottled orange and yellow and the flesh is light orange in colour. They are imported mainly from France and Holland. Season – late summer.

Care must be taken when buying. Melons should not be over- or under-ripe. This can be assessed by carefully pressing the top or bottom of the fruit. There should be a slight degree of softness to the cantaloup and charentais melons.

The stalk should be attached, otherwise the melon deteriorates quickly.

Uses: Melon is mainly used as a dessert and for hors d'œuvre and sweet dishes.

Menu examples: Melon frappé; Melon surprise

Paw-paw (papaya): green to golden skin, orangey flesh with a sweet subtle flavour and black seeds. Used as for mangoes.

Passion fruit: the name comes from the flower of the plant which is meant to represent the Passion of Christ. Size and shape of an egg with crinkled purple-brown skin when ripe. Flesh and seeds are all edible.

Persimmon: a round orange red fruit with a tough skin which can be cut when the fruit is ripe. When under ripe they have an unpleasant acidlike taste of tannin.

Pineapple: Pineapple is served as a dessert; it is also used in many sweet dishes and as a garnish to certain meat dishes.

Menu examples:

Ananas Créole	Grilled gammon and pineapple
Ananas en surprise	Pineapple fritters, apricot sauce

Rhubarb: Forced or early rhubarb is obtainable from January. The natural rhubarb from April–June.

Used for pies, puddings, fool and compôte.

Further information: Fruit Trades Journal, Market Towers, New Covent Garden. London W8.

8 Nuts *Les noix (f)*

Nuts are the reproductive kernel (seed) of the plant or tree from which they come. Nuts are perishable and may easily become rancid or infested with insects. Nuts are used extensively in pastrywork, confectionery, vegetarian cooking and the preparation of some liqueurs.

		Wholesale		Retail	
English	French	Unit	Cost	Unit	Cost
almond	l'amande (f)				
Brazil	la noix de Bresil				
chestnut	le marron				
coconut	la noix de coco				
filbert					
hazel nut	la noisette				
pecan	la pacane				
pistachio	la pistache				
walnut	la noix				

Season
Dessert nuts are in season during the autumn and winter.

Food value
Nuts are highly nutritious because of their protein, fat and mineral salts. They are of considerable importance to vegetarians, who use nuts in place of meat; it is therefore a food which builds, repairs and provides energy. Nuts are difficult to digest.

Storage
Dessert nuts, those with the shell on, are kept in a dry, ventilated store. Nuts without shells, whether ground, nibbed, flaked or whole, are kept in air-tight containers.

Quality and purchasing points

1 Nuts should be of good size.
2 They should be heavy for their size.
3 There must be no sign of mildew.

Use
Nuts are used as a dessert, as a main ingredient to vegetarian dishes, also for decorating and flavouring.

They are used whole, or halved, and almonds are used ground, nibbed and flaked.

Almonds
Salted almonds for cocktail parties and bars.

Ground, flaked, nibbed, for use in sweet dishes and for decorating cakes.

For cake mixtures, large and small, such as Dundee cake, Congress tarts, macaroons; for petits fours and large cakes.

Marzipan and frangipan for Bakewell tarts.
Praline for ice cream and gâteaux.
Coating for Pomme Berny.

Brazil nuts
Brazil nuts are served with fresh fruit as dessert and are also used in confectionery.

Chestnuts

Stuffing for turkeys.	Chestnut flour for soup.
Garnish for ice cream.	As a sweet dish (Mont Blanc aux marrons).
Petits fours	Marron glacé.

Coconut
Coconut is used in desiccated form for curry preparations, in cakes and for decorating cakes, such as madeleines.

Filberts and hazel nuts
These nuts are used as a dessert and for praline.

Pecans
Pecan nuts are used salted for dessert, various sweets and ice cream.

Peanuts and cashew nuts
These are salted and used in cocktail bars.

Pistachio nuts
These small green nuts, grown mainly in France and Italy, are used for decorating galantines, small and large cakes and petits fours. They are also used for ice cream.

Walnuts
Walnuts, imported mainly from France and Italy, are used as a dessert, in salads and for decorating cakes and sweet dishes. They are also pickled.

9 Eggs — *Les oeufs*

The term eggs applies not only to those of the hen, but also to the edible eggs of other birds, such as turkeys, geese, ducks, guinea fowl, quails and gulls.

Hens' eggs

Cost

Wholesale unit	crate 360 eggs
Retail unit	dozen

Quality points for buying

1 The eggshell should be clean, well shaped, strong and slightly rough.
2 When broken there ought to be a high proportion of thick white to thin white.
3 The yolk should be firm, round and of a good even colour.

If an egg is kept, the thick white gradually changes into thin white and water passes from the white into the yolk. The yolk loses strength and begins to flatten, water evaporates from the egg and is replaced by air, and as water is heavier than air fresh eggs are heavier than stale ones.

The professional method of testing eggs for quality is by a process known as 'candling' in which a candling lamp is used; this produces a beam of light strong enough to penetrate the shell of an egg to illuminate the contents and to show up any defects.

It is also possible to determine the freshness of an egg by placing it in a 10% solution of salt (60 g salt to $\frac{1}{2}$ litre water). A two-day-old egg will float near the bottom of the solution with its broad end upward. As the egg ages it becomes lighter and floats closer to the surface of the solution.

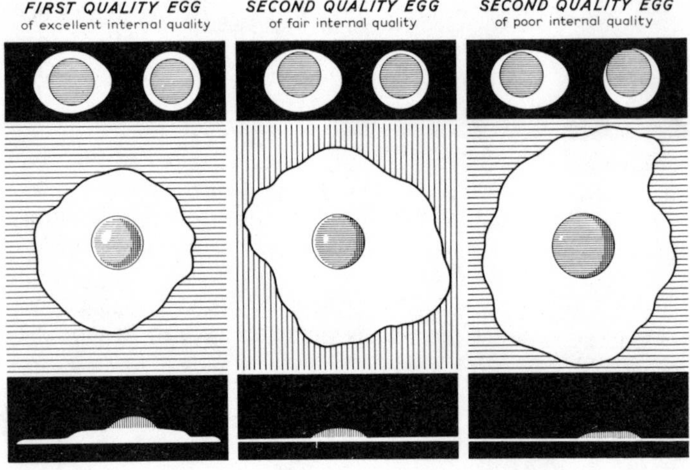

Quality of eggs

Storage

1 Eggs must be stored in their packing trays blunt end upwards, in a cool but not too dry place; a refrigerator of 0–5°C is ideal.
2 No strongly smelling foods such as cheese, onions and fish should be stored near the eggs because the eggshells are porous and the egg will absorb strong odours.

3 Eggs should not be washed before being stored as washing would remove the natural protective coating.
4 Eggs are stored point end down.

Food value

Eggs contain most nutrients and are low in Calories - two eggs containing 180 Calories. Egg protein is complete and easily digestible, therefore it is useful to balance meals. Eggs may also be used as the main dish; they are a protective food and provide energy and material for growth and repair of the body.

Production

Hens' eggs are graded in seven sizes:

Size 1	70g	Size 5	50g
2	65g	6	45g
3	60g	7	under 45g
4	55g		

Small eggs are usually more plentiful in late summer and autumn. The size of an egg does not affect the quality but does affect the price. Eggs are collected from the farmers at least once a week by the egg packing station. The eggs are tested for quality by candling, then weighed and stamped with the grade, and the packing station number. They are then packed into boxes containing 30 dozen, 360 (3 long hundreds). The wholesale price of eggs is quoted per long hundred (120).

All egg-boxes leaving the packing station are dated.

Preservation

Cold storage: Eggs are kept a little above freezing-point. The humidity of the air and the amount of carbon dioxide in the air are controlled. They will keep about 9 months under these conditions.

Frozen eggs: These are used mainly by bakers and confectioners; they are sold in 14, 16, 28 and 42 lb tins. The eggs are broken, thoroughly mixed and then frozen.

Dried eggs: The eggs are broken, well mixed and then spray dried at a temperature of approx. 71°C. These are also used mainly by bakers and confectioners.

Grease method: A pure grease free from salt, water and other impurities must be used, such as Oteg, paraffin wax or lard. The eggs are dipped into the liquid grease and then allowed to dry. The grease fills up the porous shell, forms a skin and so excludes air. They may then be stored in the same way as fresh eggs.

Waterglass - sodium silicate: A solution is made up with sodium silicate and boiling water. When used it must be quite cold; the newlaid eggs are packed point downwards in an earthenware bowl or galvanised pail and covered with the solution. A lid is placed on the container to prevent evaporation.

Uses of eggs

Hors d'œuvre: Chiefly hard-boiled egg for egg mayonnaise and for many composed salads such as fish salad.

Soups: For the clarifying of consommé, in the preparation of royales for garnishing consommé, for thickening certain soups, eg Germiny and veloutés.

Egg dishes: These dishes are very popular on breakfast and luncheon menus and include scrambled, poached, soft-boiled, hard-boiled, en cocotte, sur le plat and omelets.

Farinaceous: Eggs are used in the making of pastes for ravioli, canneloni and noodles.

Fish: In the preparation of frying batters and for coating fish prior to crumbing.

Sauces: Mayonnaise, hollandaise, béarnaise, sabayon are some of the sauces made with eggs.

Meat and poultry: For binding mixtures such as Vienna steaks and chicken cutlets, and for coating cuts of meat and poultry prior to cooking.

Salads: Usually hard-boiled and included in many composed salads.

Sweets and pastries: Eggs are used in many ways for these items.

Savouries: Scotch woodcock, cheese soufflé, savoury flans are some of the dishes in which eggs are used.

Other eggs

Turkeys' and guinea fowls' eggs may be used in place of hens' eggs.

The eggs of the goose or duck may be used only if they are thoroughly cooked.

Quails eggs are used in some establishments as a garnish.

Further information: British Egg Information, 37 Panton Street, London SW1.

10 Milk — *Le lait*

The milk most used in this country is that obtained from cows. Goats' milk and ewes' milk can also be used.

Food value

Milk is almost a perfect food as it contains all the nutrients required for growth, repair, energy, protection and regulation of the body.

Storage

Milk keeps less well than almost any other food; it readily becomes dirty and unsafe; therefore it must be stored with care.

Milk is an excellent food for human beings; it is also, unfortunately, an excellent food for bacteria. Cows are subject to tuberculosis and other infections, precautions therefore are taken and herds are tested for TB.

As milk is so easily contaminated it can be dangerous. Contamination can occur before milking - cows may be diseased; during milking - by unclean cows, premises, utensils or milker; after milking - at the dairy, in transit or in the kitchen.

Storage points

1 Fresh milk should be kept in the container in which it is delivered.
2 Milk must be stored in the refrigerator.
3 Milk should be kept covered as it absorbs strong smells such as onion or fish.
4 Fresh milk and cream should be purchased daily.
5 Tinned milk is stored in a cool, dry ventilated room.
6 Dried milk is stored in air-tight tins and kept in a dry store.
7 Imitation cream is kept in the refrigerator.

	Wholesale		Retail	
Milk	Unit	Cost	Unit	Cost
pasteurised				
UHT				
Channel Island				
sterilised				
homogenised				
dried milk	bags, tins			
evaporated	case		tin	
condensed	case		tin	
Cream				
double				
whipping				
single				
Devon – clotted cream				
UHT				

Pasteurised: Milk which has been heated for 15 seconds at 72°C to kill harmful bacteria, then cooled quickly.

U.H.T. (ultra heat treatment): Milk which has been subjected to ultra heat treatment, that is 132°C for one second. Unopened it will keep without refrigeration for 6 months.

Homogenised: Milk which is homogenised is treated so that the cream is dispersed throughout the milk. The milk and cream do not separate when left to stand.

Channel Island and South Devon milk: This is produced by Jersey, Guernsey, and South Devon herds in this country. It has a high (4%) fat content.

Sterilised milk: This is homogenised milk which is heated to 104°C–110°C for 30–40 minutes. It keeps for 2–3 months under refrigeration in the unopened bottle, but has a different taste from fresh milk.

Preservation
Evaporated milk has had 60% of the water removed by evaporation before canning.

Condensed milk is richer than evaporated because more water has been removed. It can be sweetened or unsweetened.

Dried milk is either spray or roller processed.

Uses of milk
1 Soups and sauces
2 Cooking of fish, vegetables and gnocchi
3 Making of puddings, cakes, sweet dishes
4 Cold drinks – milk, milk shakes, malts
5 Hot drinks – tea, coffee, cocoa, chocolate

Cream *La crème*

This is the concentrated milk fat which is skimmed off the top of the milk and should contain at least 18% butterfat. Cream for whipping must contain more than 30% butterfat. (See charts, pages 216–217)

Single cream (18% fat content)
Does not whip, but it can be used in its liquid state with sweet dishes and for finishing soups, fish, meat and poultry sauces and stews (Blanquette, Fricassée).

Whipping cream (35% fat content)

Double cream (48% fat content)
Owing to its higher fat content, more than 30% butterfat, it whips and is used for decorating and filling pastries, gâteaux and with sweet dishes.

Devon or clotted cream (55% fat content)
Served with Devon Teas and fresh fruit; also fruit compôtes and pies.

Imitation creams
There are several types and qualities available which are produced from an emulsion of oil, margarine or butter with milk powder and water. Imitation cream is used for filling and decorating small and large cakes and making and finishing sweet dishes.

Use of cream

1 Fresh cream must be cold when required for whipping.
2 For preference it should be whipped in china or stainless steel bowls. If any other metal is used, the cream should be transferred to china bowls as soon as possible.
3 If fresh cream is whipped too much it turns to butter. This is more likely to happen in hot conditions. To prevent this, stand the bowl of cream in a bowl of ice while whisking.

4 When adding cream to hot liquids dilute the cream with some of the liquid before adding to the main bulk. This helps to prevent the cream from separating.

Yogurt
Yogurt is a curd-like food, prepared from milk fermented by the action of bacteria feeding on the lactose (milk sugar) and producing lactic acid. Species of lactobacillus bacteria are used.

Types of yogurt

1 Fat free yogurt – contains less than 0.5% milk fat
2 Low fat yogurt – contains maximun 1.5% milk fat
3 Wholemilk yogurt – contains fat as in whole milk
4 Whole or real fruit yogurt – contains whole fruit in sugar syrup
5 Fruit flavoured yogurt – contains fruit juices or syrup

Further information: Milk Marketing Board, Thames Ditton, Surrey.

11 Fats and oils

Storage of all fats
Fats should be kept in a cold store and in warm weather in a refrigerator.

Butter — *Le beurre*

Butter must be kept away from strong-smelling foods. Butter is produced by churning the cream of milk. One litre of cream yields approximately one ½ kg butter.

Food value
Butter is an energy food as it has a very high fat content.

	Wholesale		Retail	
Fats	Unit	Cost	Unit	Cost
butter margarine lard suet cooking fat	boxes or packets cartons			

Quality

1 The taste should be creamy and pleasant.
2 The texture should be soft and smooth.
3 It must smell fresh.
4 The colour of pure butter is almost white or very pale yellow.
5 Fresh butter should be used fairly quickly, otherwise it goes rancid (acquires an unpleasant taste and smell).

Type of Cream	Legal Minimum Fat %	Processing and Packaging	Storage	Characteristics and Uses
CREAM OR SINGLE CREAM	18	Homogenised and pasteurised by heating to about 79.5°C (175°F) for 15 seconds then cooled to 4.5°C (40°F). Automatically filled into bottles and cartons after processing. Sealed with foil caps. Bulk quantities according to local suppliers.	2-3 days in summer 3-4 days in winter under refrigeration.	A pouring cream suitable for coffee, cereals, soup or fruit. A valuable addition to cooked dishes. Makes delicious sauces.
WHIPPING CREAM	35	Not homogenised, but pasteurised and packaged as above.	2-3 days in summer 3-4 days in winter under refrigeration.	The ideal whipping cream. Suitable for piping, cake and dessert decoration, ice-cream, cake and pastry fillings.
DOUBLE CREAM	48	Slightly homogenised, then pasteurised and packaged as above.	2-3 days in summer 3-4 days in winter under refrigeration.	A rich pouring cream which will also whip. The cream will float on coffee or soup.
DOUBLE CREAM 'THICK'	48	Heavily homogenised, then pasteurised and packaged. Usually only available in domestic quantities.	2-3 days in summer 3-4 days in winter under refrigeration.	A rich spoonable cream which will not whip.
DOUBLE CREAM 'EXTENDED LIFE'	48	Heated to 82°C (180°F) for 15 seconds and cooled. Homogenised and filled into bottles, vacuum sealed, then heated in bottles to 115°C (240°F) for 12 minutes. Usually only available in domestic quantities.	2-3 weeks under refrigeration.	A spoonable cream which will lightly whip.

CLOTTED CREAM	55	Heated to 82°C (180°F) and cooled for about $4\frac{1}{2}$ hours. The cream crust is then skimmed off. Usually packed in cartons by hand. Bulk quantities according to local suppliers.	2-3 days in summer 3-4 days in winter under refrigeration.	A very thick cream with its own special flavour and colour. Delicious with scones, fruit and fruit pies.
STERILISED HALF CREAM	12	Homogenised, filled into cans and sealed. Heated to 115°C (240°F) for 20 minutes, then cooled rapidly.	Up to two years if unopened.	A pouring cream with a slight caramel flavour.
STERILISED CREAM	23	Processed as above.	Up to two years if unopened.	A thicker, caramel flavoured cream which can be spooned but not whipped.
ULTRA HEAT TREATED CREAM	18	Homogenised and heated to 132°C (270°F) for one second and cooled immediately. Aseptically packed in foil lined containers. Available in bigger packs for catering purposes.	6 weeks if unopened. Needs no refrigeration. Usually date stamped.	A pouring cream.

Production
Butter consumed in this country is produced in England, New Zealand, Australia, France, Holland and Denmark. Butter is often blended and can be salted or unsalted; the salt does act as a preservative. Butter is also mixed with margarine and sold as a special blend.

Use
Butter is used for most kitchen purposes where expense does not have to be considered.

Butter can be used for: making roux for soups and sauces; finishing sauces such as sauce vin blanc for fish, sauce madère for meat; hard butter sauces (maître d'hôtel) and butter sauces (hollandaise); pot-roasting meat, poultry and game; finishing vegetables (petits pois au beurre); making of all pastes except suet, hot water and nouille paste; decorating cold dishes and cocktail savouries; making cakes and butter creams.

It can be clarified and used for shallow frying of all kinds of food.

Further information: Butter Information Council, Pantiles House, 2 Neville Street, Tunbridge Wells, Kent.

Margarine *La margarine*
Margarine is produced from milk and a vegetable oil (groundnut, palm, coconut, cotton seed or soya bean).

Food value
Margarine is an energy and protective food. With the exception of palm oil, the oils used in the manufacture of margarine do not contain vitamins A and D; these are added during production. Margarine is not inferior to butter from the nutritional point of view.

Quality
There are several grades of margarine and some are blended with butter. Taste is the best guide to quality.

Production
The vegetable oils are obtained from Commonwealth countries, West Africa and South-East Asia. Margarine is made by first extracting the oils and fats from the raw materials. These are refined, blended, flavoured and coloured, then mixed with fat-free pasteurised milk. The emulsion is then churned, cooled and packed. Cake and pastry margarines are blended in a different manner to table margarine to produce the required texture suitable for mixing.

Use
Margarine can be used in place of butter. The difference being: the smell is not so pleasant; nut brown (beurre noisette) or black butter

(beurre noir) cannot satisfactorily be produced from margarine. The flavour of margarine when used in the kitchen is inferior to butter – it is therefore not so suitable for finishing sauces and dishes.

It should be remembered that it is equally nutritious and may be cheaper than butter.

Animal fats

Lard *Le saindoux*

Lard is the rendered fat from the pig. Lard has almost 100% fat content. It may be used in hot water paste and with margarine to make short paste. It can also be used for deep or shallow frying.

Suet

Suet is the hard solid fat deposits in the kidney region of animals. Beef suet is the best and it is used for suet paste stuffing and mince-meat.

Dripping

Dripping is obtained from clarified animal fats and it is used for deep or shallow frying.

Further information: Unilever Ltd, Unilever House, Blackfriars, London, EC4.

Oil *L'huile*

Oils are fats which are liquid at room temperature.

	Wholesale		Retail	
	Unit	Cost	Unit	Cost
olive	cans			
maize	cans			
groundnut (arachide)	cans or drums			

Other varieties of oil are obtained from sunflower seed, soya bean, walnut, grape seed, sesame, almond and wheatgerm.

Food value

As oil has a very high fat content it is useful as an energy food.

Storage

1. Oil should be kept in a cool place.
2. If refrigerated some oils congeal; they return to a fluid state in a warm temperature.
3. Oils keep for a fairly long time, but they do go rancid if not kept cool.

Quality

Olive oil is the best, owing to its flavour. Better grade oils are almost without flavour, odour and colour.

Production

Olive oil is extracted from olives grown in Mediterranean countries, particularly Spain, Italy, Greece and France.

Groundnut oil is obtained from groundnuts grown in West Africa.

Maize oil is obtained from maize grown in Europe and the USA.

The oil is extracted from the raw material, refined and stored in drums.

Use

Olive oil is used for making vinaigrette and mayonnaise and in the preparation of hors d'œuvre dishes. Walnut and groundnut oils may also be used. It is also used in making farinaceous pastes and for shallow frying.

Other oils are used for deep frying. Oil is used for lubricating utensils, trays and also marble slabs to prevent cooked sugar from sticking.

Oil may also be used to preserve foods by excluding air.

Points on the use of all fats and oils

For frying purposes a fat or oil must, when heated, reach a high temperature without smoking. The food being fried will absorb the fat if the fat smokes at a low temperature.

Type	Approx. flash-point	Smoke point	Recommended frying temp
	(°C)	(°C)	(°C)
finest quality vegetable oils	324	220	180
finest vegetable fat	321	220	180
high-class vegetable oil	324	204	180
pure vegetable fat	318	215	170–182
pure vegetable oil	330	220	
finest quality maize oil	224	215	180
finest fat	321	202	180
finest quality dripping	300	165	170–180
finest natural olive oil	270–273	148–165	175

Fats and oils should be free from moisture, otherwise they splutter.

As they are combustible, fats and oils can catch fire. In some fats the margin between smoking and flash point may be narrow.

A good frying temperature is 75°C–180°C.

Further information: British Edible Oils Ltd, Knights Road, London, E16; Proctor and Gamble Catering Information Service, Groat House, Collingwood Street, Newcastle upon Tyne, NE1 1XR.

12 Cheese *Les fromages*

Cheese is made from cows', ewes' or goats' milk and it takes approximately 5 litres of milk to produce $\frac{1}{2}$ kg of cheese.

There are many hundreds of varieties; most countries manufacture their own special cheeses.

Types

There are four main types of cheese with numerous varieties of each:

1 Hard cheese.
2 Semi-hard cheese.
3 Soft or cream cheese.
4 Blue-vein cheese.

Quality

1 The skin or rind of cheese should not show spots of mildew, as this is a sign of damp storage.
2 Cheese when cut should not give off an over-strong smell or any indication of ammonia.
3 Hard, semi-hard and blue-vein cheese when cut should not appear dry.
4 Soft cheese when cut should not appear runny, but should have a delicate creamy consistency.

Production

Cheese is produced by almost every country in the world, and is usually made from cows' milk, but some cheese is made from goats' milk, eg certain types of Parmesan; and some from ewes' milk, eg Roquefort.

Rennet is the chief fermenting agent used in cheese-making and is a chemical substance found in the gastric juice of a calf or lamb.

A typical cheese-making process, briefly, is as follows:

1 One gallon of milk makes approximately one pound of cheese.
2 The milk is tested for acidity and then made sour by using a starter (bacteria which produce lactic acid).
3 Rennet is added, which causes the milk to curdle.
4 The curds are stirred, warmed and then allowed to settle.
5 The liquid (whey) is run off.
6 The curds are ground, salted and put into moulds. If a hard cheese is being made, then pressure is applied in order to squeeze out more of the whey.
7 The curds are now put into the special mould and a skin or rind is allowed to form.
8 When set, the cheese is removed from the mould and is then kept in special storage in order to mature and develop flavour.

Varieties	Country	Wholesale purchasing unit (whole cheese)	Approx price per unit	Retail purchasing unit	Approx price per unit
Hard cheeses:					
Cheddar	England				
Cheshire	England				
Double Gloucester	England				
Caerphilly	England				
Derby	England				
Lancashire	England				
Leicester	England				
White Stilton	England				
White Wensleydale	England				
Emmental	Switzerland				
Gruyère	Switzerland				
Edam	Holland				
Gouda	Holland				
Parmesan	Italy				
Semi-hard cheeses:					
St.-Paulin	France				
Pont l'Éveque	France				
		square chip boxes		square chip boxes	
Bel Paese	Italy				
Soft cheeses:					
Camembert	France	case of 24 cheeses		1 round cheese in chip box	
Camembert	France	case of 48 × $\frac{1}{2}$ cheeses		1 or half round cheese in chip box	

Varieties	Country	Wholesale purchasing unit (whole cheese)	Approx price per unit	Retail purchasing unit	Approx price per unit
Soft cheeses (cont.):					
Brie	France			large wedge: in chip box	
Brie	France			small wedge: in chip box	
Carré de l'Est	France	case of 24 cheeses each		1 square cheese in chip box	
Pommel Demi-Sel	France	boxes of 6 cheeses		individual cheeses	
Pommel Demi-Suisse	France	box of 3 cheeses		individual cheeses	
Blue-vein cheeses:					
Stilton	England				
Lymeswold	England				
Wensleydale	England				
Cheshire	England				
Roquefort	France				
Gorgonzola	Italy				
Danish Blue	Denmark				

Storage
All cheese should be kept in a cool, dry, well-ventilated store and whole cheeses should be turned occasionally if being kept for any length of time. Cheese should be kept away from other foods which may be spoilt by the smell.

Food value
Cheese is a highly concentrated form of food. Fat, protein, mineral salts and vitamins are all present. Therefore it is an excellent body-building, energy-producing, protective food.

Preservation
Certain cheeses may be further preserved by processing. A hard cheese is usually employed, ground to a fine powder, melted, mixed with pasteurised milk, poured into moulds then wrapped in lacquered tinfoil, eg processed Gruyère, Kraft, Primula.

Uses
Cheese has many uses in cookery:

Soups
Grated Parmesan cheese is served as an accompaniment to many soups, eg Minestrone.

It is also used to form a crust on top of brown onion soup, eg Soupe à l'oignon.

Farinaceous
A grated hard cheese, usually Parmesan, is mixed in with or is also served as an accompaniment to most farinaceous dishes: eg Spaghetti Italienne, Ravioli.

Egg dishes
Omelette au fromage
Œuf dur Chimay

Fish dishes
Coquille Saint Jacques Mornay
Filets de sole florentine

Vegetables
Chou-fleur au gratin
Chou de mer Mornay

Savouries
Welsh rarebit
Quiche lorraine
Soufflé au fromage

A well-ordered restaurant should always be able to offer a good selection of assorted cheeses on a 'cheese board', and this should be available after lunch and dinner. At least six varieties should be presented, eg:

Cheddar	Gorgonzola
Gruyère	Edam
Camembert	Caerphilly

Hard cheeses

Cheddar
One of the oldest English cheeses, originally made in the Cheddar district of Somerset. It is now made in Scotland, Canada, Australia, New Zealand, all using the original 'cheddaring process'. There are two main types: a) Factory Cheddar which is made whenever and wherever milk is plentiful; therefore it is made in large quantities and the price is economical. b) Farmhouse Cheddar is made from May to October when the cows are out to grass on fresh feed. The flavour of the resulting cheese is buttery, mellow and nutty.

Cheshire
This is the oldest English cheese. As with Cheddar it is produced by 'factory' and 'farmhouse' methods. It is made in two colours, red and white, and is loose, flaky cheese with a mild flavour.

Double Gloucester
A close, crumbly cheese, similar to Cheshire, which ripens slowly and takes 6 months to mature.

Caerphilly
Originally made in Wales, now produced in the south-west of England. A white, fairly firm cheese, delicate and mild in flavour, and it will keep for about 3 weeks only.

Lancashire
A hard cheese, excellent for toasting. It has a rich mellow flavour and is at its best at about 3 months old.

Gruyère and Emmental (Swiss)
These are cooked hard cheeses, pale yellow in colour, honeycombed with holes caused by rapid fermentation during manufacture. Emmental has large holes, gruyère small holes. About 10 months are needed for these cheeses to ripen.

Parmesan (Italian)
This is the hardest cheese of all; when ripe the crust is black, the cheese pale yellow. It is always used for cooking or grated as an accompaniment, never on a cheese board.

Semi-hard cheeses

St.-Paulin (French)
Originally known as Port-Salut. A mild-flavoured spongy cheese, round in shape, 14–20 cm diameter and 3–6 cm thick.

Pont L'Éveque (French)
Made in 8 cm squares, 3 cm thick. It has a rough, thick rind and elastic spongy texture and a slightly sour flavour.

Bel Paese (Italian)
Similar to St.-Paulin.

Soft cheeses

Camembert (French)
One of the most famous of the French soft cheeses. It is made round, 9 cm in diameter, $1\frac{1}{2}$ cm thick, during the summer when the cows' milk is at its creamiest. To serve this cheese at its best it must be ripe – that is, of a soft creamy consistency. When under-ripe it has a white, chalky appearance and when over-ripe it turns a dark, unpleasant colour and gives off an unpleasant smell.

Brie (French)
Another very famous cheese, which is 36–48 cm in diameter. It is traditionally served on a mat of straw. Made during the autumn and usually obtainable November–May. It should be served ripe like Camembert.

Carré de L'Est (French)
A square-shaped cream cheese similar to the Camembert.

Pommel Demi-Suisse (French)
Sometimes referred to as Petit Suisse, it is white and creamy, croquette shaped, 3 cm diameter and 5 cm high. In spite of its name it is a French Cheese.

Blue-vein cheeses

Stilton
Originally made in a village of the same name in Huntingdonshire. It is a rich double cream cheese with blue veins radiating from the centre. The blue veins are caused by inoculating the cheese with a mould. The rind should be slightly wrinkled, moist and of a drab colour. Must be eaten ripe. It takes about 6–9 months to mature.

Lymswold
The English mild blue full fat soft cheese with light white crust.

Wensleydale
Made in Yorkshire; blue-veined; similar to Stilton, but smaller.

Roquefort (French)
Made from ewes' milk in the South of France during the lambing season

and matured in limestone caves. The rennet used is taken from the lambs. Blue-veining is made by placing layers of mouldy bread-crumbs between the curds during manufacture.

Gorgonzola (Italian)
Originally made in a village of the same name near Milan in North Italy. It is a very rich, fully flavoured, blue-vein cheese.

Danish blue (Danish)
An imitation of Roquefort cheese.

Further information: The Cheese Bureau, 40 Berkeley Square, London, W 1; Comité National de Propagande des Produits Laitiers Français, 7 Rue Scribe, Paris IX.

13 Cereals

Cereals are cultivated grasses, but the term is broadened to include sago, rice and arrowroot. All cereal products contain starch. The following are the important cereals used in catering: wheat, oats, barley, maize, rice, tapioca, sago, arrowroot.

Products	Wholesale purchasing unit	Cost per unit	Retail purchasing unit	Cost per unit
flour (soft)	bags			
flour (strong)	bags			
flour (wholemeal)	bags			
semolina	bags			
macaroni	boxes			
spaghetti	boxes			
vermicelli				
noodles				
alphabets				

Wheat — *Le blé*

Source
Wheat is the most common cereal produced in the western world and is grown in most temperate regions. Large quantities are home-grown and a great deal, particularly in the form of strong flour, is imported from Canada.

Food value
Cereals are one of the best energy foods. Whole grain cereals provide vitamin B and are therefore protective foods.

Storage

1 The store room must be dry and well ventilated.

2 Flour should be removed from the sacks and kept in wheeled bins with lids.
3 Flour bins should be of a type that can be easily cleaned.

Flour is probably the most common commodity in daily use. It forms the foundation of bread, pastry and cakes and is also used in soups, sauces, batters and other foods.

Production of flour

The endosperm of the wheat grain contains all the material used by the baker. It consists of numerous large cells of net-like form in which starch grains are tightly packed. In addition, the cells contain an insoluble protein called gluten. When flour is mixed with water it is converted into a sticky dough. This characteristic is due to the gluten which becomes sticky when moistened. The relative proportion of starch and gluten varies in different wheats, and those with a low percentage of gluten are not suitable for bread-making, ie soft flour. For this reason, wheat is blended.

In milling, the whole grain is broken up, the parts separated, sifted, blended and ground into flour. Some of the outer coating of bran is removed as is also the wheat germ which contains oil and is therefore likely to become rancid and so spoil the flour. For this reason wholemeal flour should not be stored for more than 14 days.

White flour contains 70–72% of the whole grain (the endosperm only).

Wholemeal flour contains 100% of the whole grain.

Wheatmeal flour contains 90% of the whole grain.

'Self-raising flour' is white flour with the addition of cream of tartar and bicarbonate of soda.

Uses of wheat products

Soft flour: Large and small cakes, biscuits, all pastes except puff and flaky, thickening soups and sauces, batters and coating various foods.

Strong flour: Bread, puff and flaky pastry, and Italian pastes.

Wholemeal flour: Wholemeal bread and rolls.

Further information: Flour Advisory Bureau, 21 Arlington Street, London, W 1.

Semolina is granulated hard flour prepared from the central part of the wheat grain. The finest semolina is prepared from maize.
Uses: Gnocchi, milk puddings, moulds and as a dusting for certain pastes such as noodle and ravioli.

Menu example: Gnocchi Romaine

Macaroni and spaghetti: Soups, farinaceous dishes, garnishes.

Menu examples: Minestrone, Macaroni au gratin,
Escalope de veau napolitaine

Vermicelli: Garnishing soups, milk pudding.

Menu example: Purée longchamps

Noodles: Garnishing soups, farinaceous dishes, meat dishes.

Menu examples: Nouilles au beurre
Bœuf braisé aux nouilles

Alphabets: Garnishing soups.

Menu example: Consommé alphabétique

Oats — *L'avoine (f)*

Oats are either rolled into flakes or ground into three grades of oatmeal: coarse, medium and fine.

	Wholesale purchasing unit	Cost per unit	Retail purchasing unit	Cost per unit
rolled oats				
oatmeal (coarse)				
oatmeal (medium)				
oatmeal (fine)				

Source

Oats are one of the hardiest cereals, and are grown in large quantities in Scotland and the north of England.

Food value

Oats have the highest food value of any of the cereals. They contain a good proportion of protein and fat.

Storage

Because of the fat content, the keeping quality of oat products needs extra care. They should be kept in containers with tight-fitting lids, and stored in a cool, well-ventilated store room.

Uses

Rolled oats: porridge.

Oatmeal: porridge, thickening soups, coating foods, cakes and biscuits, haggis.

Patent rolled oats nowadays largely displace oatmeal and have the advantage of being already heat treated and consequently more quickly and easily cooked.

Barley — *L'orge (f)*

Barley is made into pearl barley when the grains are husked, steamed, rounded and polished. Pearl barley is also ground into a fine flour (crème d'orge). These products are used for making barley water for thickening soups and certain stews.

	Wholesale purchasing unit	Cost per unit	Retail purchasing unit	Cost per unit
Pearl barley				
Barley flour (Crème d'orge)				

Barley, when roasted, is changed into malt and as such is used extensively in the brewing and distilling of vinegar.

Barley needs the same care in storage as oats.

Maize — *Le maïs*

Maize is also known as corn, sweetcorn or corn-on-the-cob, and besides being served as a vegetable it is processed into cornflakes and cornflour. Maize yields a good oil suitable for cooking.

Cornflour

Cornflour is the crushed endosperm of the grain which has the fat and protein washed out so that it is practically pure starch.

Cornflour is used for making custard and blancmange powders, because it thickens very easily with a liquid, and sets when cold into a smooth paste that cannot be made from other starches.

Custard powder consists of cornflour, colouring and flavouring.

Cornflour is used for thickening soups, sauces, custards and also in the making of certain small and large cakes.

	Wholesale purchasing unit	Cost per unit	Retail purchasing unit	Cost per unit
cornflour				
custard powder				

Rice — *Le riz*

Rice needs a hot, wet atmosphere and is grown chiefly in India, the Far East, South America, Italy and the USA.

There are mainly two types used in this country:

a) *Long grain:* a narrow, pointed grain, best suited for savoury dishes and plain boiled rice because of its firm structure, which helps to keep the rice grains separate.

Menu examples: Pilaff aux foies de volailles; Kedgeree; Curried beef and rice; Œuf poché Bombay

b) *Short grain:* a short, rounded grain, best suited for milk puddings and sweet dishes because of its soft texture.

Menu examples: Baked rice pudding; Poire Condé

Brown rice

Brown rice is any rice that has had the outer covering removed but retains its bran and as a result is more nutritious.

Wild rice
Wild rice is the seed of an aquatic plant related to the rice family.

Ground rice
Ground rice is used for milk puddings.

Rice flour *La crème de riz*
Rice flour is used for thickening certain soups, eg cream soups.

Rice paper
This is a thin edible paper produced from rice, used in the making of macaroons and nougat.

Storage
Rice should be kept in tight-fitting containers in a cool, well-ventilated store.

	Wholesale purchasing unit	Cost per unit	Retail purchasing unit	Cost per unit
long grain				
short grain				
ground rice				
rice flour				
crème de riz				

Pre-cooked instant rice and par-boiled rice are also obtainable.

Tapioca ***Le tapioca***
Tapioca is obtained from the roots of a tropical plant called Cassava.

Types
Flake (rough); seed (fine).

Uses
Garnishing soups, milk puddings.

Menu examples: Purée lamballe
Tapioca pudding

	Wholesale purchasing unit	Cost per unit	Retail purchasing unit	Cost per unit
flake tapioca				
seed tapioca				

Storage
Tapioca should be stored as for rice.

Sago ***Le sagou***

Sago is produced from the pith of the sago palm. It is used for garnishing soups and for making milk puddings.
Menu example: Consommé au sagou

	Wholesale purchasing unit	Cost per unit	Retail purchasing unit	Cost per unit
sago				

Storage
Sago should be stored as for rice.

Arrowroot ***La marante***

Arrowroot is obtained from the roots of a West Indian plant called Maranta.

It is used for thickening sauces and is particularly suitable when a clear sauce is required as it becomes transparent when boiled. Arrowroot is also used in certain cakes and puddings, and is particularly useful for invalids as it is easily digested.

	Wholesale purchasing unit	Cost per unit	Retail purchasing unit	Cost per unit
arrowroot				

Storage
Arrowroot is easily contaminated by strong-smelling foods, therefore it must be stored in air-tight tins.

Potato flour ***Le fécule de pomme de terre***

Potato flour is a preparation from potatoes, suitable for thickening certain soups and sauces.

	Wholesale purchasing unit	Cost per unit	Retail purchasing unit	Cost per unit
potato flour (fécule)				

Breakfast cereal foods
A wide variety of cereals is processed into breakfast foods, eg barley, wheat, rice, bran, corn, etc.

14 Raising agents

The method of making mixtures light or aerated may be effected in several ways:

1 Sifting the flour.

When sifting flour air is incorporated.

2 Rubbing fat into flour.
During this process air can be incorporated.

3 Whisking or beating with:

a) eggs – for sponges, genoise, Swiss rolls,
b) egg whites for meringue,
c) butter or margarine for puff or rough puff pastry,
d) sugar and fat for creaming method of sponge puddings and rich cakes.

In all cases the whisking, beating or rolling (as with puff pastry) encloses air in the mixture.

4 Using baking powder.
5 Using yeast.
6 Layering of fat in a puff paste, (known as lamination).

During cooking, steam develops in between the layers of fat and paste in puff and flaky pastry, thus causing the pastry to rise.

1 Baking powder

Wholesale		Retail	
Unit	Cost	Unit	Cost
Tin			

Baking powder is made from 1 part sodium-bicarbonate to 2 parts of cream of tartar.

When used under the right conditions it produces carbon dioxide gas; to produce gas a liquid and heat are needed. As the acid has a delayed action, only a small amount being given off when the liquid is added, the majority of the gas is released when the mixture is heated. Therefore cakes and puddings when mixed do not lose their property of the baking powder if they are not cooked right away.

Hints on using baking powder

1 Mix the baking powder thoroughly with the flour.
2 Replace the lid tightly on the tin.
3 Measure accurately.
4 Do not slam oven doors in early stages of cooking.
5 Excess baking powder causes a cake to collapse in the middle and dumplings to break up.
6 Insufficient baking powder results in a close, heavy texture.

Use: It is used in sponge puddings, cakes and scones and in suet puddings and dumplings.

2 Yeast **La levure**

Yeast is a form of plant life.

	Wholesale		Retail	
	Unit	Cost	Unit	Cost
fresh yeast				
dried yeast				

Storage and quality points

1 Yeast should be wrapped and stored in a cold place.
2 It is ordered only as required.
3 It must be perfectly fresh and moist.
4 It should have a pleasant smell.
5 Yeast should crumble easily.
6 It is pale grey in colour.

Food value: Yeast is rich in protein and vitamin B. It is therefore a help towards building and repairing the body and provides protection.

Production: Yeast is a form of plant life consisting of minute cells; these grow and multiply at blood heat provided they are fed with sugar and liquid. The sugar causes fermentation – this is the production of gas (carbon dioxide) in the form of small bubbles in the mixture or dough. When heat is applied to the mixture or dough it causes it to rise.

Dried yeast has been dehydrated and requires creaming with a little water before use. Its main advantage is that it will keep for several months in its dry state.

Use: To use yeast these points should be remembered:

1 The yeast should be removed from the refrigerator and used at room temperature.
2 Salt retards the working of yeast.
3 The more salt used the slower the action of yeast.
4 Best temperature for yeast action is 21°–27°C.
5 The liquid for mixing the dough should be 36°C–37°C.
6 Temperatures over 52°C destroy yeast.
7 Yeast can withstand low temperatures without damage.

Hints on the use of yeast

1 *Warmth* – the flour, bowl and liquid should be warm.
2 *Kneading* – yeast doughs must be kneaded (worked) to make an elastic dough and to distribute the yeast evenly. An elastic dough is required to allow the gases to expand.
3 *Proving* – this term means that the dough is allowed to double its size. This should occur in a warm place, free from draughts. The dough must be covered. The quality of the dough is improved by 'knocking back'. This means the dough is pressed down to its original size and allowed to prove again. The dough is lightly kneaded, moulded and proved again before baking.

Overproving: The dough should not overprove, either in the bowl or in the moulded state. Excess or uneven heat or too long a proving time can cause overproving.

Uses

Bread doughs: rolls, white, brown, wholemeal loaves, etc.
Bun doughs: currant, Chelsea, Swiss, Bath, doughnuts.
Baba, savarin and marignans.
Croissants and brioche.
Danish pastry.
Frying batter.

15 Sugar *Le sucre*

Sugar is produced from sugar cane grown in a number of tropical and sub-tropical countries and from sugar beet which is grown in Great Britain and Europe.

Food value

As sugar contains 99.9% pure sugar, it is invaluable for producing energy.

	Wholesale		Retail	
	Unit	Cost	Unit	Cost
granulated				
castor				
cube				
coffee crystals				
icing				
Demerara				
Barbados				
syrup				
treacle				

Storage

Sugar should be stored in a dry, cool place. When purchased by the sack, the sugar is stored in covered bins.

Production

The sugar is extracted from the cane or beet, crystallised, refined and then sieved. The largest holed sieve produces granulated, the next size castor and fine linen sieves are used for icing sugar. Loaf or cube sugar is obtained by pressing the crystals while slightly wet, drying them in blocks and then cutting the blocks into squares.

Demerara sugar or brown sugar is unrefined.

Syrup is produced during the production of sugar. It is filtered and evaporated to become the required colour and thickness.

Use

Sugar is chiefly used for pastry, confectionery and bakery work.

Pastry uses: for pies, puddings, sweet dishes, ice creams and pastries.

Confectionery uses: decorating gateaux and celebration cakes (birthday, christening, wedding), sweets and petits fours. Sugar work (pulled, blown and spun).

Bakery uses: yeast doughs, large and small cakes.

In the kitchen it is used in certain sauces, such as mint and Robert. Sugar may be added to peas and carrots. It is used in some meat dishes, eg Carbonnade de bœuf, Baked sugar ham. Sugar is also added to the brine solution.

Sugar as well as being used to sweeten foods is also used to give colour, eg Crème caramel and the production of blackjack.

Glucose is a syrup made from potatoes, cane sugar and fruit treated and refined to a liquid or powder form. Glucose is not as sweet as sugar, but it is an important energy producer. Glucose is used extensively in confectionery work.

Further information: British Sugar Bureau, 140 Park Lane, London, W1Y 3AA.

16 Cocoa *Le cacao*

Cocoa is a powder produced from the beans of the cacao tree. It is imported mainly from West Africa.

Purchasing Units

Wholesale		Retail	
Unit	Cost	Unit	Cost
bags			
bags			
cask			

Food value

As cocoa contains some protein and a large proportion of starch it helps to provide the body with energy. Iron is also present in cocoa.

Storage

Cocoa should be kept in air-tight containers in a well-ventilated store.

Production

The cocoa beans grow in the pods of the cacao tree. The beans are dried, fermented, re-dried and roasted. The shells are cracked and removed; the nibs which are left are ground to a thick brown liquid called cocoa mass. The mass is compressed, then crushed, ground and sifted, making cocoa. The cocoa butter has been removed, because it would make the drink greasy.

Uses

For hot drinks cocoa is mixed with milk, milk and water, or water. Hot liquid is needed to cook the starch and make it more digestible.

Cocoa can be used to flavour puddings, cakes, sauces, icings and ice cream.

To make cocoa:

1 Measure the amount of cocoa and liquid carefully (30 g cocoa; 60 g sugar; 1 litre milk or milk and water)
2 Mix the cocoa with a little of the cold liquid.
3 Bring the remaining liquid to the boil, add the cocoa, stirring all the time.
4 Return to the pan and bring to the boil, stirring until it boils, then add the sugar.

Chocolate — *Le chocolat*

Chocolate is produced from cocoa mass mixed with fine sugar and cocoa butter. They are all ground well together and extra flavourings are sometimes added.

Purchasing units

	Wholesale	
	Unit	Cost
couverture (sweetened)	blocks	
couverture (unsweetened)	blocks	

Uses

Chocolate or couverture is used for icings, butter creams and sauces.

Drinking chocolate

This is ground cocoa from which less fat has been extracted and to which sugar has been added. It can be obtained in flake or powder form.

Further information: Cocoa, Chocolate and Confectionery Alliance, 11 Green Street, London, W1.

17 Coffee — *Le café*

Coffee is produced from the beans of the coffee tree.

Varieties

Coffee is grown in South America, Central America, Mexico, West Indies, East and Central Africa, Madagascar, India and Indonesia, and the varieties of coffee are named after the areas where they are grown eg: Mysore, Kenya, Brazil, Mocha, Java, Jamaican (Blue Mountain).

Purchasing unit

Coffee beans *a*) unroasted, *b*) roasted, *c*) ground, are sold by the pound and in 7-lb, 14-lb or 28-lb parcels or tins.

Coffee essence is obtained in 5½-oz, 10-oz, 25-oz and 1-gallon bottles.

Quality points for buying

1 A good quality coffee bean should be bought.
2 The beans should be freshly ground.
3 As water varies in different areas, sample brews with several kinds of coffee should be made to select the best result.

Storage

Coffee should be kept in air-tight containers in a well-ventilated store. The beans should be roasted and ground as they are required.

Food value

It is the milk and sugar served with coffee that have food value. Coffee has no value as a food by itself.

Production

The coffee tree or bush produces fruit called a cherry which contains seeds. The outer-side pulp is removed and the seeds or beans are cleaned, graded and packed into sacks. When required, the beans are blended and roasted to bring out the flavour and aroma.

French coffee usually contains chicory; the root is washed, dried, roasted and ground. The addition of chicory gives a particular flavour and appearance to the coffee.

Coffee essence is a concentrated form of liquid coffee which may contain chicory.

Instant coffee is liquid coffee which has been dried into powder form.

Uses

Coffee is mainly used as a beverage which may be served with milk, or as a flavouring for cakes, icings, bavarois and ice cream.

There are several methods for making coffee:

1 Still set
2 Jug method
3 Drip or filter method
4 Saucepan method
5 Cona
6 Espresso

Points to note when making coffee

The following are the rules for making good coffee:

1 Use good coffee which is freshly ground.

2 Use freshly drawn, freshly boiled water.
3 Measure the quantity of coffee carefully 300–360 g per 5 litres.
4 After the coffee has been made it should be strained off, otherwise it will acquire a bitter taste.
5 Milk, if served with coffee, should be hot but not boiled.
6 All coffee-making equipment must be kept scrupulously clean, washed thoroughly after each use and rinsed with clean hot water (never use soda).

The various methods for making coffee are as follows:

Instant coffee: Boiling water is added to soluble coffee solids.

Jug or saucepan method: Boiling water is poured on to the coffee grounds in a jug or a saucepan, allowed to stand for a few minutes, then strained.

Percolator: When the water boils it rises up through a tube and percolates through the coffee grounds.

Cona coffee: The water is boiled in a glass globe then it passes up a tube to a glass cup which contains the ground coffee. Here it infuses and as it cools it drops as liquid coffee into the bottom of the glass globe.

Still set: This consists of a container into which the ground coffee is placed. Boiling water is passed through the grounds and the coffee is piped into an urn at the side.

Espresso: This method involves passing steam through coffee grounds and infusing under pressure.

Filter method: Boiling water is poured into a container into which the ground coffee has been placed. The infusion takes place and the coffee drops into the cup below.

Further information: London Coffee Information Centre, 21 Berners Street, London WIP 4DD.

18 Tea — *Le thé*

Tea is the name given to the young leaves and leaf buds of the tea plant after they have been specially treated and dried.

Tea is produced in India, Pakistan, Assam, Darjeeling, Sri Lanka, Java, Sumatra, China, East Africa, Uganda, Kenya, Tanzania and Malawi.

Teas show marked differences according to the country and district in which they are produced and it is usual to blend several types.

China teas have the most delicate flavour of any, but lack ‘body’.

Types

There are a large number of teas on the market, and as water in different districts affects the flavour, the only sure way to select a tea for continual use is by trying out several blends, tasting them and then assessing the one that gives the most satisfactory flavour.

Buying

Tea may be obtained in packs from $\frac{1}{2}$ oz to 100 lb so that obviously many factors concerned with the type of business must be considered when deciding how to buy.

The cheapest way of purchasing tea is in 100-lb chests which are lined with lead or aluminium paper. This is to prevent the tea from absorbing moisture and odours.

Storage

Tea should always be stored in dry, clean, air-tight containers in a well-ventilated store room.

Food value

Tea alone has no nutritional properties, but it is a most refreshing drink. Nutritional value is only supplied by the milk and sugar in the tea.

Use

1 Use a good tea – the ideal recipe is 60 g to 5 litres of boiling water; there should be no guess-work, the tea should be weighed or measured for each brew.
2 Always use freshly drawn, freshly boiled water.
3 Heat the pot – unless this is done the water goes off the boil rapidly, thus preventing the correct infusion of the tea.
4 Take the pot to the boiling water – the water must be as near boiling point as possible to enable the leaves to infuse properly.
5 Allow the tea to brew for 4–5 minutes, and stir well before pouring.

Further information: The Tea Council, Sir John Lyon House, Upper Thames Street, London EC4.

19 Pulses

Pulses are the dried seeds of plants which form pods.

	Wholesale		Retail	
Types	Unit	Cost	Unit	Cost
beans, butter	bags			
beans, haricot	bags			
lentils (split)	bags			
lentils (whole green)	bags			
peas (blue–small round)	bags			
peas (marrowfat)	bags			
peas (green split)	bags			
peas (yellow split)	bags			

Food value

Pulses are good sources of protein and carbohydrate and therefore help to provide the body with energy. With the exception of the soya bean, they are completely deficient in fat and are therefore suitable for serving with fatty foods, eg fat bacon or pork.

Storage

All pulses should be kept in clean containers in a dry, well-ventilated store.

Use

Pulses are used mainly for soups, stews, vegetables, accompaniments to meat dishes and vegetarian cookery.

Menu examples:	Haricot-bean soup	Purée soissonnaise
	Lentil soup	Purée de lentilles
	Green-pea soup	Purée St.-Germain
	Yellow-pea soup	Purée égyptienne
	Boiled belly of pork and pease pudding, Haricot oxtail, Lentil croquettes	

20 Herbs — *Les herbes*

Of the thirty known types of herbs, approximately twelve are generally used in cookery. Herbs may be used fresh, but the majority are dried, so as to ensure a continuous supply throughout the year. The leaves of herbs contain an oil which gives the characteristic smell and flavour. They are simple to grow and where possible any well-ordered kitchen should endeavour to have its own fresh herb patch. Tubs or window-boxes can be used if no garden is available.

Dried herbs are obtainable in 1-lb bags and $\frac{1}{4}$-oz packets. Herbs have no food value but are important from a nutritive point of view in aiding digestion because they stimulate the flow of gastric juices. These are the most commonly used herbs:

Basil — *Le basilic*

Basil is a small leaf with a pungent flavour and sweet aroma. Used in raw or cooked tomato dishes or sauces, salads and lamb dishes.

Bay-leaves — *Le laurier*

Bay-leaves are the leaves of the bay laurel or sweet bay trees or shrubs. They may be fresh or dried and are used for flavouring many soups, sauces, stews, fish and vegetable dishes, in which case they are usually included in a faggot of herbs (bouquet-garni).

Celery seed — *La graine de céleri*

Celery seed is dried and used for flavouring soups, sauces, stews, eggs, fish and cheese dishes, when fresh celery is unobtainable. If used in a white soup or sauce it should be tied in a piece of muslin, otherwise it can cause discoloration. When celery seed and salt are ground together it is known as celery salt.

Chervil *Le cerfeuil*

Chervil has small, neatly shaped leaves with a delicate aromatic flavour. It is best used fresh, but may also be obtained in dried form. Because of its neat shape it is employed a great deal for decorating chaud-froid work. It is also one of the 'fines-herbes', the mixture of herbs used in many culinary preparations.

Chive *La ciboulette*

Chive is a bright green member of the onion family resembling a coarse grass. It has a delicate onion flavour. It is invaluable for flavouring salads, hors d'œuvre, fish, poultry and meat dishes, and chopped as a garnish for soups and cooked vegetables, and should be used fresh.

Marjoram *La marjolaine*

Marjoram is a sweet herb which may be used fresh in salads and pork, fish, poultry, cheese, egg and vegetable dishes, and when dried, can be used for flavouring soups, sauces, stews and certain stuffings.

Mint *La menthe*

There are many varieties of mint. Fresh sprigs of mint are used to flavour peas and new potatoes. Fresh or dried mint may be used to make mint sauce or mint jelly for serving with roast lamb. Another lesser known but excellent mint for the kitchen is apple mint.

Oregano

Oregano has a flavour and aroma similar to marjoram but stronger. Used in Italian and Greek-style cooking in meats, salads, soups, stuffings, pasta, sauces, vegetable and egg dishes.

Parsley *Le persil*

Parsley is probably the most common herb in this country and has numerous uses for flavouring, garnishing and decorating a large variety of dishes. It is generally used fresh, but may be obtained in dried form. When garnishing deep fried fish it is customary to fry whole heads of fresh parsley till crisp.

Rosemary *Le romarin*

Rosemary is a strong fragrant herb which should be used sparingly and may be used fresh or dried for flavouring sauces, stews, salads and for stuffings. Rosemary can also be sprinkled on roasts or grills of meat, poultry and fish during cooking and on roast potatoes.

Sage *La sauge*

Sage is a strong, bitter, pungent herb which aids the stomach to digest rich fatty meat and is therefore used in stuffings for duck, goose and pork.

Tarragon *L'estragon (m)*

This plant has a bright green attractive leaf. It is best used fresh, particularly when decorating chaud-froid dishes. Tarragon has a pleasant flavour and is used in sauces, one well-known example being Sauce béarnaise. It is one of the 'fines-herbes' and as such is used for omelets, salads, fish and meat dishes.

Thyme *Le thym*

Thyme is a popular herb in this country and is used fresh or dried for flavouring soups, sauces, stews, stuffings, salads and vegetables.

Fine herbs *Fines-herbes*

This is a mixture of fresh herbs, usually chervil, tarragon and parsley, which is referred to in many classical cookery recipes.

Balm, bergamot, borage, dill, fennel, savory, sorrel, tansy, lemon thyme

These herbs are used in cookery, but on a much smaller scale.

Harvesting and drying of herbs

1 The shoots and leaves should be collected from the plants just before they bloom.
2 They should be inspected to see that they are sound.
3 They are then tied in small bundles and hung up to dry in a warm but not sunny place.
4 After 24 hours paper bags should be tied over them to keep out dust and to help retain colour in the leaves.
5 When sufficiently dry they should break up easily if rubbed between forefinger and thumb.
6 The leaves have the middle vein removed and they are then passed through a sieve.
7 The sieved herbs must be kept in air-tight bottles or tins in order to conserve flavour.

Angelica *L'angélique (f)*

Although not a herb, angelica is found growing in a herb garden. It has a long bamboo-like stem growing to a height of about 1.5 metres. The stems are bleached, cut into 36-cm pieces and boiled in a green syrup, cooled, then reboiled daily in syrup for 5 days.

Further information: Heath and Heather, Herb Specialists, St. Albans, Hertfordshire.

21 Spices *Les épices*

Spices are natural products: they are fruits, seeds, roots, flowers or the bark of a number of different trees or shrubs.

Allspice or pimento *La toute épice*

This is so called because the flavour is like a blend of cloves, cinnamon and nutmeg. It is the unripe fruit of the pimento tree which grows in the West Indies. Allspice is picked when still green and dried; the colour then turns reddish brown. Allspice is ground and used as a flavouring in sauces, sausages, cakes, fruit pies and milk puddings. It is one of the spices blended for mixed spice.

Cloves *Le clou de girofle*

Cloves are the unopened flower-buds of a tree which grows in Zanzibar, Penang and Madagascar. The buds are picked when green and dried in the sun, when they turn to a rich brown colour. They are used for flavouring stocks, sauces, studding roast ham joint's and mulled wine.

The studded onion (oignon piqué or clouté) is an onion and a bay leaf studded with a clove.

When apples are cooked, in most cases cloves are used as a flavouring. Cloves may be obtained in ground form and as such they are used in mixed spice.

Cinnamon *La cannelle*

Cinnamon is the bark of the small branches of the cinnamon shrub. The inner pulp and the outer layer of the bark are removed and the remaining pieces dried. It is a pale brown colour and is obtained and used in stick or powdered form. It is mainly used in bakery and pastry work. When stewing pears a stick of cinnamon improves the flavour. Doughnuts may be passed through a mixture of sugar and ground cinnamon. Slices of apple for fritters may be sprinkled with cinnamon before being passed through the frying batter. It is another of the spices blended for mixed spice.

Nutmegs and mace *La muscade et la macis*

The tropical nutmeg tree bears a large fruit like an apricot which, when ripe, splits. Inside is a dark brown nut with a bright red net-like covering which is the part that becomes mace. Inside the nut is the kernel or seed, which is the nutmeg. Although the two spices come from the same fruit the flavour is different. Mace is more delicate and is used for flavouring sauces and certain meat and fish dishes. Nutmeg is used in sweet dishes (particularly milk puddings), sauces, soups, vegetable and cheese dishes. It is also used for mixed spice.

Coriander *La coriandre*

Coriander is a pleasant spice obtained from the seed of an annual plant grown chiefly in Morocco. It is a yellowish brown colour and tastes like a mixture of sage and lemon peel. It is used in sauces, curry powder and mixed spice.

Caraway *Le carvi*

Caraway seeds come from a plant grown in Holland. The seeds are about $\frac{1}{2}$ cm long, shaped like a new moon and brown in colour. Caraway seeds are used in seed-cake and certain breads, sauerkraut, cheese and confectionery. Also for flavouring certain liqueurs such as Kummel.

Ginger *Le gingembre*

Ginger is the rhizome or root of a reed-like plant grown in the Far East. The root is boiled in water and sugar syrup until soft. Ground ginger is used mainly for pastry and bakery work and for mixed spice. Whole root is used for curries, pickles and sauces.

Turmeric *Le curcuma*

Turmeric grows in the same way as ginger and it is the rhizome which is used. It is without any pronounced flavour, its main use being for colouring curry powder. It is ground into a fine powder, after which the colour is bright yellow. Turmeric is also used in pickles, relishes and as a colouring in cakes and rice.

Saffron *Le safran*

The stigmas from a crocus known as the saffron crocus (grown chiefly in Spain) are dried and form saffron, which is a flavouring and colouring spice. It is used in soups, sauces and particularly in rice dishes, giving them a bright yellow colour and distinctive flavour. Saffron is very expensive as it takes the stigmas from approximately 4000 crocus flowers to yield 30 g.

Chillies and capsicums *Le piment*

These are both from the same family and grow on shrubs. They are bright red and are used in pickles and for red pepper. The larger kind are called capsicums; they are not as hot and, when ground, are called paprika. This is used for a Hungarian type of stew known as goulash.

Ingredients for a typical curry powder

2 parts bay-leaves	2 parts garlic	4 parts cinnamon
3 parts ginger	3 parts caraway	4 parts mace
3 parts chillies	40 parts coriander	4 parts mustard
2 parts nutmeg	3 parts clove	4 parts pepper
3 parts saffron	3 parts allspice	20 parts turmeric

Ingredients for mixed spice

4 parts allspice	4 parts cloves	4 parts cinnamon
4 parts coriander	1 part nutmeg	1 part ginger

Other spices
Anis, dill seed, fennel seed, juniper berries, celery seed, sesame seed, poppy seed, caraway seed, cassia, cardamon, cumin, and fenugreek.

22 Condiments

Salt — Le sel

	Wholesale		Retail	
	Unit	Cost	Unit	Cost
cooking salt				
table salt				
sea salt				

Food value: Salt (sodium chloride) is essential for stabilising body fluids and preventing muscular cramp.

Storage: Salt must be stored in a cool, dry store as it readily absorbs moisture. It should be kept in air-tight packets, drums or bins.

Production: Salt occurs naturally in the form of rock salt in underground deposits, mainly in Cheshire. It may be mined or pumped out of the earth after water has been introduced into the rock salt. The salt is extracted from the brine by evaporation and it is then purified.

Use: Salt is used for curing fish such as herrings and haddocks and for cheese and butter making. Salt it also used for the pickling of foods, in the cooking of many dishes and as a condiment on the table.

Pepper — Le poivre

Pepper is obtained from the berry of a tropical shrub. These berries are black pepper corns. White pepper corns are obtained by removing the skin from the black pepper corn. White pepper is less pungent than the black, and both may be obtained in ground form.

Wholesale		Retail	
Unit	Cost	Unit	Cost

Pepper corns are used whole in stocks, court-bouillons, sauces and dishes where the liquid is passed. They are crushed for reductions for sauces and used in a pepper-mill for seasoning meats before frying or grilling. Green peppercorns are fresh unripe pepper berries, milder than dried pepper corns, available frozen or in tins. Pink peppercorns are softer and milder than green peppercorns, available preserved in vinegar.

Ground pepper is used for seasoning many dishes and as a condiment at the table.

Cayenne pepper — *Le cayenne*

Wholesale		Retail	
Unit	Cost	Unit	Cost

Cayenne is a red pepper used on savoury dishes and cheese straws. It is a hot pepper which is obtained from grinding chillies and capsicums, both of which are tropical plants related to the tomato.

Paprika — *Le paprika*

Wholesale		Retail	
Unit	Cost	Unit	Cost

Paprika is a bright red mild pepper used in goulash and for decorating hors d'œuvre dishes such as egg mayonnaise.

It is produced from capsicums grown in Hungary.

Mustard — Le moutarde

Wholesale		Retail	
Unit	Cost	Unit	Cost
tins	tins		

Mustard is obtained from the seed of the mustard plant, which is grown mainly in East Anglia. It is sold in powder form and is diluted with water, milk or vinegar for table use.

Mustard is used in the kitchen for sauces (e.g. mustard, mayonnaise, vinaigrette), for devilled dishes such as grilled leg of chicken, and in Welsh Rarebit.

A large variety of continental mustards are sold as a paste in jars, having been mixed with herbs and wine vinegar.

Vinegar — Le vinaigre

Wholesale		Retail	
Unit	Cost	Unit	Cost
casks	bottles		

Malt vinegar is made from malt, which is produced from barley. Yeast is added, which converts it to alcohol, and bacteria are then added to convert the alcohol into acetic acid. The resulting vinegar is stored for several months before being bottled or casked.

Artificial, non-brewed, pure or imitation vinegars are chemically produced solutions of acetic acid in water. They are cheaper and inferior to malt vinegar, having a pungent odour and a sharp flavour.

Spirit vinegars are produced from potatoes, grain or starchy vegetables, but they do not have the same flavour as malt vinegar.

Red or white wine vinegars are made from grapes and are more expensive and have a more delicate flavour than the other vinegars.

All vinegars can be distilled; this removes the colour. The colour of vinegar is no indication of its strength as burnt sugar is added to give colour.

To produce flavoured vinegar the required herbs, eg tarragon, are stored in a jar and covered with good quality vinegar and then stored for at least two weeks and used as required (eg Sauce béarnaise).

Uses: Vinegar is used as a preservative for pickles, rollmops and cocktail onions.

As a condiment on its own or with oil as a salad dressing.

Vinegar is used in marinades to tenderise meats (beef for braising, hare, venison) before cooking, and in court-bouillons to tenderise the flesh of fish during cooking (crab and lobster).

Vinegar is used for flavouring sauces such as mayonnaise and in reductions for sharp sauces (Sauce piquante, Sauce à la diable).

23 Colourings, flavourings, essences

A number of food colourings are obtainable in either powder or liquid form.

1 Natural colours

Cochineal
Cochineal is a red colour used in pastry and confectionery work.

Green colouring
This can be made by mixing indigo and saffron, but chlorophyll, the natural green colouring of plants, eg spinach, may also be used. This is used in pastry, confectionery and in green sauce which is sometimes served with salmon.

Indigo
Indigo is the blue colour seldom used on its own, but which, when mixed with red, produces shades of mauve.

Yellow colouring
A deep yellow colour can be obtained from turmeric roots and is prepared in the form of a powder mainly used in curry and mustard pickles.

Yellow colour is also obtained by using egg yolks or saffron.

Brown sugar
This is used to give a deep brown colour in rich fruit cakes; it also adds to the flavour.

Blackjack or browning
Blackjack is usually made in the kitchen by lightly burning sugar, then diluting it with water; this gives a dark brown, almost black, liquid. Blackjack is used for colouring soups, sauces, gravies, aspics and in pastry and confectionery.

Chocolate colour
This can be obtained in liquid or powder form, and is used in pastry and confectionery.

Coffee colour
This is usually made from coffee beans with the addition of chicory.

A large range of artificial colours are also obtainable; they are produced from coal tar and are harmless. Some mineral colours are also used in foodstuffs. All colourings must be pure and there is a list of those permitted for cookery and confectionery use.

2 Essences

Essences are generally produced from a solution of essential oils with alcohol, and are prepared for the use of cooks, bakers and confectioners.

Among the many types of essence obtainable are:

Almond	Lemon	Orange	Peppermint
Pineapple	Raspberry	Strawberry	Vanilla

Essences are available in many size bottles.

Flavouring essences are obtainable in three categories. *Natural essences* are made from:

1 Fruit juices pressed out of soft fruits, eg raspberries or strawberries.
2 Citrus fruit peel, eg lemon, orange.
3 Spices, beans, herbs, roots, nuts, eg caraway seeds, cinnamon, celery, mint, sage, thyme, clove, ginger, coffee beans, nutmeg and vanilla pod.

Artificial essences such as vanilla, pineapple, rum, banana, coconut are produced from various chemicals blended to give a close imitation of the natural flavour.

Compound essences are made by blending natural products with artificial products.

The relative costs vary considerably and it is advisable to try all types of flavouring essence before deciding on which to use for specific purposes.

24 Grocery, delicatessen

Delicatessen means table delicacies and the word covers a wide range of foods.

Anchovy essence
This is a strong, highly seasoned commodity used for flavouring certain fish sauces and fish preparations such as anchovy sauce or fish cakes.

Aspic jelly

Unit	Cost
tins	

Aspic jelly is a clear savoury jelly which may be the flavour of meat, game or fish. It may be produced from fresh ingredients (*Practical Cookery*, page 42) or obtained in a dried form.

It is used for cold larder work, mainly for coating chaud-froid dishes, and may also be chopped or cut into neat shapes to decorate finished dishes.

Boar's head
This is a combination of a brawn and galantine, prepared from the head of the boar. It is moulded back in the skin of the head in its original shape and coated with a dark reddish-brown aspic and decorated with piped butter. Boars' heads are used on cold buffets.

Bombay duck
These are dried fillets of a fish found in Southern Asia. They are lightly cooked, usually by grilling and served as an accompaniment to curry dishes of meat and poultry. Bombay duck are purchased in packets of 12 fillets.

Brawn
This is a preparation made from the boiled, well-seasoned head of a pig. After being cooked the meat is picked off the bones, roughly chopped or minced, then set in a mould with some of the cooking liquor. When cold and set it is carved in thick slices and served as a cold meat.

Caviar
Caviar is the uncooked roe of the sturgeon which is prepared by carefully separating the eggs from the membranes of the roe and gently rubbing them through sieves of coarse hemp. It is then soaked in a brine solution, sieved and packed.

Sturgeon fishing takes place in the estuaries of rivers which run into the Caspian or Black Sea, therefore caviar is Russian, Persian or Romanian in origin. The types normally obtainable in Britain are

Beluga, Osetrova and Sevruga. These names refer to the type of sturgeon from which the caviar is taken.

Caviar is extremely expensive and needs to be handled with great care and understanding. Caviar should be kept at a temperature of 0°C but no lower otherwise the extreme cold will break the eggs down. Caviar must never be deep frozen.

A red caviar (keta) is obtained from the roe of salmon. From the lumpfish a mock caviar is obtained. These are both considerably cheaper than genuine caviar.

Cèpes

A species of French mushroom obtainable as Cèpes au naturel or Cèpes à l'huile.

They are usually sold in tins or bottles or in dried form. They are used in many French-style dishes.

Chow-chow

a) A Chinese or pidgin English word for a mixture. It is the name given to oriental fruits preserved in syrup which is served with curry.
b) The name also of a fleshy fruit obtainable at Christmas time.

Cockscombs and kidneys

These are dressed and preserved in bottles and are used for garnishing certain dishes, eg Financière.

Continental sausages

A large variety of these are imported from European countries.

Salami: Salami is a popular sausage imported chiefly from Italy and Hungary. It is usually made from pork, beef and bacon; highly seasoned and coloured with red wine it is then well dried and cured so as to keep for years.

It is thinly sliced and eaten cold; usually as part of an hors d'œuvre.

Mortadella: This large, oval-shaped, cooked sausage is imported from Italy. Mortadella is made from pork and veal and has pieces of pork fat showing. It is thinly sliced and eaten in the same way as salami.

Cervelat beef or pork sausage: These sausages are chiefly imported from Germany; they are dried, smoked and eaten raw.

Frankfurt or Vienna sausage: There are several varieties of these small sausages which are made from ham or pork. They are dried, then smoked, and are boiled before being used. Frankfurters are obtainable in tins. They should be served as part of the garnish to Sauerkraut, eg Choucroûte garni.

Liver sausage: Liver sausage is produced in large quantities in this country and may be made from pigs' or calves' liver mixed with lean and fat pork and highly seasoned. It is sliced and served cold, usually as part of an hors d'œuvre. Liver sausage is also used in sandwiches, and it may be served with other cold meats. This needs to be kept in a refrigerator.

Extracts (meat and vegetable)

Unit	Cost
tins	

Extracts are highly concentrated forms of flavouring used in some kitchens to strengthen stocks and sauces. They are also used for making hot drinks, eg Bovril, Marmite, Maggi, etc.

Foie gras
This expensive delicacy is obtained from the livers of specially fattened geese and is produced mainly in Strasbourg. It is obtainable either plain or with truffles in tins of various sizes and at certain times of the year is also obtained in round pastry cases (foie gras en croûte) and in earthenware terrines. Foie gras is a classic first course for any lunch, dinner or supper menu. It is also used as a garnish, eg Tournedos Rossini, and is included in the rice stuffing for certain chicken dishes. A purée or mousse of foie gras is obtainable and is suitable for sandwiches and to help the flavour of certain stuffings.

Frogs' legs — *Les cuisse de grenouilles (f)*
The flesh of the hindquarters of a certain species of green frog are esteemed as a delicacy in certain continental restaurants. They are cooked in various ways, eg fried, braised, grilled.

Galantine
This is a cooked meat preparation made from well-seasoned finely minced chicken, veal or other white meat. A first-class galantine is stuffed with strips of fat pork, tongue, chicken or veal, truffles and pistachio nuts, then rolled in thin fat pork, tied in a cloth and boiled. When cold, galantines are coated with chaud-froid and masked with aspic and served on cold buffets (see page 248, *Practical Cookery*).

Gelatine

	Unit	Cost
leaf		
powdered		

Gelatine is obtained from the bones and connective tissue collagen of certain animals; it is manufactured in leaf or powdered form in varying qualities.

Gelatine is used for setting foods such as jellies and aspics, and sweets, eg bavarois.

Haggis
This traditional Scottish dish is made from the heart, lungs and liver of

the sheep, mixed with suet, onion and oatmeal and sewn up in a stomach bag. It is boiled and served with mashed potatoes.

Hams

A ham is the hind leg of a pig cured by a special process which varies according to the type of ham. One of the most famous English hams is the York ham (6–7 kg) which is cured by salting, drying and sometimes smoking. The Bradenham ham is of coal-black colour and is a sweet-cured ham from Chippenham in Wiltshire. Hams are also imported from Northern Ireland and Denmark.

All the above hams should be soaked in cold water for several hours before being boiled or braised. Ham may be eaten hot or cold in a variety of ways. Continental raw hams, Bayonne and Ardenne from France and Parma from Italy, are cut in thin slices and served raw, usually as part of hors d'œuvre.

Horseradish — *Le raifort*

Horseradish is a plant of which only the root is used. The root is washed, peeled, grated and used for horseradish sauce and horseradish cream. It is obtainable in sauce or cream form in jars or bottles; either may be served with hot or cold roast beef and smoked eel.

Paté maison

Paté is a well seasoned cooked mixture of various combinations of meat, poultry, game fish or vegetable, usually served cold as a first course. There are numerous recipes; two are on page 38 of *Practical Cookery*.

Pickles

These are vegetables and/or fruits preserved in vinegar or sauce and include:

Red cabbage, which can be served as part of hors d'œuvre and may also be offered as an accompaniment to Irish Stew.

Gherkins are a small, rough-skinned variety of cucumber, the size of which should not exceed that of the small finger. Gherkins are used for hors d'œuvre, tartare sauce, charcutière sauce, certain salads and for garnishing some cold dishes, and as an accompaniment to Bœuf bouilli à la française.

Olives are the fruit of the olive tree and there are three main varieties:

a) *Manzanilla* – the small green olive used for cocktail savouries, hors d'œuvre and garnishing many dishes such as Escalope de veau viennoise. These olives may also be obtained stuffed with pimento.
b) *Spanish queens* – the large green olives used for hors d'œuvre and cocktail savouries.
c) *Black olives* – used for hors d'œuvre and certain salads.

Cocktail onions: These are the small queen or silver-skin onion used for cocktail savouries and hors d'œuvre.

Walnuts are pickled when green and tender before the shell hardens. They are used for hors d'œuvre, salads and garnishing certain dishes such as Canapé Ivanhoe.

Capers are the pickled flower buds of the caper plant. They are used in Caper sauce, Tartare sauce, Piquant sauce and for garnishing many hot and cold fish dishes such as Truite grenobloise, Mayonnaise de homard.

Mango chutney: This is a sweet chutney which is served as an accompaniment to curried dishes.

Poppadums

These are thin round biscuits prepared in India from a mixture of finely ground pigeon peas (dhal) and other ingredients. When lightly cooked, either by frying or grilling, they are also served as an accompaniment to curry dishes. Poppadums are obtainable in tins of 50 pieces.

Potted shrimps

These are the peeled tails of cooked shrimps, which are preserved in butter and are usually served as an hors d'œuvre. They must be refrigerated.

Rollmops – Bismark herrings

These are fillets of herring which are rolled, well spiced and pickled, then served cold, usually as an hors d'œuvre.

Saltpetre

This is a natural product (nitrate of potash) which may also be produced artificially. It is used for pickling, and is one of the chief ingredients in a brine-tub for pickling meats.

Sauerkraut

Sauerkraut is a pickled product made by finely cutting white cabbage. Salt is added in a ratio of 1 kg salt to 40 kg cabbage, the mixture is packed tightly in containers and heavy weights placed on top. A liquid soon forms to cover the cabbage and fermentation begins. Sauerkraut can be kept in a cool place for 4–6 months.

Smoked herring fillet

These are preserved in oil and used as hors d'œuvre.

Smoked salmon

London is a world-famous centre for this very popular food. British, Scandinavian or Canadian salmon weighing between 6–8 kg are used for smoking. The salmon are cleaned, split into two sides, salted, rinsed, dried, then smoked. A good quality side of smoked salmon should have a bright deep colour and be moist when lightly pressed with the finger tip at the thickest part of the flesh. A perfectly smoked side of salmon will remain in good condition for not more than 7 days when stored at a temperature of 18°C. This versatile food is used for canapés, hors d'œuvre, sandwiches, and as a first course for lunch, dinner or supper.

Snails *Les escargots (m)*
These edible snails are raised on the foliage of the vine. They are obtainable in boxes which include the tinned snails and the cleaned shells. The snails are replaced in the shells with a mixture of butter, garlic, lemon juice and parsley, then heated in the oven and served in special dishes as a first course.

Tomatoes
These are obtainable: *a*) peeled whole in tins of various sizes; *b*) as a purée in tins of various sizes and of different strengths; *c*) in paste form in tubes and tins.

All types are used a great deal in the preparation of many soups, sauces, egg, fish, and farinaceous, meat and poultry dishes.

Truffles *La truffe*
The truffles chiefly used in this country are imported in tins of varying sizes. Truffles are a fungus and many varieties are found in many parts of the world. The black truffle found in the Périgord region of France is the most famous. White truffles are found in Italy.

Because of the jet black colour truffles are used a great deal in the decorating of cold buffet dishes, particularly on chaud-froid work. Slices of truffle are used in the garnishing of many classical dishes such as Sole Cubat, Tournedos Maréchale, Poulet Sauté Archiduc. Truffles are considered to be a delicacy and are extremely expensive.

A truffle substitute suitable for cold buffet decorative work is available at a much lower price.

Worcestershire sauce
This is a thin, highly seasoned, strong-flavoured sauce used as an accompaniment at table and in flavouring certain sauces, meat puddings and pies.

25 Confectionery goods

Marrons glacés
Peeled and cooked chestnuts preserved in syrup. They are used in certain large and small cakes, sweet dishes and as a variety of petit fours.

Fondant
A soft white preparation of sugar. It is made by boiling sugar and

glucose to a temperature of about 102°C, allowing it to cool slightly, then working it to a soft cream. Fondant has many uses in pastry and confectionery work, chiefly for coating petit fours, pastries and gâteaux. It may also be obtained as a ready-made preparation.

Cocktail cherries
Bright red cherries preserved in a syrup often flavoured with a liqueur known as Maraschino. They are obtainable in jars and bottles of various sizes. In addition to being used for cocktails they are also used to give colour to grapefruit and grapefruit cocktails.

Piping jelly
A thick jelly of piping consistency obtainable in different colours and flavours. It is used for decorating pastries and gâteaux and cold sweets. Piping jelly is obtainable in large tins.

Honey
A natural sugar produced by bees working upon the nectar of flowers. It is generally used in the form of a preserve and as such it may be offered on breakfast and tea menus. Honey is obtainable in 1-lb jars and 7-lb tins.

Jam
A preserve of fruit and sugar which is obtainable in 1-oz, 1-lb and 2-lb jars and 7-lb tins. Raspberry and apricot jams are those mostly used in the pastry.

Marmalade
A preserve of citrus fruits and sugar, which is used mainly for breakfast menus and for certain sweets.

Red-currant jelly
A clear preserve of red currants used as a jam and also in the preparation of savoury sweet sauces such as Réforme and Cumberland sauce. Red-currant jelly is also used as an accompaniment to Roast saddle of mutton and Jugged hare (Civet de lièvre).

Mincemeat
A mixture of dried fruit, fresh fruit, sugar, spices, nuts, etc., chiefly used for mince-pies. A recipe for mincemeat is on page 352 in *Practical Cookery*. It can also be obtained in 1-lb and 2-lb jars and 6-lb tins.

Gum tragacanth
A soluble gum used for stiffening pastillage; only a very clear white type of gum tragacanth should be used.

Pastillage
A mixture of icing sugar and gum tragacanth which may be moulded

into shapes for set pieces for cold buffets and also for making baskets, caskets, etc., for the serving of petits fours.

Rennet

A substance obtained from the stomach of calves, pigs and lambs. Rennet is prepared in powder, extract or essence form and is used in the production of cheese and for making junket. See page 332 *Practical Cookery*.

Chocolate vermicelli

A ready-made preparation of small fine chocolate pieces used in the decorating of small and large cakes and some chocolate-flavoured sweets. Chocolate vermicelli is obtainable in 7-lb boxes.

Marzipan

A preparation of ground almonds, sugar and egg yolks used in the making of petits fours, pastries and large cakes. Marzipan is freshly made by pastry cooks; it is also obtainable as a ready-prepared commodity.

Cape gooseberries (Physalis)

A tasty, yellow-berried fruit resembling a large cherry. Cape gooseberries are often dipped into fondant and served as a petit four.

Ice cream — *La glace*

A frozen preparation of a well-flavoured, sweetened mixture which can be made in many ways and in many flavours. Ice cream may be bought ready prepared, usually in 5-litre cans which are suitable for deep-freeze storage. The storage temperature for ice cream should not exceed - 19°C. A large number of sweets can be prepared using ice cream as a base mixed with various fruits, nuts, sauces and cream. Many variations of semi-hot sweets of the 'omelet surprise' type have ice cream as one of the chief ingredients. Other sweets are made from an enriched ice-cream mixture and frozen in specially shaped moulds which give their names to the sweets, eg a heart-shaped mould - cœur glacé; a bomb-shaped mould - bombe glacé. All caterers must comply with the Ice Cream Regulations which govern the production and labelling of ice cream.

Wafers — *La gaufrette*

Thin crisp biscuits of various shapes and sizes usually served with ice-cream. They are obtainable in large tins of approximately 1000 and half-tins of approximately 500 wafers.

Further information: *Food Commodities*, B. Davis, Heinemann, *The Book of Ingredients*, Dowell and Bailey, Michael Joseph.

8
Preservation of foods

Having read the chapter, these learning objectives, both general and specific, should be achieved.

General objective Understand how foods are preserved and be aware of the various methods used. Appreciate the value of preserving foods and know which foods are preserved and by which method.

Specific objective Explain in a simple manner the methods of preserving food. State which foods are preserved by which method. Give examples of the uses of preserved foods.

Food spoilage

Unless foods are preserved they deteriorate; therefore, to keep them in an edible condition it is necessary to know what causes food spoilage. In the air there are certain micro-organisms called moulds, yeasts and bacteria which cause foods to go bad.

Moulds
These are simple plants which appear like whiskers on foods, particularly sweet foods, meat and cheese. To grow, they require warmth, air, moisture, darkness and food; they are killed by heat and sunlight. Moulds can grow where there is too little moisture for yeasts and bacteria to grow, and will be found on jams and pickles.

Although not harmful they do cause foods to taste musty and to be wasted. Example: the top layer of a jar of jam would be removed if it had mould on it.

Correct storage in a dry cold store prevents moulds from forming.

Not all moulds are destructive. Some are used to flavour cheese (Stilton, Roquefort) or to produce antibiotics (penicillin, streptomycin).

Yeasts
These are single-cell plants, or organisms larger than bacteria, which grow on foods containing moisture and sugar. Foods containing only a small percentage of sugar and large percentage of liquid, such as fruit

juices and syrups, are liable to ferment because of yeasts. Although they seldom cause disease yeasts do increase food spoilage; foodstuffs should be kept under refrigeration if they may be spoiled by yeasts. Yeasts are also destroyed by heat. Yeast's ability to feed on sugar and produce alcohol is the basis of the beer and wine-making industry.

Bacteria

Bacteria are minute plants, or organisms, which require moist, warm conditions and a suitable food to multiply. They spoil food by attacking it, leaving waste products, or by producing poisons in the food.

Their growth is checked by refrigeration and they are killed by heat. Certain bacterial forms (spores) are more resistant to heat than others and require higher temperatures to kill them.

Pressure cooking destroys heat-resistant bacterial spores provided the food is cooked for a sufficient length of time, because increased pressure increases the temperature; therefore heat-resistant bacterial spores do not affect canned foods. Acids are generally capable of destroying bacteria – for example, vinegar in pickles.

Dehydrated foods and dry foods do not contain much moisture and, provided they are kept dry, spoilage from bacteria will not occur. If they become moist then bacteria can multiply – for example, if dried peas are soaked and not cooked the bacteria present can begin to multiply.

Other causes

Food spoilage can occur from other causes, such as by chemical substances called enzymes, which are produced by living cells. Fruits are ripened by the action of enzymes; they do not remain edible indefinitely because other enzymes cause the fruit to become over-ripe and spoil.

When meat and game are hung they become tender; this is caused by the enzymes. To prevent enzyme activity going too far, foods must be refrigerated or heated to a temperature high enough to destroy the enzymes. Acid retards the enzyme action; for example, lemon juice prevents the browning of bananas or apples when they are cut into slices.

The acidity and alkalinity of foods

The level of acidity or alkalinity of a food is measured by its pH. The pH can range from 1–14, with pH 7 denoting neutral (neither acid nor alkaline).

1 2 3	4 5	6 7	8 9	10 11	12 13 14
strong acid	medium acid	weak acid	weak alkali	medium alkali	strong alkaline
lemons	pears		some mineral waters	egg white	
vinegar	bananas		hard water	bicarbonate of soda	
rhubarb	carrots				
	tomatoes				

Most micro-organisms grow best at near neutral pH. Bacteria (particularly harmful ones) are less acid tolerant than fungi, and no bacteria will grow at pH less than 3.5. Spoilage of high acid foods such as fruit is usually caused by yeasts and moulds. Meat and fish are more susceptible to bacterial spoilage, since their pH is nearer neutral.

The pH may be lowered so that the food becomes too acidic (less than ph 1.5) for any micro-organisms to grow, for example use of vinegar in pickling. In the manufacture of yoghurt and cheese bacteria produce lactic acid, this lowers the pH, and retards the growth of food poisoning and spoilage organisms.

Methods of preservation

This may be achieved by several methods:

a) By removing the moisture from the food, eg drying, dehydration.
b) By making the food cold, eg chilling, freezing.
c) By applying heat, eg canning, bottling.
d) By radiation, using X- or gamma (γ) rays.
e) By chemical means, eg salting, pickling, crystallising.

Food may be preserved by the following methods:

1	Drying	5	Canning	9	Smoking (curing)
2	Dehydration	6	Bottling	10	Chemical
3	Chilling	7	Pickling	11	Gas storage
4	Freezing	8	Salting	12	Radiation

Drying and dehydration

This method of preserving is achieved by extracting the moisture from the food, thus preventing moulds, yeasts and bacteria from growing. This was done by drying foods, such as fruits, in the sun; today many types of equipment are used, and the food is dried by the use of air at a regulated temperature and humidity.

Accelerated freeze drying

This is a process of dehydration whereby food requires no preservation or refrigeration yet, when soaked in water, regains its original size and flavour. It can be applied to every kind of food. The food is frozen in a cabinet, the air is pumped out and the ice vaporised. This is called Accelerated Freeze Drying and it is the drying of frozen foods by sublimation under conditions of very low pressure. Sublimation is the action of turning from solid to gas without passing through a liquid stage; in this case it is ice to steam without firstly turning to water.

The food when processed in this way does not lose a great deal of its bulk, but it is very much lighter in weight. When water is added the food gives off its natural smell.

Advantages of drying

a) If kept dry, food keeps indefinitely.
b) Food preserved by this method occupies less space than food preserved by other methods. Some dried foods occupy 10% of the space required when fresh.
c) Dried foods are easily transported and stored.
d) The cost of drying and the expenses incurred in storing are not as high as other methods of preservation.
e) There is no waste after purchase, therefore portion control and costing are simplified.

Foods preserved by drying

Vegetables: peas, onions, beetroot, beans, carrots, lentils, cabbage, mixed vegetables, potatoes.

Herbs

Eggs

Milk

Fruits: apples, pears, plums (prunes), apricots, figs, grapes (sultanas, raisins, currants).

Meat

Fish

Vegetables

Many vegetables are dried; those most used are the pulse vegetables: beans, peas and lentils, which are used for soups and vegetable purée. Usually potatoes are cooked, mashed and then dried. The other dried vegetables are used as a vegetable, eg cabbage, onions.

Pulse vegetables are generally soaked in water before use, then washed before being cooked. Vegetables which are dehydrated (having a lower content of water as more moisture has been extracted) are soaked in water.

Potatoes which are dehydrated are in powder form and are reconstituted with water, or milk, or milk and water. They usually have manufactured Vitamin C added as dehydration results in loss of Vitamin C.

Herbs

Herbs are tied into bundles and allowed to dry out in a dry place.

Fruits

Sultanas, currants and raisins are dried grapes which have been dried in the sun or by hot air. Figs, plums, apricots, apples and pears are also dried by hot air. Apples are usually peeled and cut in rings or diced and then dried.

All dried fruits must be washed before use, and fruits such as prunes, figs, apricots, apples and pears are cooked in the water in which they are soaked.

Little flavour or food value is lost in the drying of fruits with the exception of loss of Vitamin C.

Milk

Milk is dried either by the roller or spray process. With the roller method the milk is poured on to heated rollers which cause the water to evaporate; the resulting powder is then scraped off. With the spray process the milk is sent through a fine jet as a spray into hot air, the water evaporates and the powder drops down. The temperature is controlled so that the protein in the milk is not cooked.

Use: Milk powder may be used in place of fresh milk mainly for economic reasons (especially skimmed milk powder) and is used for cooking purposes such as custard, white sauce, etc.

Eggs

Eggs are dried in the same way as milk, and although they have a food value similar to fresh eggs, dried eggs do not have the same aerating quality. When reconstituted the eggs should be used at once; if left in this state in the warm atmosphere of the kitchen bacteria can multiply and food poisoning may result because the heat used in the drying process is insufficient to kill the bacteria present in the original egg.

Use: Dried eggs are mainly used in the bakery trade.

Chilling and freezing

Refrigeration is a method of preservation where the micro-organisms are not killed; they are only prevented from multiplying. The lower the temperature the longer foods will keep. Refrigerators kept at a temperature between 0°C–7°C prevent foods from spoiling for only a short time; most frozen foods can be kept at −17°C for a year and at −28°C for two years. Foods must be kept in a deep freeze until required for use.

Cold chilled storage of fresh foods merely retards the decay of the food; it does not prevent it from eventually going bad. The aim of chilling is to slow down the rate of spoilage; the lower the chill temperature within the range −1 C and +8 C the slower the growth of micro-organisms and the bio-chemical changes which spoil the flavour, colour, texture and nutritional value of foods. Lowering the temperature to this range also reduces food poisoning hazards although it is important to remember that the food should not be contaminated before chilling.

If frozen slowly, large uneven crystals are formed in the cells of food. The water in each cell contains the minerals which give flavour and goodness to food; if frozen slowly, the minerals are separated from the ice crystals which break through the cells; on thawing, the goodness and flavour drain away. Quick-freezing is satisfactory because small ice crystals are formed in the cells of food; on thawing, the goodness and flavour are retained in the cells.

The following graph shows how only small ice crystals are formed when the temperature falls rapidly through zero, whereas in the second curve, the slower fall results in larger ice crystals.

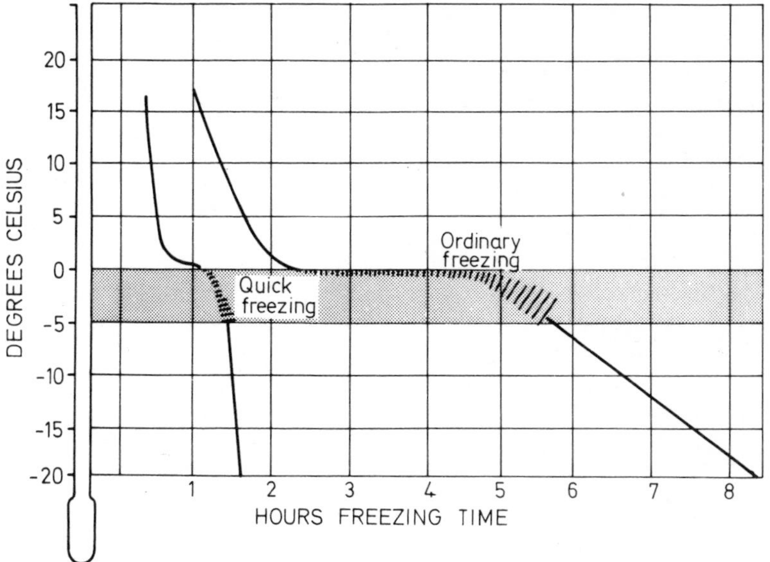

Speeds of freezing

Meat

Chilling: Meat which is chilled is kept at a temperature just above freezing-point and will keep for up to one month. If the atmosphere is controlled with carbon dioxide the time can be extended to 10 weeks. Meat from Australia is frozen because the journey may take about 12 weeks.

Freezing: Mutton and lamb carcasses are frozen; beef carcasses are not usually frozen because owing to the size of the carcass it takes a long time to freeze and this causes ice crystals to form which, when thawed, affect the texture of the meat. Frozen meat must be thawed before it is cooked.

Quick-freezing

During the cooking and freezing process, foods undergo physical and/or chemical changes. If it is found that these changes are detrimental to the product, then recipe modification is required. The following products require some modification: sauces, casseroles, stews, cold desserts, batters, vegetables, egg dishes.

Conventional recipes normally use wheat flour for thickening, but in the cook-freeze system this will not give an acceptable final product because separation of the solids from the liquids in the sauce will occur if the product is kept in frozen storage for over a period of several weeks. To overcome this problem it is necessary to use wheat flour in conjunction with any of a number of classically modified starches, eg tapioca starch, waxy maize starch. Many recipes prove successful with a ratio of 50% wheat flour with 50% modified starch.

To quick-freeze, trays of foods are placed on hollow shelves through which passes a refrigerant having a very low temperature. The foods stay in the freezing chamber for 60–90 minutes according to size.

Rapid freezing of foodstuffs can be achieved by a variety of methods using different types of equipment, for example:

a) Plate freezer.
b) Blast freezer.
c) Low-temperature immersion freezer.
d) Still-air cold room.
e) Spray freezer (using liquid nitrogen or carbon dioxide) known as *cryogenic* freezing which is a method of freezing food by very low temperature. It also freezes food more quickly than any other method. The food to be frozen is placed on a conveyor belt and passes into an insulated freezing tunnel. The liquefied nitrogen or carbon dioxide is injected into the tunnel through a spray, and vaporises, resulting in a very rapid freezing process.
f) Freeze flow – a system which freezes food without hardening it.

Foods which are frozen

A very wide variety of foods are frozen, either cooked or in an uncooked state.

Examples of cooked foods: Whole cooked meals; braised meat; vol-au-vents; éclairs; cream sponges; puff pastry items.

Examples of raw foods: Fillets of fish; fish fingers; poultry; peas; French beans; broad beans; spinach; sprouts; broccoli; strawberries; raspberries; blackcurrants.

With most frozen foods, cooking instructions are given; these should be followed to obtain the best results.

Fillets of fish may be thawed out before cooking; vegetables are cooked in their frozen state. Fruit is thawed before use and as it is usually frozen with sugar the fruit is served with the liquor.

Advantages of using frozen foods

a) Frozen foods are ready-prepared, therefore there is a saving of time and labour.
b) Portion control and costing are easily assessed.
c) Foods are always ‘in season’.
d) Compact storage.
e) Additional stocks to hand.
f) Guaranteed quality.

g) Very little loss of Vitamin C from fruits and vegetables even after several months in a deep freeze.

Pre-cooked frozen foods

In addition to the ever-growing range of pre-cooked frozen food products offered by large-scale food manufacturers, numerous catering organisations are continuously conducting research and development into pre-cooked frozen food systems.

Systems have been developed for the catering services of hospitals, school-meals, industrial dining rooms and almost every type of commercial establishment irrespective of the price range.

It is claimed by research workers developing these systems that the increased use of pre-cooked frozen foods is the likely solution to many of the present and future problems of large-scale catering operations. The advantages offered by a catering system based on frozen foods are: standardised food quality, greater staff and equipment utilisation, reduced building space requirements and closer control of food costs.

All cooking for these systems must be carried out correctly and hygienically in a properly planned and operated installation. The maintenance of high food standards depends on continuous and strict supervision.

Pre-cooked frozen foods are frozen rapidly by plate or blast freezer and are stored at a temperature of minus 18°C or below.

Freezing (or pre-cooling where needed) must immediately follow cooking.

Food that has been issued for consumption and has thawed either partially or completely must not be allowed back to the freezer or deep-freeze store. The food must be eaten as soon as possible after the final heating.

Further information: Freeze Production, Electricity Council, 30 Millbank, London SW1.

Pre-cooked chilled foods – the regethermic food service system

This system which is in use for numbers up to 2,500 meals per day is less expensive than cook freeze. Food is cooked using standard cooking procedures and is then placed in a blast chiller, reduced to a temperature of 3°C after which it is kept in refrigerated storage at 3°C. The food is then removed in bulk from storage, plated or put in bulk dishes and brought to serving temperature in 12 minutes (plated), 20 minutes (bulk) with no re-cooking, evaporation or dehydration in special regethermic ovens which are specifically designed to regenerate - not to cook. The system has been used in hospitals in many countries over a number of years and can also be used for other types of catering.

Further information: Regethermic (UK), 2 Devonshire Gdns, Chiswick, London W4 3TW.

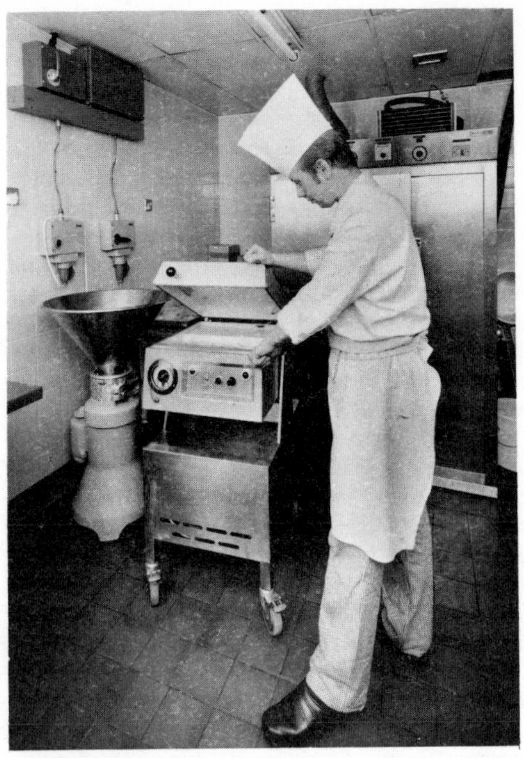

A hotel ready-food kitchen
Left to right: High-speed mixer, vacuum packer, blast freezer

Canning and bottling

Bottled and canned foods are sealed in air-tight bottles or tins and heated at a high enough temperature for a sufficient period of time to destroy harmful organisms.

Dented cans which do not leak are safe to use, but blown cans, that is those with bulges at either end, must not be used. See page 270.

Tinned hams are canned at a low temperature to retain flavour and avoid excessive shrinkage in the can and therefore should be stored in a refrigerator and consumed soon after purchase. Other tinned foods are kept in a dry, cool place and following table indicates the advised storage time:

Fruit	up to 12 months
Milk	up to 12 months
Vegetables	up to 2 years
Meat	up to 5 years
Fish in oil	up to 5 years
Fish in tomato sauce	up to 1 year

Foods are canned in tins of various sizes, eg:

Size	Approx. weight	Use
	142 g	baked beans, peas
	227 g	fruits, meats, vegetables
A 1	284 g	baked beans, soups, vegetables, meats, pilchards
14Z	397 g	fruits, vegetables
A 2	567 g	fruits, vegetables, fruit and vegetable juices
A 2½	794 g	fruits, vegetables
A 10	3079 g	fruit, vegetables, tongues

The advantages of canned foods are similar to those of frozen foods, but a disadvantage is that due to the heat processing a proportion of the Vitamin C and B1 (thiamine) may be lost.

Preservation by salt and by salting and smoking

Micro-organisms cannot grow in high concentrations of salt. This method of preservation is used mainly to preserve meat and fish, and the advantage lies chiefly in the fact that a wider variety of dishes with different flavours can be put on the menu.

Meats

Meats which are salted or 'pickled' in a salt solution (brine) are brisket, silverside of beef, ox tongues, legs of pork.

Fish

Fish are usually smoked as well as being salted and include: salmon, trout, haddock, herrings, cods' roes.

The amount of salting varies. Bloaters are salted more than kippers and red herrings more than bloaters.

The salt added to butter and margarine and also to cheese acts as a preservative.

Preservation by sugar

A high concentration of sugar prevents the growth of moulds, yeasts and bacteria. This method of preservation is applied to fruits in these forms: jams, marmalades, jellies, candied, glacé and crystallised.

Jams are prepared by cooking fruit and sugar together in the correct quantities to prevent the jam from spoiling. Too little sugar means the jam will not keep.

Jellies, such as red-currant jelly, are prepared by cooking the juice of the fruit with the sugar.

Marmalade is similar to jam in preparation and preservation, citrus fruits being used in place of other fruits.

Candied: The peel of such fruit as orange, lemon, grapefrit and lime and also the flesh of pineapple are covered with hot syrup, the syrup's sugar content is increased each day until the fruit is saturated in a very heavy syrup, then it is allowed to dry slowly.

Crystallised: After the fruit has been candied it is left in fresh syrup for 24 hours. It is then allowed to dry slowly until crystals form on the fruit. Angelica, ginger, violet and rose petals are prepared in this way.

Glacé: The fruit, usually cherries, is first candied, then dipped in fresh syrup to give a clear finish.

Preservation by acids (see page 259 for explanation of pH)

Foods may be preserved in vinegar, which is acetic acid diluted with water. In this country malt vinegar is the one most used, although distilled or white wine vinegar is used for pickling white vegetables such as cocktail onions and also for rollmops (herrings).

Foods usually pickled in vinegar are: gherkins, capers, onions, shallots, walnuts, red cabbage, mixed pickles, chutneys.

Preservation by chemicals

Certain chemicals are permitted by law to be used to preserve certain foods such as sausages, fruit pulp, jam, etc. For domestic fruit bottling, Campden preserving tablets can be used.

Preservation by radiation

Foods subjected to certain rays, for example X-rays or gamma (γ) rays, are preserved and research is being carried out on this method of preserving food.

Preservation by gas storage

Gas storage is used in conjunction with refrigerators to preserve meat, eggs and fruit. Extra carbon dioxide added to the atmosphere surrounding the foods increases the length of time they can be stored. Without the addition of gas these foods would dry out more quickly.

Further information: Accelerated Freeze Drying, HMSO; Freeze Production Electricity Council, 30 Millbank, London SW1;
Education Department, Unilever Ltd., Unilever House, Blackfriars EC4.

9
Storekeeping

Having read the chapter, these learning objectives, both general and specific, should be achieved.

General objective Understand the principles of storekeeping and know why stores control is essential.

Specific objective Explain why efficient storekeeping is necessary. List the features which make a well planned store. Specify how each commodity is stored. Explain the correct use of cold room, chill room and refrigerator. Identify the responsibilities of the storekeeper. Accurately use bin cards, stores ledger, requisition book, order book and stock sheets. Operate a control system using delivery notes, invoices, credit notes and statements. Define and calculate gross profit, nett profit and percentage discounts.

A clean, orderly food store, run efficiently, is essential in any catering establishment for the following reasons:

1 Stocks of food can be kept at a suitable level, so eliminating the risk of running out of any commodity.
2 All food entering and leaving the stores can be properly checked; this helps to prevent wastage.
3 A check can be kept on the percentage profit of each department of the establishment.

This control may be assisted by computer application, see chapter 15 page 396.

A well-planned store should include the following features:

a) It should be cool and face the north so that it does not have the sun shining into it.
b) It must be well ventilated and free from dampness (dampness in a dry store causes it to be musty, and encourages bacteria to grow and tins to rust).
c) It should be in a convenient position to receive goods being delivered by suppliers and also in a suitable position to issue goods to the various departments.
d) A good standard of hygiene is essential, therefore the walls and ceilings should be free from cracks, and either painted or tiled so as

to be easily cleaned. The floor should be free from cracks and easy to wash. The junction between the wall and floor should be rounded to prevent the accumulation of dirt.

e) Shelves should be easy to clean.

f) Good lighting, both natural and artificial, is very necessary.

g) A counter should be provided to keep out unauthorised persons, thus reducing the risk of pilfering.

h) The storekeeper should be provided with a suitable desk.

i) There should be ample well-arranged storage space, with shelves of varying depths and separate sections for each type of food. These sections may include deep-freeze cabinets, cold rooms, refrigerators, chill rooms, vegetable bins and container store. Space should also be proivided for empty containers.

j) Efficient, easy-to-clean weighing machines for large and small scale work should be supplied.

Store containers

Foods delivered in flimsy bags or containers should be transferred to suitable store containers. These should be easy to wash and have tight-fitting lids. Glass or plastic containers are suitable for many foods, such as spices and herbs, as they have the advantage of being transparent; therefore it is easy to see at a glance how much of the commodity is in stock.

Bulk dry goods (pulses, sugar, salt, etc) should be stored in suitable bins with tight-fitting lids. These bins should have wheels so that they can be easily moved for cleaning. All bins should be clearly labelled or numbered.

Sacks or cases or commodities should not be stored on the floor; they should be raised on duckboards so as to permit a free circulation of air.

Some goods are delivered in containers suitable for storage and these need not be transferred. Heavy cases and jars should be stored at a convenient height to prevent any strain in lifting.

Special storage points

1 All old stock should be brought forward with each new delivery.

2 Commodities with strong smells or flavours should be stored as far away as possible from those foods which readily absorb flavour. For example, strong-smelling cheese should not be stored near eggs.

3 Bread should be kept in a well-ventilated container with a lid. Lack of ventilation causes condensation and encourages moulds. Cakes and biscuits should be stored in airtight tins.

4 Tinned goods should be unpacked, inspected and stacked on shelves. When inspecting tins, these points should be looked for:

a) Blown tins – this is where the ends of the tins bulge owing to the formation of gases either by bacteria growing on the food and producing gases or by the food attacking the tin-plate and producing gases. All blown tins should be thrown away as the contents are dangerous and the use of the contents may cause food-poisoning.

b) Dented tins – these should be used as soon as possible not because the dent is an indication of inferior quality but because dented tins if left will rust and a rusty tin will eventually puncture.

c) Storage life of tins varies considerably and depends mainly on how the contents attack the internal coating of the tin which may corrode and lay bare the steel.

d) Cleaning materials often have a strong smell, therefore they should be kept in a separate store. Cleaning powders should never be stored near food.

Storage accommodation

Foods are divided into two groups for the purpose of storage: dry foods and perishable foods.

a) *Dry stores* include: cereals, pulses, sugar, flour, etc; bread, cakes; jams, pickles and other bottled foods; canned foods; cleaning materials.

b) *Perishable foods* include: meat, poultry, game, fish; dairy produce and fats; vegetables and fruit.

The cold room (see also Refrigeration, page 262)

A large catering establishment may have a cold room for meat, with possibly a deep-freeze compartment where supplies can be kept in a frozen stage for long periods. Poultry is also stored in a cold room. Fish should have a cold room of its own so that it does not affect other foods. Game, when plucked, is also kept in a cold room.

Chill room

A chill room keeps foods cold without freezing, and is particularly suitable for those foods requiring a consistent, not too cold, temperature, such as dessert fruits, salads, cheese, etc.

Refrigerator

The refrigerator gives cold-room and chill-room conditions and is ideal for storing fats.

Use of refrigeration

1 All refrigerators, cold rooms, chill rooms and deep-freeze units should be regularly inspected and maintained by qualified refrigeration engineers.

2 Defrosting should occur regularly, according to the instructions issued from the manufacturers. Refrigerators usually require to be defrosted weekly; if this not done, then the efficiency of the refrigerator is lessened.

3 While a cold unit is being defrosted it should be thoroughly cleaned, including all the shelves.

4 Hot foods should never be placed in a refrigerator or cold room because the steam given off can affect nearby foods.
5 Peeled onions should never be kept in a cold room because the smell can taint other foods.

Vegetable store

This should be designed to store all vegetables in a cool, dry, well-ventilated room with bins for root vegetables and racking for other vegetables. Care should be taken to see that old stocks of vegetables are used before the new ones. This is important as fresh vegetables and fruits deteriorate quickly. If it is not convenient to empty root vegetables into bins they should be kept in the sack on racks off the ground.

Ordering of goods within the establishment

In a large catering establishment the stores carry a stock which for variety and quantity often equals a large grocery store. Its operation is similar in many respects, the main difference being that requisitions take the place of cash. The system of internal and external accountancy must be simple but precise.

The storekeeper

The essentials which go to making a good storekeeper are:

1 Experience.
2 Knowledge of how to handle, care for and organise the stock in his or her charge.
3 A tidy mind and sense of detail.
4 A quick grasp of figures.
5 Clear handwriting.
6 A liking for his or her job.
7 Honesty.

There are many departments which draw supplies from these stores – kitchen, still room, restaurant, grill room, banqueting, floor service. A list of these departments should be given to the storekeeper, together with the signatures of the heads of departments or those who have the right to sign the requisition forms.

All requisitions must be handed to the storekeeper in time to allow the ordering and delivery of the goods on the appropriate day. Different coloured requisitions may be used for the various departments if desired.

Types of books used in stores control

a) *Bin card:* There should be an individual bin card for each item held in stock. The following details are found on the bin card:

BIN CARD

PRICE ....

UNIT (lbs., Tins, etc.)............ MAX. STOCK........

COMMODITY.............................. MIN. STOCK........

DATE	RECEIVED	ISSUED	STOCK IN HAND	

1 Name of the commodity
2 Issuing unit
3 Date goods are received or issued
4 From whom they are received and to whom issued
5 Maximum stock
6 Minimum stock
7 The quantity received
8 The quantity issued
9 The balance held in stock

b) *Stores ledger:* This is usually found in the form of a loose-leaf file giving one ledger sheet to each item held in stock. The following details are found on a stores ledger sheet:

BIN No DESCRIPTION CLASSIFICATION CODE UNIT MAXIMUM MINIMUM

Date	DETAIL	Invoice or Req No	QUANTITY			UNIT PRICE	VALUE		
			Received	Balance	Issued		Received	Balance	Issued

1 Name of commodity
2 Classification
3 Unit
4 Maximum stock
5 Minimum stock
6 Date of goods received or issued
7 From whom they are received and to whom issued
8 Invoice or requisition number
9 The quantity received or issued and the remaining balance held in stock
10 Unit price
11 The cash value of goods received and issued and the balancing cash total of goods held in stock

Every time goods are received or issued the appropriate entries should be made on the necessary stores ledger sheets and bin cards. In this way the balance on your bin card should always be the same as the balance shown on the stores ledger sheet.

c) *Departmental requisition book:* One of these books should be issued to each department in the catering establishment which finds it necessary to draw goods from the store. These books can either be of different colours or have serial numbers denoting to which department they belong. Every time goods have to be drawn from the store a requisition must be filled out and signed by the necessary head of department. This applies whether one item or twenty items are needed from the store. When the storekeeper issues the goods he will check them against the requisition and tick them off. At the same time he fills in the cost of each item. In this way the total expenditure over a period for a certain department can be quickly found. The following details are found on the requisition sheet:

DEPARTMENTAL REQUISITION BOOK 267

Date .... Class ....

Description	Quan.	Unit	Price per Unit	Issued if Different	Quan.	Unit	Price per Unit	Code	£	

1 Serial number
2 Name of department
3 Date
4 Description of goods required
5 Quantity of goods required
6 Unit
7 Price per unit
8 Issue, if different
9 Quantity of goods issued
10 Unit
11 Price per unit
12 Cash column
13 Signature

d) *Order book:* This is in duplicate and has to be filled in by the storekeeper every time he or she wishes to have goods delivered. Whenever goods are ordered, an order sheet must be filled in and sent to the supplier. On receipt of the goods they should be checked against both delivery note and duplicate order sheet. All order sheets must be signed by the storekeeper. Details found on an order sheet are as follows:

1 Name and address of catering establishment
2 Name and address of supplier
3 Serial number of order sheet
4 Quantity of goods
5 Description of goods to be ordered
6 Date
7 Signature
8 Date of delivery, if specific day required

e) *Stock sheets.* Stock should be taken at regular intervals of either one week or one month. Spot checks are advisable about every 3 months. The stock check should be taken where possible by an independent person, thus preventing the chance of 'pilfering' and 'fiddling' taking place. The details found on the stock sheets are as follows:

1 Description of goods
2 Quantity received and issued and balance
3 Price per unit
4 Cash columns

The stock sheets will normally be printed in alphabetical order.

All fresh foodstuffs such as meat, fish, vegetables, etc, will be entered in the stock sheet in the normal manner, but as they are purchased and used up daily a NIL stock will always be shown on their respective ledger sheets.

Commercial documents

Essential parts of a control system of any catering establishment are delivery notes, invoices, credit notes and statements.

Delivery notes are sent with goods supplied as a means of checking that everything ordered has been delivered. The delivery note should also be checked against the duplicate order sheet.

Invoices are bills sent to clients, setting out the cost of goods supplied or services rendered. An invoice should be sent on the day the goods are despatched or the services are rendered or as soon as possible afterwards. At least one copy of each invoice is made, and used for posting up the books of account, stock records and so on.

Invoices contain the following information:

a) The name, address, telephone numbers, etc (as a printed heading), of the firm supplying the goods or services.
b) The name and address of the firm to whom the goods or services have been supplied.
c) The word INVOICE.
d) The date on which the goods or services were supplied.
e) Particulars of the goods or services supplied together with the prices.
f) A note concerning the terms of settlement, eg 'Terms 5% one month', which means that if the person receiving the invoice settles his account within one month he may deduct 5% as discount.

Credit notes are advices to clients, setting out allowances made for goods returned or adjustments made through errors of overcharging on invoices. They should also be issued when chargeable containers such as crates, boxes or sacks are returned. Credit notes are exactly the same in form as invoices except that the word CREDIT NOTE appears in place of the word INVOICE. To make them more easily distinguishable they are usually printed in red, whereas invoices are always printed in black. A credit note should be sent as soon as it is known that a client is entitled to the credit of a sum with which he has been previously charged by invoice.

Statements are summaries of all invoices and credit notes sent to clients during the previous accounting period, usually one month. They also show any sums owing or paid from previous accounting periods and the total amount due. A statement is usually a copy of a client's ledger account and does not contain more information than is necessary to check invoices and credit notes.

When a client makes payment he usually sends a cheque, together with the statement he has received. The cheque is paid into the bank and the statement may be returned to the client duly receipted.

Cash discount is a discount allowed in consideration of prompt payment.

At the end of any length of time chosen as an accounting period (eg one month) there will be some outstanding debts. In order to encourage customers to pay within a stipulated time, sellers of goods frequently offer a discount. This is called cash discount. By offering cash discount the seller may induce his customer to pay more quickly, so turning debts into ready money. Cash discount varies from $1\frac{1}{4}$ % to 10%, depending on the seller and the time, eg $2\frac{1}{2}$ % if paid in 10 days; $1\frac{1}{4}$ % if paid in 28 days.

Simple discount table

Percentage	Part of £1
$2\frac{1}{2}$	$2\frac{1}{2}$ p
5	5 p
10	10 p
$12\frac{1}{2}$	$12\frac{1}{2}$ p
15	15 p
20	20 p
25	25 p

Trade discount is discount allowed by one trader to another.

This is a deduction from the catalogue price of goods made before arriving at the invoice price. The amount of trade discount does not therefore appear in the accounts. For example, in a catalogue of kitchen equipment a machine listed at £250 less 20% trade discount shows:

Catalogue price	£250
Less 20% trade discount	50
Invoice price	£200

The £200 is the amount entered in the appropriate accounts.

In the case of purchase tax on articles, discount is taken off *after* the tax has been deducted from list price.

Gross price is the price of an article before discount has been deducted.

Nett price is the price after discount has been deducted; in some cases a price on which no discount will be allowed.

Stores control

In conclusion, the following list gives the duties of a storekeeper.

Duties of a storekeeper

1 To keep a good standard of tidiness and cleanliness.
2 To arrange proper storage space for all incoming foodstuffs.
3 To keep up-to-date price lists of all commodities.
4 To ensure that an ample supply of all important foodstuffs is always available.
5 To check that all orders are correctly made out, and despatched in good time.
6 To check all incoming stores – quantity, quality and price.
7 To keep all delivery notes, invoices, credit notes, receipts and statements efficiently filed.
8 To keep a daily stores issue sheet.
9 To keep a set of bin cards.
10 To issue nothing without receiving a signed chit in exchange.
11 To check all stock at frequent intervals.
12 To see that all chargeable containers are properly kept, returned and credited – that is, all money charged for sacks, boxes, etc, is deducted from the account.
13 To obtain the best value at the lowest buying price.
14 To know when foods are in or out of season.

Cash a/c

The following are the essentials for the keeping of a simple cash account:

1 All entries must be dated.
2 All moneys received must be clearly named and entered on the left-hand or debit side of the book.
3 All moneys paid out must also be clearly shown and entered on the right-hand or credit side of the book.
4 At the end of a given period – either a day, week or month or at the end of each page – the book must be balanced – that is, both sides are totalled and the difference between the two is known at the balance. If, for example, the debit side (money received) is greater than the credit side (money paid out), then a credit or right-hand side balance is shown, so that the two totals are then equal. A credit balance then means cash in hand.
5 A debit balance cannot occur because it is impossible to pay out more than is received.

INVOICE

Phone: 574 1133 Telegrams: SOUT	No. 03957 Vegetable Suppliers Ltd., 5 Warwick Road, Southall, Middlesex
Messrs. L. Moriarty & Co., 597 High Street, Ealing, London, W.5	Terms: 5% One month

Your order No. 67 Dated 3rd September, 19....			£
Sep. 26th	25 kg	Potatoes at 20p per kg	5·00
	2 kg	Sprouts at 40p per kg	00·80
			5·80

STATEMENT

Phone: 574 1133 Telegrams: SOUT	Vegetable Suppliers Ltd., 5 Warwick Road, Southall, Middlesex
Messrs. L. Moriarty & Co., 597 High Street, Ealing, London, W.5.	Terms: 5% One month

19....		£
Sept. 10th	Goods	45·90
17th	Goods	32·41
20th	Goods	41·30
26th	Goods	16·15
		135·76
28th	Returns credited	4·80
		130·96

General rule:
Debit moneys coming in
Credit moneys going out

Daily Stores Issues Sheet

Commodity	Unit	Stock in hand	Monday In	Monday Out	Tuesday In	Tuesday Out	Wednesday In	Wednesday Out	Thursday In	Thursday Out	Friday In	Friday Out	Total pur-chases	Total issues	Total stock
Butter	kg	27		2						3				5	22
Flour	Sacks	2		1			1						1	1	2
Olive oil	Litres	8		1						$\frac{1}{2}$				$1\frac{1}{2}$	$6\frac{1}{2}$
Spices	30 g packs	8		4			8						8	4	12
Peas, tin	A10	30		6						3				9	21

.... CANTEEN Week ending .... No. meals served .... Cost per meal ....

Commodity	hand B/F	Stock received during week						Stock used during week									in hand C/F
		M.	Tu.	W.	Th.	F.	Total	M.	Tu.	W.	Th.	F.	S.	Total	@*	Cost*	
Apples, canned																	
Apples, dried Apricots, etc – dried																	
Baking powder																	
Baked beans																	

*The cost of stock used can also be checked by using two extra columns.

Example:
Make out a cash account and enter the following transactions:

Oct.	1	Paid for repair to stove	£54·50
	2	Paid to grocer	36·85
	3	Received for lunches	65·00
	4	Received for teas	25·15
	5	Received tax rebate	48·92
	6	Paid to butcher	25·64
Oct.	8	Paid to fishmonger	16·30
	9	Received for sale of pastries	32·45
	10	Paid for fuel	25·00
	11	Paid tax	30·00
	11	Received for goods	65·64
	12	Paid to greengrocer	30·16
Oct.	15	Received for teas	35·10
	17	Received for pastries	30·00
	19	Paid to butcher	15·42
	21	Paid to grocer	14·65
	24	Received for goods	31·40
	26	Received for goods	32·16
	29	Received for goods	25·10

An example of a well-laid out store

Dr.			*First week*			Cr.
Date	*Receipts*	£	*Date*	*Payment*	£
Oct. 3	To Lunches	65·00	Oct. 1	By Repairs	54·50
4	,, Teas	25·15	2	,, Grocer	36·85
5	,, Tax Rebate	48·92	6	,, Butcher	25·64
				,, Balance c/fwd	22·08
		139·07			116·99

Dr.			*Second week*			Cr.
Date	*Receipts*	£	*Date*	*Payment*	£
Oct.	To Balance b/fwd	22·08	Oct. 8	By Fishmonger	16·30
9	,, Sale of Pastries	32·45	10	,, Fuel	25·00
11	,, Goods	65·64	11	,, Tax	30·00
			12	,, Greengrocer	30·16
				,, Balance c/fwd	18·71
		120·17			101·46

Dr.			*Third week*			Cr.
Date	*Receipts*	£	*Date*	*Payments*	£
Oct.	To Balance b/fwd	18·71	Oct. 19	By Butcher	15·42
15	,, Teas	35·10	21	,, Grocer	14·65
17	,, Pastries	30·00		,, Balance c/fwd	142·40
24	,, Goods	31·40			
26	,, Goods	32·16			
29	,, Goods	25·10			
		172·47			172·47

10 Kitchen organisation and supervision

Having read the chapter these learning objectives both general and specific, should be achieved.

General objectives Be aware of the history of catering so as to appreciate the existing situation. Know how kitchens operate and understand how they are organised thus enabling the theories of organisation and supervision to be applied.

Specific objectives Define the personnel structure and function of individuals in establishments of various sizes. Explain how the system enables the customer to receive the food that is ordered. Specify the aspects of supervision and explain their application in catering.

Historical development

The history of the development of the kitchen is fascinating, and to look at pictures of baronial kitchens and banquets gives a glimpse of past glories. To visit some of the preserved or restored kitchens in palaces, castles, mansions, or humble cottages today indicates the place of importance they had in their own times. Books have been written about the era of affluence and about those who peopled the palaces both in the mirrored and chandeliered banqueting halls and the kitchens and sculleries downstairs. Many of these people have been immortalised by having soups, garnishes, sweets and other dishes named after them; some have given their names to culinary equipment, to hotels or restaurants. Catering students need to be aware of the past, so as to have a perspective of the industry they have chosen to enter.

Effects of travel

It should be borne in mind that until the advent of the railways, the speed and distance human beings could travel was limited to the capabilities of the horse. Therefore coaching inns were the 'tourist hotels' of the day until the expansion of railways and the increased movement of people from England to places like Switzerland and the

South of France. César Ritz and Escoffier became illustrious names in the hotel world as Mrs Beeton had been in private houses and homes before them. Hotels were built which subsequently became famous for excellence and whose names have become part of the English language. A comparatively short span of time has seen the birth, growth and, in some cases, decline of the luxurious hotels for the few who could afford them in Europe and the USA.

Effects of World Wars

Two World Wars affected both the economy and our society, so that catering has expanded and become the servant to many, rather than the slave to the affluent few. Because of World War Two, industrial catering came into being, although Robert Owen tried to provide meals for workers in Lanark in the early 1800s. It was necessary to provide food for factory workers in addition to the basic ration and school meals had to be available for all children so a School Meals Service came into being. The hospitals became nationalised and the feeding of patients became the responsibility of catering officers instead of the nursing staff.

Catering had reached a peak on the Atlantic liners and Pullman Cars provided food on trains, but with the advent of the aeroplane a new concept – 'Airline Catering' – came on the scene. In recent years American influences such as fast-food operations and franchise units as well as an explosion of small eating places from every corner of the globe have become commonplace, due mainly to the increased amount of travel, more people having greater spending power and also many people from overseas settling in Great Britain and opening eating places.

Provision has always been made for the needs of residential institutions such as halls of residence of universities and colleges of education. Catering students will be interested to know that the industry was served by only one catering college, namely Westminster Technical Institute, from 1910 to the 1930s. In 1983 the industry is served by over 200 colleges with catering training facilities.

This very brief historical introduction leads into the organisation which has developed and is developing in kitchens today.

Understanding the organisation which has existed in many establishments since this pattern was implemented by Escoffier, it should be clear that when labour is relatively cheap, plentiful and skilled, and when there is a demand for elaborate and extensive menus, this system is very effective. Every cook knows what he or she has to do, how to do it and by when. Sometimes certain cooks at varying periods were overworked, at other times they had little to do. There could also develop a specialisation which restricted movement of staff, eg the pastry chef was not employed to his best if placed in the larder, and vice versa.

Effects of deep freezing

With the arrival of deep freezing, the changes in public demands regarding choice of menu, and the need to be more cost conscious on the part of the establishment, changes have occurred, and these are some of the reasons for adapting the traditional system or producing a different system. For example, a very extensive menu can be offered if much of the *mise en place* is prepared throughout the day and kept refrigerated until required at service time. By having a finishing kitchen for final preparation and presentation by a small number of skilled cooks with adequate *mise en place* then fish, meat, vegetables, potatoes, pastas, eggs, etc, cooked by sautéing, grilling, deep frying and so on can be completed quickly and efficiently to the benefit of the customer. This system, which has been operated for many years very effectively in some establishments, means that all staff are fully used. The design of the finishing kitchen needs to include refrigerated cabinets for holding perishable foods, adequate cooking facilities and *bain-marie* space for holding sauces, etc.

The restaurants which provide a limited menu, eg steak houses, are able to organise very few staff to cope with large numbers of customers, to quite a high degree of skill. The required standard can be produced because few skills are needed. Nevertheless an employee producing grilled steaks, pancakes or whatever has to be organised in a systematic way and the flow of the work should be smooth.

Other kinds of establishments are required to produce large amounts of food to be served at the same time, for example schools, hospitals, industrial establishments, airlines and departmental stores. Staff have to be organised with large-scale preparation and production equipment and the means of finishing dishes quickly. To enable this to happen satisfactorily the preparation - production - freezing or chilling - reheat cycle has been developed, enabling staff to be fully used producing, and in many cases other staff to be involved in reheating or finishing the foods. Needless to say very high standards of hygiene must be practised in situations using a system of deep freezing or chilling and reheating.

Effects of kitchen systems

The principle of kitchen organisation is to produce the right quantity of food at the best standard for the correct number of people at the correct time by the most effective use of staff, equipment and materials. As can be seen from the previous examples the system of organisation can be complex, it can be simple, but the factor which will have the greatest effect on the organisation will be the menu, which in turn is decided by the customer or the prevailing economic situation.

The system of kitchen organisation will vary, due mainly to the size and the type of the establishment. Obviously, where a kitchen has 100 chefs preparing for banquets of up to 1000 people and a lunch and dinner service of 300–400 customers with an *à la carte* menu, the organisation will be quite different from a small restaurant doing thirty table d'hôte lunches, a full-view coffee shop, a speciality restaurant with a busy turnover, or a hospital diet kitchen preparing diets. As costs of space, equipment, fuel, maintenance and labour are continually increasing, considerable time, thought and planning have had to be given to the organisation and lay-out systems of kitchens.

The requirements of the kitchen have to be clearly identified with regard to the type of food that is to be prepared, cooked and served. All space and equipment must be fully justified and the kitchen organisation planned at the same time.

An organised kitchen

Given a good understanding of the total market availability of food of all kinds, fresh, part-prepared and ready-prepared, together with sound knowledge of kitchen equipment both traditional and labour-saving, kitchen plans and organisation can be economically and efficiently prepared.

Even when there are two kitchens of a similar nature the internal organisation may vary as each person in charge will have their own way of running the kitchen. It has been found most satisfactory in organising the work of a kitchen to divide it into 'parties' or corners, according to the person responsible for the kitchen organisation, eg chef de cuisine, kitchen superintendent, catering manager.

The number of parties required will depend on the size of the establishment. The organisation of the kitchen will be greatly affected by the type of food that is being used. If a great deal of frozen and prepared foods are in use then obviously the kitchen organisation can be streamlined and simplified.

Many speciality restaurants with simple menus are able to operate efficiently in an extremely small kitchen area with a small number of kitchen staff serving a large number of customers.

A trend which is to be commended is where the entire kitchen operation is on full view to the customer. This type of operation not only gives the customer the opportunity of seeing the standard of hygiene and cleanliness practised both by the establishment and the kitchen staff, but also makes the employees more aware of the importance of first-class standards of hygiene, both kitchen and personal. The organisation chart on page 288 could be for a large hotel, that on page 287 for a medium-sized establishment.

Role and function of personnel

The Head Chef — *le Chef de cuisine*

In large establishments the duties of the Chef de cuisine, Head Chef or person in charge are mainly administrative; only in small establishments would it be necessary for the chef to be engaged in handling the food. His function is –

to organise the kitchen,
to compile the menus,
to order the foodstuffs,
to show the required profit,
to engage the staff,
to supervise the kitchen (particularly at service time),
to advise on purchase of equipment,
and in many cases to be responsible wholly or partially for the stores, still room and the washing up of silver, crockery, etc.

The Second Chef — *le Sous-Chef*

The Second Chef relieves the head chef when he is off duty. He is the Chef's right-hand man, whose main function is to supervise the work in the kitchen so that it runs smoothly and according to the chef's wishes. In large kitchens there may be several sous-chefs with specific responsibility for separate services such as banquets, grill room, etc.

The Chefs de Partie

The Chefs de Partie is in charge of a section of the work in the kitchen. Usually the chef in charge of the sauce 'partie' is next in status to the sous-chefs; and the larder chef, being responsible for the perishable foods, is often considered of a higher status than the other chefs except the pastry chef. This is the job of the specialist. The Chef de partie organises his own section, delegates the work to his assistants and is in fact the 'backbone' of the kitchen.

Chef de cuisine
- Le Saucier (Sauce)
- Le Rôtisseur (Roast)
- Le Poissonnier (Fish)
- L'Entremettier (Vegetable)
- Le Garde-Manger (Larder)
- Le Pâtissier (Pastry)

Organisation of medium-sized kitchen

The Assistant Cooks *Les Commis Chefs*

The Chefs de partie is assisted by commis or assistants, the number varying with the amount of work done by the party, eg the vegetable party is larger than the fish party, due to the quantity of work to be prepared, so there are more assistants on that party. The first commis is usually capable of taking over a great deal of the responsibility, and in some cases will take charge of the party when the chef is off duty.

The Apprentice *l'Apprenti*

The apprentice is learning the trade and is moved to each of the parties to gain knowledge of all the sections in the kitchen.

The work of the chefs and their parties

The Sauce Party (head: le Saucier)

The sauce cook prepares the entrées; that is to say all the meat, poultry and game dishes which are not roasted or grilled. This includes all made-up dishes such as, vol-au-vents, stews, braised, boiled, poêled and sautéd dishes. The sauce cook will prepare certain garnishes for these dishes and make the meat, poultry and game sauces.

The Roast Party (head: le Rôtisseur)

All roasted and grilled meat, poultry and game are cooked by the roast cooks. All grilled and deep-fried fish and other deep-fried foods, including potatoes, are cooked by this party, as well as many savouries. The only deep-fried foods which may not be cooked by the roast party are cooked in the pastry. The work of the Rôtisseur includes the garnishing of the grills and roasts; he therefore grills the mushrooms and tomatoes and makes the Yorkshire pudding and roast gravy.

The Fish Party (head: le Poissonnier)

Except for grilled and deep-fried fish, all the fish dishes and fish sauces and garnishes are cooked by this party, as well as béchamel, sauce hollandaise and melted butter. The preparation of the fish is usually done by a fishmonger in the larder.

The Vegetable Party (head: l'Entremettier)

All the vegetables and potatoes, other than deep fried, and the egg and farinaceous dishes are the responsibility of the vegetable party as well as the vegetable garnishes to the main dishes. Such things as savoury soufflés and in some places pancakes will be cooked by this party.

The organisation of large traditional hotel kitchen

(Very few establishments operate this system in its entirety today.)

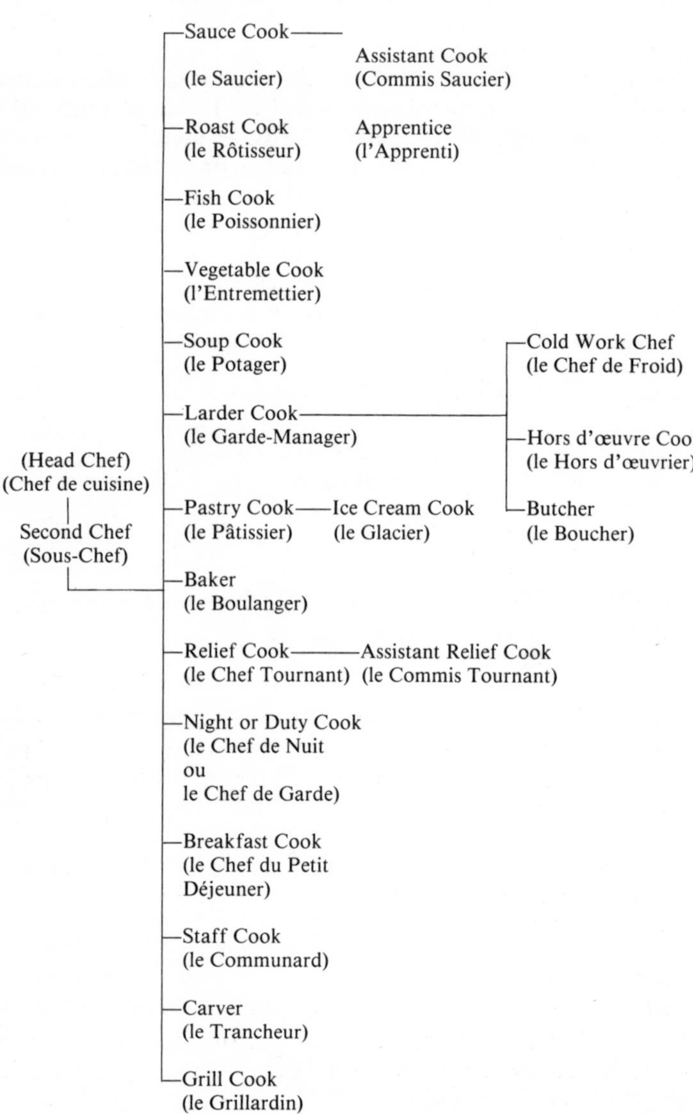

The Soup Party (head: le Potager)

In large establishments there was a separate party to make the soups and their garnishes. In some brigades, the eggs and farinaceous dishes were the responsibility of this party.

The Larder Party (head: le Garde-manger)

The larder is mainly concerned with the preparation of food which is cooked by the other parties. This includes the preparation of poultry and game and in smaller establishments the preparation of meat. The fish is prepared by a fishmonger in the larder by cleaning, filleting and portioning.

All the cold soup, egg, fish, meat, poultry and game dishes are decorated and served by this party. This work is done by the Chef de Froid and will entail a certain amount of cooking.

The sandwiches and certain work for cocktail parties such as canapés and the filling to bouchées is done here.

The hors d'œuvre and salads are made up by the Hors d'œuvrier in his own place which is near to the larder.

All cold sauces are prepared in the larder.

The oysters, cheeses and the dessert fruits may also be served by a person in the larder.

The Pastry Party (head: le Pâtissier)

All the sweets and pastries are made by the pastry cooks, also items required by other parties, such as vol-au-vent, bouchées, noodles, etc as well as the covering for meat and poultry pies.

Ice cream and petits fours are also made here. Formerly, a Glacier was employed to make all the ice creams, but the majority of ice cream is now produced in factories.

The bakery goods such as croissants, brioche, etc, may be made by the pastry when there is no separate bakery.

The Butcher — *le Boucher*

Usually the butcher worked under the direct control of the Chef or sous-chef and dissected the carcasses and prepared all the joints and cuts ready for cooking. The majority of establishments now order meat pre-jointed or pre-cut.

The Baker — *le Boulanger*

The baker would make all the bread, rolls, croissants, etc, but few hotels today employ their own bakers.

The Relief Cook — *le Chef Tournant*

The Chef Tournant usually relieves the chefs of the sauce, roast, fish, and vegetable parties on their day off. The first commis in the larder and pastry usually relieves his own chef. In some places a commis tournant will also be employed.

The Duty Cook — *le Chef de Garde*

The Chef de Garde was employed where split duty was involved and this chef was on guard to do any orders in the kitchen during that time when most of the staff were off duty and also for the late period when the other staff had gone home. Split duty hours involved a break from approximately 2.30 pm to 5.30 pm. Usually a commis would be on guard in the larder and pastry.

The Night Cook — *le Chef de Nuit*

A Chef de Nuit was employed to be on duty part of the night and all night if necessary to provide late meals.

The Breakfast Cook — *le Chef du Petit Déjeuner*

The breakfast cook will prepare all the breakfasts and often in smaller establishments will do additional duties, sometimes working until after lunch.

Staff Cook — *le Communard*

The staff cook provides the meals for the employees who use the staff room; these are the wage-earning staff and include uniformed and maintenance staff, chamber maids, etc. This applies in large hotels.

The Grill Cook — *le Grillardin*

The Carver — *le Trancheur*

In places where there is call for a large number of grills and roast joints a grill cook and a carver will be employed in many cases in the dining room or grill room.

Subsidiary departments under the control of the Chef

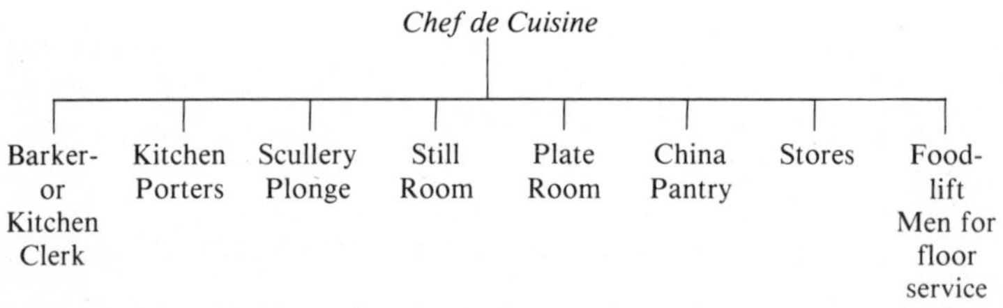

Kitchen Clerk and Barker — *le Contrôler/l'Aboyeur*

These two duties are usually performed by one person.

The kitchen clerk is responsible for much of the chef's routine clerical work and is in fact secretary to the chef.

During service time he will often call out the orders from the hot plate. Aboyeur means barker, caller or announcer.

Kitchen Porters — *les Garçons de cuisine*

Kitchen porters are responsible for general cleaning duties.

Larger parties such as the pastry, larder and vegetable parties may have one or more porters to assist the chefs. They may prepare breadcrumbs, chop parsley, peel vegetables and carry food from one section to another.

When several porters are employed one is usually appointed head porter and he may be responsible for extra duties, eg changing laundry, etc.

Vegetable preparer — *le Légumier*

Staff responsible for preparing vegetables.

Scullery — *le Plongeur de batterie*

The sculleryman or 'plongeur' is responsible for collecting, washing and returning to the appropriate place in the kitchen all the pots and pans.

Stillroom — *le Garçon/la Fille d'office*

In the stillroom are prepared and served all the beverages, eg tea, coffee, chocolate, etc, also the bread and butter, rolls and toast. Simple afternoon teas are also served from the stillroom.

Plate Room (Silver) — *l'Argentier*

All silver dishes and cutlery are cleaned and polished in the plate room. The kitchen is supplied from here with clean silver ready for service.

China Pantry — *le Vaisselier*

Here the used crockery and glass are returned, washed and stored ready for service.

Food-lift Men — *le Room Service*

Where food is served to customers in their rooms, lift men are employed to send it to the floors by a food lift.

Stores — *l'Economat*

The storekeeper is in charge of the stores, and is responsible for checking all inward delivery of goods. In some places the storekeeper will not be responsible for checking perishable foodstuffs; these will go direct to the kitchen and be checked by the chef or sous chef.

The storekeeper will be responsible for the issuing of food to the separate parties.

More details of the stores and the storekeepers' duties will be found in the chapter on Storekeeping, p 269.

Kitchen service – traditional large kitchen

To help understand the traditional organisation of the kitchen and food checking system, it is best to follow what happens from the time the customer gives the order to the waiter to the time he or she receives the meal. This will vary from place to place.

If the customer orders:

Cauliflower soup
Grilled rump steak
Fried potatoes   Stuffed tomatoes
Lemon pancakes
Coffee

then the waiter makes out the order in triplicate and this is the procedure:

1 One copy goes to the cashier.
2 A copy is taken to the kitchen hot plate.
3 It may then be stamped with the time, so that if the customer reads on the menu that he or she will have to wait, for example, 10 minutes for the steak to be cooked, then if he or she is kept waiting longer or if the waiter wishes to collect the steak too soon, the kitchen have proof of how long and at what time the steak should be ready.
4 The aboyeur will call out for the soup and the main course and vegetables.
5 The soup will be sent immediately to the hot plate by the soup cook.
6 The aboyeur will tick the check and give it to the waiter in order to confirm that the meal has gone to the customer.
7 The roast cook on hearing the order for a rump steak orders it from the larder, or the butcher, and it is sent out to be cooked. The roast cook puts it on to cook.
8 When the waiter returns, the roast cook will be asked by the aboyeur for the steak.
9 The vegetable cook will be asked for the vegetables.
10 The aboyeur passes them on to the waiter and ticks off the check. If he keeps a note of the various main dishes served he will mark off the chart so that he knows how many portions have been sold.
11 When the sweet is required, the waiter takes a separate check to the pastry in return for the pancakes.
12 When the waiter is ready for the coffee he takes another check to the stillroom.

When the cashier receives the cashier's copy the cashier records the information on a food checking cash sheet or register. At the end of the meal the kitchen copies can be compared with the cashier's copies to make sure all the food issued from the kitchen has been charged for.

Because this system is labour intensive, it has had to be considerably stream-lined and modified. Many more simpler methods using less movement and labour now operate as a result of work study.

Further information: *Food and Beverage Service*, Lillicrap (Arnold).

Kitchen supervision

Introduction to supervision

The organisation of different industries varies according to their specific requirements and the names given to people doing similar jobs in different industries may also vary. Some firms will require operatives, technicians, technologists; others need craftsmen, supervisors, managers, and the supervisory function of the charge hand, foreman, chef de parti, supervisor may be similar.

The structure of the catering industry is made up of people with craft skills, the craftsmen who are involved with production; the supervisors such as chefs de parti; and those who use managerial skills and determine policy. The group of people who have a supervisory function are those with whom we are now concerned.

Good supervision is the effective deployment of money, material and manpower.

Supervisory function

For a supervisor to be effective leadership qualities are needed to enable the supervisor to carry out the role. These qualities include the ability to:

communicate	mediate
co-ordinate	inspire
motivate	make decisions
initiate	organise

Those supervised expect from the supervisor:

consideration	understanding
respect	consistency

and in return the supervisor expects from those supervised:

loyalty	co-operation
respect	

The good supervisor is one able to obtain the best from those for whom he or she responsibility and who is able to give full satisfaction to the management of the establishment.

Supervision

The function of the supervisor is to be an overseer, and in the catering industry the name given to the supervisor may vary – for example, sous chef, chef de parti, kitchen supervisor or corner chef. In hospital catering the name would be sous chef or chef de parti or kitchen supervisor. The kitchen supervisor will be responsible to the catering manager, while in hotels and restaurants the chef de parti will be responsible to the chef de cuisine. The exact details of the job will vary with different aspects of the industry and the size of the various units, but generally the supervisor performs three functions:

a) Technical
b) Administrative
c) Social

Technical
Culinary skills and the ability to use kitchen equipment are essential for the kitchen supervisor. Most kitchen supervisors will have worked their way through the section or sections before reaching supervisory responsibility. The supervisor needs to be able 'to do' as well as knowing 'what to do' and 'how to do it'. He or she also needs to be able to do it well and to be able to impart some of his or her skill to others.

Administrative
The supervisor or chef de parti will, in many kitchens, be involved with the menu planning, sometimes with complete responsibility for the whole menu but more usually for part of the menu as occurs with the larder chef and pastry chef. This includes the ordering of foodstuffs

Terrace Restaurant kitchen, Dorchester Hotel, London

(which is an important aspect of the supervisor's job in a catering establishment) and of course accounting and recording materials used. The administrative function includes allocation of duties and in all instances basic work-study knowledge is needed to enable the supervisor to operate effectively. The supervisor's job may include the writing of reports, particularly in situations where comparisons are carried out and when developments are being tried.

Social

Perhaps the role of the supervisor is most clearly seen in staff relationships because the supervisor's purpose is to motivate the staff for whom he is responsible. To motivate could be described as getting movement and action; and having got the staff moving, then the supervisor needs to exert control. To achieve the required result the staff need to be organised. From this it is clear that the supervisor has a threefold function regarding the handling of staff, namely: to organise, to motivate, to control; in other words this is the essence of staff supervision.

Elements of supervision

The accepted areas of supervision include:

Forecasting and planning
Organising
Commanding
Co-ordinating
Controlling

Each of these will be considered within the sphere of catering.

Forecasting

Before one can plan it is necessary to look ahead, to foresee possible and probable outcomes and to plan for them. If the chef de parti knows that on the following day it is his assistant's day off, he looks ahead and plans accordingly. When the catering supervisor in the hospital knows that there is a 'flu epidemic and two of his cooks are 'under the weather' he plans for their possible absence. If there is a spell of fine hot weather and the cook in charge of the larder foresees a continued demand for cold foods or when she anticipates an end to the hot spell, then the plans are moderated. Forecasting for the supervisor is the using of judgement acquired from previous knowledge and experience. Because many people are on holiday in August fewer meals will be needed in the office restaurant; no students are in residence at the college hostel, but a conference is in for the day needing 60 meals; because of the Motor Show, the Spring Bank Holiday, the effects of a rail strike or a wet day, as well as less predictable situations such as the number of customers anticipated on the opening day of a new restaurant – all need forecasting.

Planning

From the forecasting comes the planning: how many meals to prepare; how much to have in stock (should the forecast not have been accurate); how many staff will be needed; which staff and when. Are the staff capable of what is required of them? If not, the supervisor needs to plan some training. This of course is particularly important if new equipment is installed. Imagine an expensive item such as a microwave oven ruined on the day it is installed because the staff have not been instructed in its proper use; or what is more likely to occur, equipment lying idle because the supervisor may not like it, consider it is sited wrongly, does not train staff to use it, or because of some similar reason.

As can be seen from these examples the necessity is for forecasting to precede planning, and from planning we now move to organising.

Organising

Organising consists of ensuring that what is wanted is where it is wanted, when it is wanted, in the right amount, at the right time, and this applies to food, to equipment and to staff.

Arranging for this to happen entails the supervisor in the production of duty rotas, maybe training programmes and cleaning schedules. Consider the supervisor's part in organising an outdoor function where a wedding reception is to be held in a church hall: 250 guests require a hot meal to be served at 2 pm and in the evening a dance will be held for the guests at which a buffet will be provided at 9 pm. The supervisor would need to organise his staff to be available when wanted, to have their own meals, maybe to see that they have get their transport home. Calor gas stoves may be needed, and the supervisor would have to arrange for the stoves to be serviced and for the equipment used to be cleaned after the function. The food would need to be ordered so that it arrived in time to be prepared. If decorated hams are to be used on the buffet then they would need to be ordered in time so that they could be prepared, cooked and decorated over the required period of time. If the staff have never carved hams before, then instruction would need to be given – this entails organising training. Needless to say the correct amounts of food, equipment and cleaning materials would have to be at the right place when wanted, and if the situation were not organised properly problems could occur.

Commanding

The supervisor has to give instructions to his or her staff on *how, what, when* and *where*; this means that orders have to be given and a certain degree of order and discipline maintained. The successful supervisor is one who is able to do this effectively and it entails a certain amount of decision making, which in turn necessitates, very often, deciding on priorities. Explanations of *why* a food is prepared in a certain manner, *why* this amount of time is needed to dress up food, say for a buffet, *why* this decision is taken and not that decision, and how these explanations and orders are given, determine the effectiveness of the supervisor.

What do you learn from this picture?

Co-ordinating

Co-ordinating is the skill the supervisor needs to get staff to co-operate and work together. To achieve this, the supervisor has to be interested in the staff, to deal with their queries, to listen to their problems and to be helpful. Particular attention should be paid to new staff, inducting them into the situation so that they become part of the team, or *partie*; the other area for which the supervisor has particular responsibility is maintaining good relations with other departments. The important persons to consider are the customers, the patients, the school children, etc, who are to receive the service, and good service is dependent on co-operation between waiters and cooks; nurses and catering staff; store-staff, caretakers, teachers, suppliers and so on. The supervisor has a crucial role to play.

Controlling

This includes the controlling of people and products and preventing pilfering as well as improving performance. Checking that staff arrive on time, and do not leave before time, and do not misuse time in between. Checking that the product, in this case the food, is of the right standard – that is to say the correct quantity and quality. Checking to prevent waste, and also to prevent stealing and to ensure that staff operate the portion control system.

This aspect of the supervisor's function involves inspecting and requires tact; controlling may include the inspecting of the swill-bin to observe the amount of waste, checking the disappearance of a quantity of food, supervising the cooking of the meat so that shrinkage is minimised and reprimanding the unpunctual member of the team.

Standards of any catering establishment are dependent on the supervisor doing his job efficiently, and standards are set and maintained by effective control, which is the function of the supervisor.

Responsibilities of the supervisor

Delegation

It is recognised that delegation is the root of successful supervision - in other words, by giving a certain amount of responsibility to others the supervisor can be more effective himself. To do this there needs to be a desire to develop the potential of those under his control, a recognition of the abilities of those subordinated to him, and, having delegated, the wisdom of allowing the person entrusted with the job to get on with it. Naturally the supervisor has judged the person to be capable before he delegates responsibility.

Motivation

Since not everyone is capable, or wants responsibility, the supervisor still needs to motivate those less ambitious, and since most people like to be occupied with other people so as to have companionship and because it is the accepted thing to do, they are prepared to work so as to improve their standard of living. There is, however, another very important motivating factor: most people desire to get *satisfaction* from the work they do. The supervisor must be aware of why people work and how different people achieve job satisfaction and so be able to act upon this knowledge.

Welfare

People work best in good working conditions and good working conditions include freedom from fear: fear of becoming unemployed, fear of failure at work, fear of discrimination. Security of the job, and incentives such as opportunities for promotion, bonuses, profit sharing and time for further study, etc, encourage a good attitude to work, but as well as these tangible factors people need to feel wanted, to feel important, and the supervisor is in an excellent position to do this. Personal worries affect the individuals' performance and can have a very strong influence on how well or how badly they work. The physical environment naturally causes problems if the atmosphere is humid, the working situation ill-lit, too hot or too noisy, and there is constant rush and tear and frequent major problems to be overcome. In these circumstances staff are more liable to be quick-tempered, angry and aggressive.

Understanding

The supervisor needs to understand both men and women, to anticipate problems, and build up a team spirit so as to overcome the problems. This entails always being fair when dealing with staff and giving them encouragement. It also means that work needs to be allocated according to ability and to keeping everyone occupied as well as checking that the environment is the most conducive to work.

Job satisfaction

Communication

Finally, and most important of all, the supervisor must be able to communicate. To convey orders, instructions, information and manual skills requires the supervisor to possess the right attitude to those with whom he or she needs to communicate. The ability to convey orders and instructions in a manner which is acceptable to the one receiving the orders is dependent not only on the words but the emphasis given to the words, the tone of voice, the time selected to give them and who is also present when they are given. This is a skill which supervisors need to develop. Instructions and orders can be given with authority without being authoritative.

Therefore the supervisor needs technical knowledge and the ability to direct staff and to carry responsibility so as to achieve specified targets and standards as set out by the policy of the organisation; this he is able to do by organising, co-ordinating, controlling and planning through effective communication.

Further information: *Catering Supervision* (*with 21 Case Studies*), A. E. Bevan (Arnold 1974). *The Supervisor's Handbook*, Julia Reay, (Northwood).

11
Industrial relations

General objective To have a knowledge and understanding of what is meant by industrial relations in the catering industry and an awareness of the basic legal requirements in the working environment. To create an appropriate attitude towards the responsibilities and the rights of people at work.

Specific objectives To relate recent legislation to specific examples in the catering industry. To explain the common terms used in committee procedure in industrial relations. To define the specific problems associated with the industry. State the role and outline the purpose of the trade unions.

Industrial relations

Effective relationships in industry depend upon:

1 co-operation between employee and employer,
2 knowing the Laws passed by Parliament (legislation),
3 having the right attitude towards those laws.

This brief introduction to industrial relations is intended to produce just such an attitude so that students and employees will work harmoniously and efficiently and thus obtain job satisfaction. Good industrial relations are created when employees and employers know both their responsibilities and their rights and use this knowledge to their mutual benefit and to the advantage of those they serve in the catering industry.

However, because of the nature of the catering industry good industrial relationships are not easy to create for the following reasons:

1 labour turnover is often high and considered to be acceptable by many people,
2 a vast majority of establishments are small,
3 many employers take pride in the relationship they have with their employees and do not favour unions,
4 few employees are 'union-minded',
5 no one union is solely concerned with the industry,
6 many employees are foreign and do not speak English,
7 many establishments are seasonal and operate for only part of the year,
8 many part-time workers are employed,

9 un-social hours are worked,
10 a history of poor conditions, poor wages and low status.

Legal aspects

Recent legislation which affects employees and employers should improve the situation in the industry. This legislation includes:

Equal Pay Act 1970,
Trade Union & Labour Relations Act 1974,
Sex Discrimination Act 1975,
Employment Protection Act 1975,
Race Relations Act 1976 and the
Health and Safety at Work Act 1974 (see page 1).

Industrial tribunals

To administer and enforce the implementation of the Acts, *Industrial tribunals* are set up to settle disputes over unfair dismissal, redundancy and cases of alleged discrimination. The intention has been to protect the individual employee and to strengthen and encourage good relationships between employer and the trade unions.

Communication

In effect communication channels between staff and management should exist with opportunities and facilities for Trade Union meetings so that employees have greater participation in the establishments organisation. To be effective communication often needs to be written and by law, staff working 16 hours a week or more are entitled to a written statement of terms and conditions of employment. An example of such a statement follows:

Name of establishment
Address

Employees Name .... **Address** ....
Job title ....
....
Date of commencement .... **Renumeration** ....
Terms & conditions of holidays and holiday pay
Sick pay arrangements
Pensions & pension schemes
Disciplinary rules **Grievance procedure**
Previous service
Length of notice to terminate contract
Other specific conditions

Trade unions and the catering industry

Those people belonging to trade unions who are employed in the catering industry mainly belong to the Transport and General Workers Union or the General and Municipal Workers Union. The aim of the union is to look after the interests of its members at work, in particular, hours of work, conditions and pay. Their function is to negotiate on

Cooperation

behalf of their members with the employers through their local officials. In the event of disagreement between the unions and the employers the Advisory Conciliation and Arbitration Service (ACAS) could be used.

For example a shop steward in a kitchen should listen to any employee who is a member of the union if he or she has a grievance which seems justified. On his/her behalf the shop steward could approach the chef or catering manager to resolve the issue. The shop steward may obtain the advice of the branch or district union officer before taking up the case. The role of the trade union is to look after members' interests and this includes the smooth running of a hygienic and safe kitchen.

Women in employment

A woman employee who is pregnant and who has been working for a stated period of time for the same employer may not be dismissed because of her pregnancy unless her condition makes it impossible to work competently. An employee who leaves work to have a baby is entitled to her job back after the baby is born, provided she has the required minimum service, has continued to work up to the specified time before confinement and informs her employer of her intention to return before her absence. Maternity pay is paid to employees complying to the stated conditions.

Details are obtainable from the Social Services and Citizens Advice Bureaux.

Equal Pay Act 1970

This Act specifies that pay must be the same for men and for women doing like work or work that is graded as similar. For example, a man or a woman employed to wash-up are both, by law, entitled to the same pay; this would also apply if a female chef de partie for the pastry was employed. The pay would be the same as for a male chef de partie of equal competence and experience.

Sex Discrimination Act 1975

Within the terms of this Act it is unlawful to discriminate on the grounds of a person being male or female, or being married when selection, appointment, transfer, dismissal or promotion are being considered. This means that both men and women should be assessed and decisions made according to their ability, competence, reliability etc, and not because a person is male or female. For example if a kitchen supervisor and a storekeeper are required, both positions should be available to men and women; it is against the law to restrict the posts to male or to female.

Race Relations Act 1976

This Act makes it illegal to discriminate against a person because of their race. However factors which could affect issues relate, for example, to language and religion. The potential waiter unable to make himself or herself understood or to understand the language of the customers, or the cook whose religion restricts the handling of certain foods may well jeopardise the opportunity for obtaining and keeping employment. But a person could not be discriminated against solely because of their race.

Employment Protection Act 1975

The purpose of this Act is to prevent the unfair dismissal of a member of staff. It is essential to know what are considered to be fair reasons for dismissal of staff as well as some reasons considered to be unfair.

Examples of fair reasons for dismissal:

a) incompetence – such as lack of skill, lack of technical or academic qualifications,
b) misconduct – eg stealing or flagrant breach of hygiene rules,
c) illegality – eg a foreign worker without a work permit,
d) genuine redundancy.

Unfair dismissal would include:

a) for being a member of, or proposing to join a trade union;
b) unfair selection for redundancy.

Industrial relations terms

A list of some of the terms used in negotiations and committees:

Abstain	Voting neither for nor against, effectively not voting.
Abstention	This is a vote which is cast neither for nor against.
Address the Chair	When speaking at a meeting members must speak to the Chair and not to other members.
Adjourn	When the meeting or part of the business is delayed until later
Ad hoc committee	A committee formed to do a specific job which ceases to exist when this special work is complete.
Against	A vote in disagreement with the motion.
Agenda	The list of items to be discussed by the committee.
Amendment	This is a proposition to change the wording of a motion. The amendment needs a proposer and a seconder; it is then put to the vote and if carried (has a majority vote for it,) the motion as amended replaces its original form. If the amendment is defeated then the original motion stands. More than one amendment can be put forward; each is discussed separately and then voted on in the order received.
AGM	Annual General Meeting
AOB	Any other business. Usually the last item on the agenda for the matters not included on the agenda. Should be used as little as possible since matters need to be considered before discussion
Arbitration	When in a dispute agreement cannot be reached and the decision of an independent person or persons is accepted.
Check-off	Union dues deducted by the employer who then pays the dues to the union.
Closed Shop	This is the situation where individuals must be a member of the same union to which the other workers belong, or when new employees must join the union. Both the employer and the trade union's agreement are required to operate a closed shop
Collective agreement	Agreements made between several trade unions and several employers or employers associations.
Collective bargaining	Negotiations to result in collective agreement regarding conditions of work, allocation of work, discipline, facilities for trade union officials and procedures for consultation.
'Chair'	The person acting as chairperson.
Consensus	A general feeling of the meeting.
Conciliation	A process whereby an outside party brings opposing parties in dispute together to settle the dispute.
Constitution	A formal document which states the aims of the organisation, its membership and how it should be managed.
Co-opted member	A person invited to join a committee by its members because of special assistance he or she can bring to the committee.
Correction of minutes	When minutes of the previous meeting are due to be signed as a correct record it is sometimes found necessary to correct them if the meeting agrees to the correction.
Defeated	Said of a motion which has been voted and decided against.
Delegate	A person authorised by a committee or organisation to represent it on its behalf.
Ex officio	Means 'because of his or her office' for example the Chairperson of an organisation is usually a member (ex officio) of all its committees because of being Chairperson
For	A vote in favour of the motion.
Job evaluation	A term used to determine the value of jobs in a way acceptable to those in the job.

Joint consultation	When employees and employers discuss before decisions are taken.
Legislation	The law, having legal authority.
Lobby	Recognised way of persuading people to support an idea.
Lost	Said of a motion when it has been voted and decided against
Minutes	The written record of a committees decisions. All resolutions, amendments and decisions, including votes must be recorded. Discussions may be summarised.
Motion	A statement which will be discussed and then may be put to the vote.
Move	The proposer of a motion is said 'to move it' when he or she asks that it be put to the vote.
Nem con	'No one against' A motion is passed 'nem con' when some vote for, none against, but some abstain.
Nomination	The giving of the name of a person for office or to be a member of a committee.
Out of order	Not conforming to the rules of procedure or the proper conduct of the meeting.
Point of order	When a member requests the Chairperson to correct or bring the running of the meeting to comply with the Rules of Procedure.
Proposer	The principal speaker for the motion.
Proxy	A person given authority to exercise a member's vote, to vote instead of that member.
Quorum	The minimum number of members required to be present for business. The number should be stated in the Rules of Procedure. If insufficient members attend then the situation is described as inquorate.
Resolution	The actual wording of a motion.
Ruling	The chairperson's decision on a matter of procedure which must be followed without discussion.
Shop steward	The trade union member elected by the department or section he or she represents. He or she communicates management proposals to his or her colleagues and represents their views to management.
Status Quo	If a change causes a dispute then the original situation is reverted to until agreement is reached.
Seconder	The next person to support a motion following the proposer.
Substantive agreement	An agreement which determines for example rates of pay, hours of work, holiday arrangements.
Trade dispute	A dispute between employers and employees or between employees and employees.
Unanimous	When all members present vote for the motion.

Further information: 'Industrial Relations' Guide No 7 published by HCITB,
The Acts of Parliament published by HMSO.

12
Kitchen French

Having read the chapter, these learning objectives, both general and specific should be achieved.

General objectives Know basic culinary French and understand when and how to use it as appropriate.

Specific objectives Translate basic culinary French terms into English and English into French. Compile menus in French so that the menu terms agree. Correctly spell French words used and include the accents.

French has been the traditional international language of professional chefs and the kitchen just as Italian has been the traditional language for musicians and in music. This results in a situation whereby many French words cannot be translated with a finely correct meaning into English; this applies particularly to kitchen and menu terms. Examples such as hors d'oeuvre, roux, canapé and fleuriste make the point (see also the chapter on menu planning).

It may or may not be considered worth while to maintain this French tradition; however, what is certain is that if French is used it must be correctly used. Not only should the spelling, grammar and pronunciation be accurate but if a dish is named traditionally then it must be authentic, for example, if meunière is stated then it should be shallow fried, served with lemon, chopped parsley and nut brown butter.

As a rule it is better to use correct English than incorrect French and it should be acknowledged that the function of language on the menu is to inform both the customers, the kitchen and the serving staffs. If follows therefore that whether French or English is used, will depend upon the establishment. If many nationalities work there or if the clientele is sophisticated or international this will have a bearing on the language used.

Every trade or industry has its own peculiar terminology, the student of catering would be wise to become acquainted with the basic French words and phrases since students do not know what opportunities may arise when some knowledge of French would be invaluable.

Lists of words in common use

Numerals

one	*un*	five	*cinq*	eight	*huit*
two	*deux*	six	*six*	nine	*neuf*
three	*trois*	seven	*sept*	ten	*dix*
four	*quatre*				

Days of the week

Monday	*lundi*	Thursday	*jeudi*	Saturday	*samedi*
Tuesday	*mardi*	Friday	*vendredi*	Sunday	*dimanche*
Wednesday	*mercredi*				

Months of the year

January	*janvier*	May	*mai*	September	*septembre*
February	*février*	June	*juin*	October	*octobre*
March	*mars*	July	*juillet*	November	*novembre*
April	*avril*	August	*août*	December	*décembre*

Seasons of the year

Spring	*le printemps*	Autumn	*l' automne*
Summer	*l' été*	Winter	*l' hiver*
Christmas	*Noël*	Easter	*Pâques*

(but at Christmas is *à Noël*, short for *à la fête de Noël*)

The grammar of kitchen French terms

There are few rules applying to the writing of menus but it requires accuracy of gender, agreement and spelling. 'The' in French is translated by different words according to the gender and number (ie singular or plural).

	Singular	*Plural*
Masculine	*le melon*	*les melons*
	l'abricot	*les abricots*
	l'hotel	*les hotels*
Feminine	*la tomate*	*les tomates*
	l'orange	*les oranges*

If the singular noun begins with a vowel then the le or *la* change to *l'*.

eau (f)	*l'eau*	the water
ail (m)	*l'ail*	the garlic

In some cases the noun begins with an *h* and the same rule applies:

hotel (m) *l'hôtel* the hotel

Note: Words that begin with an 'h' and do not take an *l'* are usually marked with an asterisk in the dictionary eg le homard.

When the noun is plural – that is, when more than one item is considered – the French for 'the' is *les*:

les saumons *les homards* *les hôtels*

When using *le, la* or *les* on the menu before a cut, part or joint then the gender used is that of the cut, part or joint and not that of the food of which it is a part:

la darne de	*la selle de*	*le troncon de*
la coeur de	*le suprême de*	*le ris de*
le délice de	*les goujons de*	*la paupiette de*

The method of cooking the cut or piece must also agree with the cut or piece:

les filets de sole frits *la selle d'agneau rôtie* *le gigot d'agneau rôti*

Verbs are often used to describe the cooking method. In the dictionary the verb is given in the infinitive, eg *brouiller* – to scramble; *sauter* – to toss. On the menu they are given in the past tense (past participles) and follow the noun and agree with it, like an adjective, in gender and number:

sauter, to toss	*sauté*, tossed	*Pommes sautées*
flamber, to set alight	*flambé*, set alight	*Pêche flambée*
concasser, to cut roughly	*concassé*, cut roughly	*Tomates concassées*

With a masculine singular noun the verb is left unaltered:
eg *Saumon fumé*.
With a masculine plural noun add *s*,
eg *Champignons grillés*
With a feminine singular noun add *e*,
Anguille fumée
With feminine plural noun add *es*,
Pommes persillées.

Most verbs to do with cooking will end in *er* but some end in *ir, re* or *ire* and these take different spelling in the past tense, for example:

frire	to fry	*frit*	fried
fondre	to melt	*fondu*	melted
bouillir	to boil	*bouilli*	boiled
farcir	to stuff	*farci*	stuffed

The agreement rule still applies:

Cabillaud (m, sing)	*Cabillaud frit*
Pommes (f, pl)	*Pommes frites*
Beurre (m, sing)	*Beurre fondu*
Boeuf (m, sing)	*Boeuf bouilli*
Tomates (f, pl)	*Tomates farcies*

à has several meanings, but on the menu the usual meanings are 'with' and 'in'. Changes in the translation of 'with the' and 'in the' take place according to the gender and number of the noun used.

à le becomes *au*
à la is not changed
à l' is not changed
à les becomes *aux*

Therefore 'tart with apples' (*tarte à les pommes*) must be written:
Tarte aux pommes
'Skate with black butter' (*raie à le beurre noir*) must be written:
Raie au beurre noir

Homard à la Americaine becomes *Homard à l'americaine*
Crêpes à le citron becomes *Crêpes au citron*
Chou-fleur à la grecque remains *Chou-fleur à la grecque*

Nouns – eg as used below: apple, butter, chocolate – as following the name of the main ingredient of a dish can be singular or plural:
Raie au beurre noir *Profiteroles au chocolat*
Tarte aux pommes

de means 'of', and the *e* is replaced by an apostrophe before words beginning with a vowel or *h*, as with *le* and *la*:

Gigot de agneau becomes *Gigot d'agneau*
Tablė de hôte becomes *Table d'hôte*
Rable de lièvre is unchanged
Longe de veau is unchanged
Filet de boeuf is unchanged
Bisque de homard is unchanged.

When translating 'of' between two nouns *de* is the only correct form:

Darne de saumon *Tronçon de barbue* *Suprème de turbot*
Delice de sole *Paupiette de boeuf* *Beignets de pommes*
Raviers de hors d'oeuvre *Purée de marrons* *Salade de tomates*

On the menu the term 'of' = *de* or 'with' = *au* is used to describe the composition of a dish. If the main ingredient of the dish is described *de* is used:

beignets de pommes
salade de tomates
If the term is only part of the dish *au* is usually used:

crêpes au citron *dinde aux marrons*
omelette au jambon

à la mode means 'in the style of' or 'fashion of' and is often left out of the name of a dish, but the word following which describes the fashion must agree with the noun *la mode*, which is feminine:

Petits pois à la mode française becomes *Petits pois à la française*

The shortening of the phrase is even taken to the stage that it is not even written, but understood:
Chou-fleur à la mode polonaise becomes *Chou-fleur polonaise.*
Polonaise is feminine because it is describing mode, which is feminine; it is not describing chou-fleur, which is masculine.

When a dish is dedicated to a particular place or person the *à la* is omitted:
Peche Melba *Poire Condé*
à la can also describe a method of cooking:
Bifteck à la poêle *Pommes à la vapeur* *Haddock au four*

en used on the menu means 'in'.:

en tasse – Consommé en tasse	consomme in a cup
en branche – Epinards en branche	leaf spinach
en goujons – Sole en goujons	sole in small strips
en colère – Merlan en colère	curled whiting (with tail in mouth)

Pommes purée is written in this way because the full name would be *Pommes de terre en purée*. The *purée* remains singular because it is not an adjective but a noun.

Care should be taken to ensure that 'various hors d'oeuvre' is translated by '*hors d'oeuvre variés*'.

Capital letters in menus

Use capital letters for:

a) the first letter of a menu item:
Darne de saumon grillée
Merlan en colère

b) proper names (places, people)
Bombe Monte Carlo
Poulet Colbert
Omelette Arnold Bennett
Oeufs Balmoral

Verbs used in the kitchen

abaisser	to roll out pastry
acidifier	to add lemon juice or vinegar
ajouter	to add
apporter	to bring
aromatiser	to flavour pastry, or cream or a dish
arrêter	to stop
arroser	to baste as with roast joints
assaisonner	to season
attendrir	to tenderise
barder	to cover with a slice of fat bacon (on game)
beurer	to add or spread butter
blanchir	to blanch

bouillir	to boil
braiser	to braise
brider	to truss poultry and game
casser	to break
chauffer	to warm
chaufroiter	to mask with sauce as in cold larder work
chemiser	to line a mould, eg with jelly
ciseler	to shred
clarifier	to clarify
concasser	to chop roughly (as for tomatoes and parsley)
couvrir	to cover
cuire	to cook
débarrasser	to clear away
déglacer	to swill out as with the roast gravy
dégraisser	to skim off the fat
délayer	to dilute
ébarber	to remove the beard (oysters, mussels)
écumer	to skim
émincer	to cut into scallops or neat slices
éplucher	to peel or discard outside leaves
etuver	to stew, to braise, to steam
farcir	to stuff
flamber	to singe as with game and poultry
fouetter	to whisk
fraiser	to finally mix sweet pastry
frire	to fry
fumer	to smoke, eg salmon, ham, etc.
garnir	to garnish
glacer	to freeze, to pass rapidly under the grill
griller	to grill
hacher	to chop
larder	to insert lardons of fat bacon, eg with roast fillet of beef
laver	to wash
lier	to bind or thicken
mariner	to souse or pickle
marquer	to prepare for cooking
mélanger	to mix
méringuer	to cover with meringue
mijoter	to simmer
monter	to beat, to add
napper	to mask with a sauce
nettoyer	to clean
paner	to crumb
parer	to trim
passer	to pass
peler	to peel
pendre	to hang game or meat

peser	to weigh
pocher	to poach
rafraîchir	to refresh
râper	to grate
réduire	to reduce
revenir	to colour meat, game, vegetables
rissoler	to brown or colour
rôtir	to roast
saler	to add salt
sauter	to toss in butter or oil
servir	to serve
sucrer	to sweeten
suer	to sweat, eg vegetables for soup
tamiser	to sieve
verser	to pour
vider	to empty (or draw poultry)

The kitchen staff

The kitchen staff	*La brigade de cuisine*
head chef	le Chef de cuisine
second chef	le Sous-Chef
party chef	le Chef de partie
assistant chef	le Commis
apprentice	l'Apprenti
the announcer or barker	l'Aboyeur
kitchen porter	le Garçon de cuisine
kitchen maid	la Fille de cuisine
stillroom maid	la Fille d'office
stillroom man	le Garçon d'office
potman or sculleryman	le Plongeur
crockery washer-up	le Vaisselier
silver-plate cleaner	l'Argentier
store-room clerk	l'Économe
kitchen clerk	le Contrôleur

Party chefs	*Les Chefs de partie*
sauce cook	le Saucier
fish cook	le Poissonnier
roast cook	le Rôtisseur
vegetable cook	l'Entremettier
soup cook	le Potager
larder cook	le Garde-Manger
hors d'œuvre cook	le Hors d'œuvrier
butcher	le Boucher
cold-work cook	le Chef du froid
pastry cook	le Pâtissier

ice-cream cook	le Glacier
confectioner	le Confiseur
baker	le Boulanger
relief cook	le Chef Tournant
duty cook	le Chef de garde
night cook	le Chef de nuit
breakfast cook	le Chef du petit déjeuner
the carver	le Trancheur
the grill cook	le Grillardin
the staff cook	le Communard

Small kitchen utensils

*In the following lists those words marked with * are those which are in more general use.*

Small kitchen utensils, etc	*Les petits utensils de cuisine*
*vegetable knife	le couteau d'office
filleting knife	le couteau à filet
chopping knife	le couteau à hacher
boning knife	le couteau à désosser
*carving knife	le tranche lard
palette	la palette
kitchen fork	la fourchette de cuisine
vegetable peeler	l'économe
scissors	les ciseaux (m)
trussing needle	l'aiguille à brider (f)
grooving knife	le couteau à canneler
steel	le fusil de boucher
sharpening stone	la pierre à aiguiser
vegetable scoop	la cuillère à pommes et à légumes
sardine tin key	la clef à sardine
tin-opener	l'ouvre-boîte (m)
corkscrew	le tire-bouchon
pastry cutters	les emporte-pièces ou découpoirs (m)
pastry cutters, plain	les emporte-pièces, uni
pastry cutters, grooved	les emporte-pièces, cannelé
column cutters	les emporte-pièces à colonne
*piping bag	la poche
*piping tube	la douille
string-box	la boîte à ficelle
salt-box	la boîte à sel
mushroom	le champignon
*skimmer	l'écumoire
*tammy cloth	l'étamine
*pepper-mill	le moulin à poivre
grater	la râpe
*steel bat	la batte
chopper	le couperet
saw	la scie
pastry pincers	la pince à pâte
pastry cutting wheel	la roulette à pâte
rolling-pin	le rouleau
pastry brush	le pinceau
*ladle	la louche
*whisk	le fouet

*sieve	le tamis
*conical strainer	le chinois
*mandolin	la mandoline
frying basket	le panier à friture
*friture (deep-fat frying-pan)	la bassine à friture
*frying-pan	la poêle
*saucepan	la russe
*stockpot	la marmite
*braising-pan	la braisière
*fish kettle for salmon	la poissonnière ou la saumonière
*fish kettle for turbot	la turbotière
the steamer	le vapeur
*the cold room	la chambre froide
*the floors	les étages
*the hot plate	la table chaude
*the bain-marie	le bain-marie
*the oven	le four
*the stove	le fourneau
*the grill	le gril

Groceries

almond	l'amande (f)
*salted almonds	les amandes salées
ground almonds	les amandes pilées
angelica	l'angélique (f)
arrowroot	la marante (l'arrowroot (m))
beer	la bière
*bay leaf	le laurier
*bread	le pain
baking powder	la poudre à lever
pearl barley	l'orge perlé
*butter	le beurre
*bacon	le lard
cochineal	la cochenille
caper	le câpre
*cheese	le fromage
cinnamon	la cannelle
cloves	le clou de girofle
*coffee	le café
cornflour	la farine de maïs
*cream	la crème
*curry	le kari
currants	les raisins de Corinthe
coriander	le coriandre
cocoa	le cacao
date	la datte
essence of almond	l'essence d'amande
essence of anchovy	l'essence d'anchois
essence of lemon	l'essence de citron
essence of orange	l'essence d'orange
essence of rum	l'essence de rhum
essence of vanilla	l'essence de vanille
*egg	l'œuf (m)
*flour	la farine
*fondant	le fondant
fig	la figue sèche

flageolet *or* dwarf kidney beans	le flageolet
gherkin	le cornichon
ginger	le gingembre
gelatine	la gélatine
glacé cherry	la cerise glacée
*haricot beans	les haricots secs (m)
honey	le miel
*ham	le jambon
*jam	la confiture
lard	le saindoux
*lentils	les lentilles (f)
macaroni	le macaroni
margarine	la margarine
*mustard	la moutarde
*milk	le lait
marmalade	la marmelade, la confiture d'oranges
nutmeg	la muscade
*olive	l'olive (f)
*olive oil	l'huile d'olive (f)
*peel (candied)	le zeste (confit)
pepper	le poivre
cayenne pepper	le poivre de Cayenne
yellow split peas	les pois cassés jaunes (m)
green split peas	les pois cassés verts (m)
prune	le pruneau
raisin	les raisins secs
*rice	le riz
sago	le sagou
*semolina	la semoule
*salt	le sel
suet	la graisse de rognon
*sugar	le sucre
castor sugar	le sucre en poudre
granulated sugar	le sucre cristallisé
*spice	l'épice (f)
sultanas	les raisins de Smyrne (m)
*tea	le thé
tomatoe purée	la purée de tomate
*vermicelli	le vermicelle
*vinegar	le vinaigre
yeast	la levure

Fish

Fish	*Le poisson*	*Typical menu examples*
*anchovy	l'anchois (m)	Anchois sur croûte
*brill	la barbue	Suprême de barbue bonne-femme
bream	la brème	
*cod	le cabillaud	Darne de cabillaud pochée, sauce persil
cod (dried *or* salt)	la morue	
carp	la carpe	
dab	le carrelet	Filet de carrelet Dugléré
eel	l'anguille (f)	
eel, conger	le congre	Anguille fumé, sauce raifort
*gudgeon	le goujon	

*haddock	l'aigrefin *or* l'aiglefin (m)	Filet d'aigrefin bretonne
*haddock (smoked)	le haddock *or* l'aigrefin fumé	Haddock Monte Carlo
hake	la merluche, le merlu *or* le colin	
*halibut	le flétan	Suprême de flétan d'Antin
*herring	le hareng	Hareng grillé sauce moutarde
John Dory	le St Pierre	
*mullet	le rouget	Rouget grenobloise
*mackerel	le maquereau	Maquereau grillé beurre d'anchois
perch	la perche	
*plaice	la plie	Filets de plie frits sauce tartare
pike	le brochet	
*salmon	le saumon	Darne de saumon grillée sauce verte
*salmon (smoked)	le saumon fumé	Cornets de saumon fumé
*salmon trout	la truite saumonée	Suprême de truite saumonée Doria
*smelt	l'éperlan (m)	Éperlans grillés beurre maître d'hôtel
sprat	le sprat	Sprats frits au citron
*skate	la raie	Raie au beurre noir
*sardine	la sardine	Sardines sur croûte
*sole	la sole	Sole meunière
sole, lemon	la limande	Limande grillée, sauce diable
sturgeon	l'esturgeon (m)	
trout	la truite	Truite au bleu
*turbot	le turbot	Tronçon de turbot poché sauce hollandaise
*young *or* chicken turbot	le turbotin	Suprême de turbotin Véronique
tunny	le thon	Thon à l'huile
*whiting	le merlan	Merlan en colère
*whitebait	la blanchaille	Blanchailles diablées
*roe	la laitance	Laitances Méphisto
*caviar	le caviar	

Shellfish	*Les crustacés (m)*	*Typical menu examples*
clam	la praire, la palourde	
cockle	la coque	
*crayfish	l'écrevisse (f)	
*crawfish	la langouste	
*crab	le crabe	Cocktail de crabe
*lobster	le homard	Salade de homard
*prawns	la crevette rose	Mayonnaise de crevettes roses
*prawns (Dublin Bay)	la langoustine	Langoustines frites sauce tartare
*shrimps	la crevette grise	
*mussel	la moule	Moules marinière
*oyster	l'huître (f)	Huîtres florentine
*scallop	la coquille St Jacques	Coquille St Jacques Mornay
coral	le corail	
turtle	la tortue	Potage tortue claire
frog	la grenouille	
snail	l'escargot (m)	

Meat

Meat	*La viande*	*Typical menu examples*
*mutton	le mouton	
lamb	l'agneau (m)	
*leg	le gigot d'agneau *or* de mouton	Gigot d'agneau rôti sauce menthe
*saddle	la selle d'agneau *or* de mouton	Selle de mouton rôti sauce aux oignons
*loin	la longe d'agneau *or* de mouton	Longe d'agneau farcie
*best-end	le carré d'agneau *or* de mouton	Carré d'agneau boulangère
*breast	la poitrine d'agneau *or* de mouton	
*shoulder	l'épaule (f) d'agneau *or* de mouton	Epaule d'agneau froide, salade française
*neck	le cou	
*liver	le foie d'agneau *or* de mouton	Foie d'agneau au lard
*kidney	le rognon d'agneau *or* de mouton	Rognons sautés Turbigo
*tongue	la langue d'agneau *or* de mouton	Langues d'agneau à la sauce poulette
*sweetbread	le ris d'agneau *or* de mouton	
*brain	la cervelle d'agneau *or* de mouton	
*heart	le coeur d'agneau *or* de mouton	Cœur de mouton braisé aux légumes
*chop	la chop d'agneau *or* de mouton	Chop de mouton grillée
*cutlet	la côtelette d'agneau *or* de mouton	Côtelettes d'agneau sauce Réforme
*middle-neck cutlets	la basse côte	
*noisette	la noisette d'agneau *or* de mouton	Noisettes de mouton fleuriste
*fillet mignon	le filet mignon	

Pork	*Le porc*	*Typical menu examples*
*leg	le cuissot de porc	Cuissot de porc rôti sauce pommes
*chop	la côte *or* chop de porc	Côte de porc à la sauce charcutière
*loin	la longe de porc	Longe de porc rôtie à l'anglaise
*belly	la poitrine de porc	
*shoulder (hand)	l'épaule (f) de porc	
trotters	le pied de porc	
head	la tête de porc	
liver	le foie de porc	

Beef	*Le bœuf*	*Typical menu examples*
*sirloin	l'aloyau (m) de bœuf	Aloyau de bœuf rôti à l'anglaise
*boned-out	le contrefilet de bœuf	Contrefilet de bœuf Dubarry

sirloin		
*fillet	le filet de bœuf	Filet de bœuf bouquetière
*sirloin steak (boned out)	l'entrecôte de bœuf	Entrecôte à la bordelaise
*head of the fillet	le châteaubriand	Chateaubriand grillé sauce béarnaise
*cut from the middle without fat	le tournedos	Tournedos à la portugaise
*ribs	la côte de bœuf	Côte de bœuf froide, salade panachée
*thin flank	la bavette	Bœuf bouilli à la française
thick flank	la tranche grasse	Carbonnade de bœuf à la flamande
*rump	la culotte de bœuf	Rump steak vert pré
topside	la tranche tendre	Paupiettes de bœuf
silverside	le gîte à la noix	Bœuf bouilli à l'anglaise
shin *or* shank	la jambe	Consommé julienne
*liver	le foie de bœuf	Foie de bœuf lyonnaise
*tail	la queue de bœuf	Queue de bœuf aux primeurs
*tripe	les tripes (f)	Tripes à l'anglaise
*marrow	la moelle	Moelle sur croûte
*tongue	la langue de bœuf	Langue braisée au madère
*heart	le cœur de bœuf	Cœur de bœuf braisé

Veal	*Le veau*	*Typical menu examples*
*cutlet	la côte de veau	Côtelette (Côte) de veau milanaise
leg	le cuissot de veau	Cuissot de veau rôti
*head	la tête de veau	Tête de veau vinaigrette
saddle	la selle de veau	Selle de veau Mercédès
loin	la longe de veau	
shoulder	l'épaule (f) de veau	
*cushion	la noix de veau	Escalope de veau viennoise
*under cushion	la sous-noix de veau	Fricassée de veau à l'ancienne
shin	le jarret	Osso buco
liver	le foie de veau	Foie de veau au lard
sweetbread	le ris de veau	Ris de veau bonne-maman
*brain	la cervelle de veau	Cervelle au beurre noir
*sausage	la saucisse	Saucisse au vin blanc
bone	l'os (m)	
suet	la graisse	

Poultry

Poultry	*La volaille*	*Typical menu examples*
poultry	la volaille	Vol-au-vent de volaille
*chicken	le poulet	Poulet sauté chasseur
*turkey	la dinde	Dinde rôtie
*young turkey	le dindonneau	Dindonneau farci aux marrons
guinea fowl	la pintade	Pintade en casserole
*duck	le canard	
*duckling	le caneton	Caneton braisé aux petits pois
goose	l'oie (f)	Oie braisée aux navets
gosling	l'oison (m)	Oison rôti, sauce pommes
pigeon	le pigeon	Pigeon braisé aux olives

Game

Game	*Le gibier*	*Typical menu examples*
*pheasant	le faisan	Faisan braisé au céleri
*partridge	le perdreau, la perdrix	Perdreau aux choux
*woodcock	la bécasse	Bécasse rôtie
*snipe	la bécassine	Bécassine rôtie
ptarmigan	la poule de neige	Poule de neige rôtie
grouse	la grouse	Grouse rôtie
*wild duck	le canard sauvage	Canard sauvage à l'orange
teal	la sarcelle	Sarcelle à l'orange
plover	le pluvier	Pluvier au porto
quail	la caille	Cailles aux raisins
ortolan	l'ortolan (m)	Ortolan à la périgourdine
venison	la venaison	
deer	le chevreuil	Selle de chevreuil sauce Grand Veneur
*hare	le lièvre	Civet de lièvre
*rabbit	le lapin	Blanquette de lapin

Vegetables

Vegetables	*Les légumes (m)*	*Typical menu examples*
*artichoke (globe)	l'artichaut (m)	Artichaut en branche sauce hollandaise
*artichokes (Jerusalem)	les topinambours (m)	Topinambours en purée
*beetroot	la betterave	Salade de betterave
*broad beans	les fèves (f)	Fèves persillées
*broccoli	le brocoli	Brocoli au beurre
*Brussels sprouts	les choux de Bruxelles (m)	Choux de Bruxelles sautés au beurre
*cabbage	le chou vert	Choux braisés
*red cabbage	le chou rouge	Choux à la flamande
*carrots	la carotte	Carottes Vichy
*cauliflower	le chou-fleur	Chou-fleur à la polonaise
celeriac	le céleri rave	Salade de céleri rave
*celery	le céleri	Céleri braisé
*chicory	l'endive	Endive belge à l'étuvée
*cucumber	le concombre	Concombre farci
*egg plant	l'aubergine (f)	Aubergine frite
*curly kale	le chou frisé	Chou frisé nature
fennel	le fenouil	Fenouil au beurre
*French beans	les haricots verts (m)	Haricots verts
*leek	le poireau	Poireaux sauce crème
*lettuce	la laitue	Laitue braisée
mange-tout	le mange-tout	Mange-tout nature
*mushroom	le champignon	Champignons sur croûte
*marrow	la courge	Courge à la provençale
*young marrow	la courgette	
*onion	l'oignon (m)	Oignons frits à la française
*pimento	le piment	Piment farci
*parsnip	le panais	Panais rôti
*peas	les petits pois (m)	Petits pois à la menthe
*potato	la pomme de terre	Pommes duchesse

salsify	le salsifi	Salsifis à la crème
sorrel	l'oseille (f)	
*spinach	l'épinard (m)	Epinards en branche
sea-kale	le chou marin	Chou marin, beurre fondu
shallot	l'échalote (f)	
*sweetcorn	le maïs	Maïs à la crème
*swede	le rutabaga	Purée de rutabaga
*tomato	la tomate	Tomate farcie
*turnip	le navet	Navets au beurre
*truffle	la truffe	
*lettuce	la laitue	
*cos lettuce	la laitue romaine	
*radishes	le radis	
*watercress	le cresson	
corn salad	la mâche	

Herbs

Herbs, spices and aromates	*Herbes, épices et aromates*
*bay leaves	les feuilles de laurier
basil	le basilic
cinnamon	la cannelle
*chervil	le cerfeuil
*chives	la ciboulette
coriander	le coriandre
cumin	le cumin
clove	le clou de girofle
*curry	le kari
fennel	le fenouil
*garlic	l'ail (m)
ginger	le gingembre
juniper	le genièvre
mace	le macis
marjoram	la marjolaine
*mint	la menthe
mignonette	la mignonette
*mustard	la moutarde
*horseradish	le raifort
nutmeg	la muscade
*parsley	le persil
*pepper (white)	le poivre blanc
pepper (cayenne)	le poivre de Cayenne
pepper (paprika)	le paprika
rosemary	le romarin
saffron	le safran
sage	la sauge
*thyme	le thym
*salt	le sel

Fruits and nuts

Fruits and nuts	*Les fruits (m) et les noix*	*Typical menu examples*
*almond	l'amande (f)	
*apple	la pomme	Tarte aux pommes
*apricot	l'abricot (m)	Bande aux abricots
*banana	la banane	Beignets de bananes

blackberry	la mûre	
blackcurrants	le cassis (m)	
black grapes	le raisin noir (m)	
*cherry	la cerise	Tartelettes aux cerises
*chestnut	le marron, la chataigne	Marrons glacés
*cranberries	les airelles (f)	Sauce airelles
damson	la prune de Damas	
*date	la datte	
fig	la figue	
gooseberry	la groseille à maquereau	
*grapefruit	le/la pamplemousse (m or f)	Pamplemousse cerisette
grapes	le raisin (m)	
greengage	le reine-claude	
hazelnut	la noisette	
hops	le houblon	
*lemon	le citron	Crêpes au citron
*melon	le melon	Melon frappé
*orange	l'orange (f)	Pouding soufflé à l'orange
*peach	la pêche	Pêche Melba
*pear	la poire	Poire Belle-Hélène
*pineapple	l'ananas (m)	Ananas Condé
pistachio	la pistache	
plum	la prune	
pomegranate	la grenade	
red currants	les groseilles rouges	

Kitchen terms

Kitchen terms, etc

l'abats (m)	offal
l'abbatis de volaille (m)	poultry offal (giblets)
l'aile (f)	wing of poultry or game birds
à la	in the style of (*à la mode*)
à la française	dishes prepared in the French way
à l'anglaise	in the English style
à la broche	cooked on a spit
à la diable	devilled, a highly seasoned dish
à la carte	dishes on a menu prepared to order and individually priced
l'aloyau (m)	sirloin of beef (on the bone)
anglaise	beaten egg with oil and seasoning
appareil (m)	a preparation of one or more ingredients mixed prior to making certain dishes
arroser	to baste as in roasting
aspic (m)	a savoury jelly
assaisonner	to season
assorti	an assortment, eg *Fromages assortis*
au bleu	when applied to meat it means very underdone. When applied to trout it is a specific dish – *Truite au bleu*
au four	cooked in the oven, eg *Pomme au four*
au gratin	sprinkled with breadcrumbs and/or cheese and browned
au vin blanc	with white wine
le bain-marie	*a)* a container to keep foods hot so as to prevent burning *b)* a shallow pan of water for cooking foods to prevent them burning or boiling

	c) a deep narrow container for storing hot soup, sauces and gravies
le blanc	a cooking liquor of water, lemon juice and salt slightly thickened with flour and used for cooking artichoke bottoms, calf's head, etc.
la blanquette	a white stew cooked in stock from which the sauce is made, eg *Blanquette de veau*
les blancs d'œufs	egg whites
blond	very slightly coloured, eg *Blond roux*
la bombe	an ice cream of different flavours in the shape of a bomb
la bordure	a ring, sometimes of rice or potatoes
bouchées	small puff-pastry cases
bouillir	to boil
le bouillon	unclarified stock
le bouquet garni	a faggot or bundle of herbs, usually parsley stalks, thyme and bay leaf, tied inside pieces of celery and leek
braiser	to braise
la braisière	braising pan
en branche	a term denoting vegetables, such as spinach, cooked and served as whole leaves
brider	to truss (poultry)
la brioche	a light yeast cake
la broche	a roasting spit
la brochette	a skewer
brouillé	scrambled, eg *œufs brouillés*
brun	brown
brunoise	small neat dice
le buffet	a sideboard of food, or a self-service table.
le canapé	a cushion of toasted or fried bread on which are served various foods. It is used as a base for savouries. When served cold as Canapé moscovite the base may be toast, biscuit or short or puff-paste pieces with the food on top and finished with aspic
le caramel	sugar cooked until coloured to a certain degree
la carte du jour	menu, or bill of fare for the day
la casserole	a fireproof dish with a lid
Chantilly	sweetened, whipped vanilla-flavoured cream
la chapelure	breadcrumbs made from dried stale crusts of bread
la charlotte	name given to various hot and cold sweet dishes which have a case of biscuits, bread, sponge, etc
le chateaubriand	the head of the fillet of beef
chaud	hot
le chaud-froid	a creamed velouté or demi-glace with gelatine or aspic added, used for masking cold dishes
le chinois	a conical strainer
le civet	a brown stew of game, usually hare
clair	clear
la cloche	a bell-shaped cover, used for special *à la carte* dishes, eg *Suprême de volaille sous cloche*
clouté	studded, as with a clove inserted into an onion (*Oignon clouté*)
la cocotte	procelain fireproof dish
la compote	stewed fruit, eg *Compote des poires*
concassé	coarsely chopped, eg parsley and tomatoes
le consommé	basic clear soup
le contrefilet	boned sirloin of beef
le cordon	a thread or thin line of sauce
la côte	rib, eg *Côte de bœuf*

la côtelette	cutlet
coupe	cut; also ice-cream dish
le court-bouillon	a cooking liquor for certain foods, eg oily fish, calf's brains, etc. It is water containing vinegar, sliced onions, carrots, herbs and seasoning
la crème fouettée	whipped cream
la crêpe	pancake
la croquette	cooked foods moulded cylinder shape, egg and crumbled and deep fried, eg *Croquette de volaille*
la croûte	a cushion of fried or toasted bread on which are served various hot foods, eg savouries, game stuffing, etc.
le croûton	cubes of fried bread served with soup, also triangular pieces which may be served with spinach and heart-shaped ones which may be served with certain braised vegetables and entrées
les crustacés (m)	shellfish
la cuisine	the kitchen
la haute cuisine	high-class cooking
le cuisinier	male cook
la cuisse de poulet	leg of chicken
cuit	cooked
culinaire	to do with cooking
dariole	a type of mould
la darne	a slice of round fish cut through with the bone, eg *Darne de saumon*
les dés	dice, *coupé en dés* – cut in dice
déglacer	to swill out a pan in which food has been cooked with wine or stock, to use the sediment
dégraisser	to skim fat off liquid
demi-glace	half glaze – reduced *espagnole*
le déjeuner	lunch
le petit-déjeuner	breakfast
désosser	to bone out
le dîner	dinner
doré	golden
du jour (plat du jour)	special dish of the day
duxelles	finely chopped mushrooms cooked with chopped shallots
l'eau	water
ébarber	to remove the beard (from mussels, etc)
écumer	to skim
émincer	to slice
l'entrecôte (f)	a steak from a boned sirloin
l'escalope (f)	thin slice of meat
espagnole	brown sauce
estouffade	brown stock
étuver	to cook in its own juice
la farce	stuffing
farci	stuffed
la farine	flour
le feuilletage	puff paste
les fines herbes	chopped parsley, tarragon and chervil
flambé	flamed or lit, eg *Poire flambée*
le flan	open fruit tart
le fleuron	small crescent pieces of puff paste used for garnishing certain dishes, eg fish and vegetable

le foie	liver
le foie gras	fat goose liver
le fondant	a kind of icing
les fonds (m)	the foundation stocks
fondu	melted
le four	oven
frappé	chilled, eg *Melon frappé*
les friandises	petits fours, sweetmeats, etc
la fricassée	a white stew in which the poultry or meat is cooked in the sauce
frisé	curled
fumé	smoked, eg *Saumon fumé*
le fumet	a concentrated stock or essence
la garniture	the trimmings on the dish
le gâteau	cake
la gelée	jelly
le gibier	game
la glace	ice or ice cream
glacé	iced
glacer	to glaze under the *salamandre*
le goût	taste
gratiner	to colour or gratinate under the *salamandre* or in a hot oven using grated cheese or breadcrumbs
le gril	grill
grillé	grilled
hacher	to chop finely or very finely dice – to mince
les herbes (f)	herbs
hors d'œuvre	preliminary dishes of an appetising nature, served hot or cold
l'huile (f)	oil
le jambon	ham
le jambon froid	cold ham
jardinière	cut into batons
julienne	cut into fine strips
le jus lié	gravy thickened with arrowroof or cornflour or *fécule*
le jus de rôti	roast gravy
le kari	curry
le lait	milk
larder	to insert strips of fat bacon into lean meat.
le lardon	small strip of bacon
la levure	yeast
liaison	name given to yolks of eggs and cream when used as a thickening
lier	to thicken
la longe	loin
losange	diamond shaped
macédoine	*a)* A mixture of fruit or vegetables, eg *Macédoine de fruits* – fruit salad *b)* Cut into 6-mm dice
macérer	to steep, to soak, to macerate
la marinade	a richly spiced pickling liquid for enriching the flavour and tenderness of meats before braising
mariné	pickled
le madère	madeira wine
le maître	master

manié	kneaded, eg *Beurre manié*
manger	to eat
marbré	marbled
la marmite	stock pot
marquer	to prepare foods for cooking
masquer	to mask or coast
médaillon	foodstuffs prepared in a round, flat shape
le menu	bill of fare
la mie de pain	fresh white breadcrumbs
mignonette	coarse ground or crushed pepper
mijoter	to simmer
les mille-feuilles (f)	'thousand leaves', a puff-pastry cream slice
à la minute	cooked to order
mirepoix	roughly cut onions, carrots, celery and a sprig of thyme and bay leaf
mise-en-place	preparations prior to service
mollet	soft, eg *Oeuf mollet*
monter	to mount a sauce, to whip egg whites or yolks or to add butter to a sauce, to add oil to mayonnaise
le moule	mould
le moulin à poivre	pepper-mill
la mousse	a hot or cold dish of light consistency, sweet or savoury
la moutarde	mustard
moutarder	to smear with mustard, or to add mustard to a sauce
le mouton	mutton or sheep
mûr	ripe, mature
la mûre	blackberry
la muscade	nutmeg
napper	to mask
natives (f)	a menu term denoting English oysters, eg *Les huîtres natives*
navarin	a brown lamb or mutton stew
le nid	nest – imitation nest made from potatoes or sugar etc.
la noisette *or* noisette	*a)*A hazelnut used in confectionery *b)*Small round potatoes cut with a special scoop *c)*As for noisette butter *d)*A cut of loin of lamb
la noix	nut, also the name given to the cushion piece of the leg of veal *(Noix de veau)*
les nouilles (f)	noodles, a flat Italian paste
l'œuf (m)	egg
œuf brouillé	a scrambled egg
œuf en cocotte	egg cooked in an egg cocotte
œuf à la coque	egg boiled and served in its shell
œuf dur	a hard-boiled egg
œuf mollet	a soft-boiled shelled egg
œuf poché	a poached egg
œuf sur le plat	egg cooked in an egg dish
œuf à la poêle	a fried egg
l'oignon	onion
l'olive farcie	stoned and stuffed olive
l'olive tournée (f)	a stoned olive
l'orge (f)	barley
l'os (m)	bone
oseille (f)	sorrel
les pailles (f)	straws (*Pommes pailles*)
les paillettes de fromage (f)	cheese straws

le pain	bread
panaché	mixed
la panade	a flour composition used for binding
pané	crumbed
panier	basket
papillote	foods cooked *en papillote* are cooked in greased greaseproof paper in their own steam in the oven
parer	to trim
passer	to strain or pass
la pâte	a dough, paste, batter, pie or pastie
la pâte à frire	frying batter
la pâtisserie	pastry department
la paupiette	a strip of fish, meat or poultry stuffed and rolled
paysanne	to cut into even thin triangles, round or square pieces
le pilon	pestle for use with a mortar, name given to the drumstick of a leg of chicken
le piment	pimento
piquant	sharp flavour
piqué	studded
poêler	a method of cookery (pot roasting)
la pointe d'asperge	asparagus tip
la praline	chopped grilled almonds or hazelnuts or crushed almond toffee
primeur	early vegetables
printanière	garnish of spring vegetables
profiteroles	small balls of *choux* paste for garnishing soups or as a sweet course
la purée	a smooth mixture obtained by passing food through a sieve
la quenelle	forcemeat of poultry, fish, game or meat, pounded, sieved and shaped, then usually poached
le râble	the back, eg *Râble de lièvre* – the back of a hare
le ragoût	stew
le ravioli	an Italian paste, stuffed with various ingredients including meat, spinach and brains
réchauffer	to reheat
réduire	to reduce, to concentrate a liquor by evaporating
le risotto	Italian rice stewed in stock
le ris	sweetbread
rissoler	to fry to a golden brown
rôtir	to roast
roux	flour and fat cooked together for thickening
le sabayon	yolks of eggs and a little water cooked till creamy
la salamandre	a gas or electric grill under which food is cooked or completed
sauter, sauté	to toss in fat or to turn in fat; also a specific meat dish
soubise	an onion purée
le soufflé	a light dish, sweet or savoury, hot or cold. Whites of egg are added to the hot basic preparation and whipped cream to the cold
le suprême	when applied to poultry it means whole wing and half the breast of the bird (there are two *suprêmes* to a bird). For other foods it is applied to a choice cut
table d'hôte	a meal of several courses, which may have a limited choice, served at a fixed price
le tamis	sieve
la terrine	an earthenware utensil with a lid. A terrine also indicates a pâté cooked and served in one
tomaté	preparations to which tomato purée has been added to dominate the flavour and colour

tourner	to turn, to shape (barrel or olive shape)
la tranche	a slice
le tronçon	a slice of flat fish cut with the bone, eg *Tronçon de turbot*
le velouté	*a)* basic sauce *b)* a soup of velvety or creamy consistency
vert	green, eg *Sauce verte*
la viande	meat
voiler	to veil or cover with spun sugar
la volaille	poultry
le vol-au-vent	a puff-pastry case

Some terms used during service (see page 291)

Le service commence *or* le service va commencer	Service begins
Ça marche, deux couverts	Warning order that an order for two covers is to be announced
Faites marcher	Begin to cook
Envoyez	Send up
Supplément	Another portion
En vitesse or vite	Quickly, with all speed
Annulez	Cancel the order
Arrêtez	Stop the order
Soigné	Carefully done (for a special customer)
Très soigné	Very carefully (for a very special customer)
Soignez la commande	The order to be done with care
Ils passent à table	The customers are now going to the table
Couvert	Cover or person
Un couvert	One person
Deux couverts	Two persons
Dix couverts	Ten persons
Deux fois deux couverts	Twice two covers
Trois et deux fois deux de soles frites	Three and twice two of fried soles
Poissonnier, faites marcher, deux soles meunière	Fish cook, start to cook two sole meunière
Saucier, envoyez quatre escalopes de veau à la viennoise, soignées, pour les étages	Sauce cook, send up four carefully done escalopes of veal viennoise for the floors (this would be for floor or room service in a hotel)
Entremettier, envoyez une omelette aux champignons en vitesse	Vegetable cook, sent up one mushroom omelet as soon as possible
La partie de vingt-cinq couverts passe à table	The party of twenty-five customers is going to the table
Rôtisseur, faites marcher deux entre-côtes doubles, une bien cuite et une saignante	Roast cook, prepare two double entrecote steaks, one well cooked and one underdone
Pâtissier, envoyez deux fois deux crème caramel	Pastry cook, send up twice two cream caramels
Garde-manger, envoyez un saumon froid supplément	Larder cook, send another portion of cold salmon
Envoyez les pommes et légumes	Send the potatoes and vegetables
Arrêtez le merlan frit	Stop the fried whiting
Argentier, envoyez les saucières	Silverman, send some sauceboats
Plongeur, envoyez les sauteuses	Sculleryman, more sauteuse wanted

Further information: Schneider and Pownie, *Le Français dans l'Hotellerie* (Arnold), Atkinson; *Menu French*, (Oxford Polytechnic Press.)

13
Menu planning

Having read the chapter these learning objectives, both general and specific, should be achieved.

General objectives Understand the principles of menu planning and know how to compile menus suitable for a wide variety of establishments.

Specific objectives Apply the principles of menu planning to menus for special parties, table d'hôte and à la carte menus. Specify suitable dishes for various courses. Compile menus taking account of all the factors – costs, season, staff capabilities, equipment, facilities, suppliers, nutritional value, types of customer. Demonstrate by correctly writing menus in English and where appropriate in French.

A menu or a bill of fare is a list of prepared dishes of food which are available to a customer, and by its content and presentation should attract the customer and represent value for money.

The compiling of a menu is one of the most important jobs of a caterer and there are a number of factors that must be taken into con-

1 The kind of menu
2 The season of year
3 The capabilities of kitchen staff
4 Size and equipment of kitchen
5 Capabilities of serving staff
6 Price of the menu
7 Type of customers
8 Supplies
9 Balance
10 Colour
11 Wording of menus.

sideration before any menu is written. Some establishments are using computers to assist in the compilation of menus, see Chapter 15, page 397.

The aim is to give the customer what he wants and not what the caterer thinks the customer wants. In general it is better to offer fewer dishes of a good standard rather than having a wide choice of dishes of mediocre quality.

Traditional cookery methods and recipes form a sound foundation of knowledge for the craftsman and the caterer. However, fashions in food change over the years and some customers tend to look for new dishes, different combinations of food and fresh ideas on menus.

The so-called Cuisine Nouvelle would probably be more correctly named as Contemporary Cooking as it represents new thinking, new food combinations and a lighter style of cooking with the accent on creativity, presentation and frequently, simplicity.

It follows therefore that students should be aware of changes in contemporary cooking and should be prepared to experiment and create new and original recipes in order that the caterer can continue to offer the customer fresh interest, variety and pleasure in his or her menus.

It is necessary to make certain that menu terms are expressed accurately so that the customer receives exactly what is stated on the menu; otherwise it may mean that the Trade Descriptions Act is being contravened. In the words of the act 'any person who in the course of a trade or business –

a) applies a false trade description to any goods

or

b) supplies or offers to supply any goods to which a false trade description is applied shall be guilty of an offence.'

For example, Paté maison must really be home-made paté, not factory-made. If Fried fillets of sole are offered on the menu, then more than one must be offered and the fish must be sole, and if an 8-oz rump steak is stated on the menu as the portion size, then it must be 8 oz raw weight.

If the sole is advertised as 'fried' and the steak as 'grilled' then these processes of cooking should be applied; if the soles are stated to be *Dover* soles and the steak as *rump* steak then the named food must be served. Likewise if the sole is stated to be served with a sauce tartare and the steak with a béarnaise sauce then the sauce should be correct and accurate.

The description on the menu should give an indication as appropriate of the quality, size, preparation and composition of the dish.

1 The kind of menu

It must be clearly understood what kind of menu is required, whether a special party menu, table d'hôte or à la carte.

a) *Special party menu* – these are menus for banquets and parties of all kinds.

b) *Table d'hôte* – this is a set menu forming a complete meal at a set price. A choice of dishes may be offered at all courses; the choice and number of dishes will usually be limited.

c) *A la carte* – this is a menu with all the dishes individually priced. The customer can therefore compile his own menu. A true à la carte dish should be cooked to order and the customer should be prepared to wait for this service.

Examples of luncheon menus suitable for a medium-sized hotel:

A Special party (at a set price)

Luncheon menu
Hors d'œuvre
Poached Scotch salmon Hollandaise sauce
Roast saddle of mutton
Garden peas New potatoes
Strawberries and cream

B Table d'hôte (at a set price)

Déjeuner	*Lunch*
Melon frappé	Chilled Melon
Hors d'œuvre	Hors d'oeuvre
Crème de volaille	Cream of chicken soup
Raviolis au jus	Raviolis
Filets de sole frits, sauce tartare	Fried fillets of sole, tartare sauce
Epaule d'agneau boulangère	Roast shoulder of lamb with savoury potatoes
Buffet froid	Cold Buffet
Petis pois — *Pommes:* Persillées	Peas — Parsley potatoes
Carottes Vichy — frites	Vichy carrots — Fried potatoes
Salade de saison	Salad in season
Gâteau praliné	Hazelnut caramel cake
Glace panachée	Mixed ice cream
ou	or
Fromage	Cheese

C A la carte (all dishes would be individually priced)

When preparing an à la carte luncheon menu, selection may be made from the following, offering only the number of course and dishes that can be well cooked and served:

A la carte luncheon menu

Hors d'œuvre:
Oysters; Smoked salmon; Pâté maison; Parma ham; Prawn cocktail; Smoked trout; Potted shrimps; Assorted sausage; Hors d'œuvre; Fruit cocktail; Melon; Avocado.

Soups:
Pea; Tomato; Onion; Vegetable; Minestrone; Consommé.

Eggs
Omelettes (Ham, Mushroom, Tomato, Cheese);
Poached egg Washington; Scrambled eggs with shrimps; Eggs in cocotte with cream.

Pastes:
Spaghetti bolognaise; Risotto with mushrooms; Noodles with cheese, Raviolis.

Fish:
Sole Colbert; Sole meunière; Fried sole; Grilled sole; Sole bonne-femme; Boiled turbot with hollandaise sauce; Devilled whitebait; Lobster Mornay; Grilled herrings with Mustard sauce; Fried or grilled plaice; Poached halibut.

Entrées:
Calf's head vinaigrette; Chicken sauté chasseur; Braised tongue with sherry; Curried beef and rice; Pork chop normande; Veal escalope viennoise.

Roasts:
Best end of lamb; Aylesbury duckling; Surrey chicken; Game (in season).

Grills:
Rump steak; Entrecôte steak; Minute steak; Fillet steak; Lamb cutlets; Lamb chop; Mixed grill; Sheep's kidneys.

Cold buffet:
Chicken; Turkey; Ham; Tongue; Galantine; Pressed beef; Pork; Dressed crab; Cold lobster.

Salads:
Lettuce, Tomato, Beetroot, French, Japanese.

Vegetables:
Peas, French beans, Leaf spinach, Vichy carrots, Cauliflower, Hollandaise sauce, Brussels sprouts, Braised celery;
Potatoes (Creamed, Allumettes, Lyonnaise, Marquise).

Sweets:
Peach Melba; Fruit salad; Lemon pancakes; Cream caramel, Jam omelet; Stewed fruits; French pastries; Assorted gâteaux; Assorted ices.

Savouries:
Welsh rarebit; Welsh rarebit with mushrooms; Mushrooms on toast; Scotch woodcock; Soft roes on toast; Sardines on toast.

Cheese:
A good selection.

Coffee:
Cona; Turkish; Irish.

Kind of meal

It is necessary to know if the menu is required for breakfast, luncheon, tea, dinner, supper or for a special function.

Kind of establishment

The type of establishment will have to be considered. Menus will vary for the following examples:

Luxury-class hotel; first-class restaurant
Medium-price hotel; or restaurant, country hotel
Licenced house; wine bar
School meal; speciality eating houses
Directors' room in an industrial canteen
Female workers' dining-room in an industrial canteen
Heavy manual workers' dining-room in an industrial canteen
Hospital; transport café

2 The season of the year

If menus have to be compiled a long time in advance of the actual date of production, the season of the year should be considered –

a) Because of the weather; it may be hot, cold or mild and certain dishes which would be acceptable in hot weather would be most unsuitable in cold weather and vice versa.
b) Because foods in season, eg strawberries, should be included in menus where possible, as they are usually in good supply, of good quality and a reasonable price.

c) Special dishes for certain days or a time of the year should be considered. For example, pancakes on Shrove Tuesday and turkey at Christmas.

3 The capabilities of the kitchen staff

The capabilities of the cooks should be considered. There are many excellent cooks, male and female, whose training is such that they can cope with simple dishes but who might be at a loss if asked to produce highly complicated foreign fare. On the other hand, if a staff of well-trained cooks, capable of a good standard of international cookery, is available, they should be given a chance to produce dishes that can express their skill and pride of craftsmanship.

4 Size and equipment of kitchen

The type of kitchen should be considered – how large? how small? And what large-scale equipment is available such as stoves, steamers, hot plates, etc. A good cook will usually manage to produce a required meal somehow in spite of any shortcomings of space or equipment. Nevertheless, the writer of menus should be aware of such deficiencies and be wary of putting on dishes that might be difficult to produce because of shortages. Care should be taken to see that the method of cooking is not repeated, otherwise certain pieces of equipment can be overloaded; for example, steamer, friture.

5 Capabilities of serving staff

The ability of person who serves the meal should be considered. If the standard of the waiter or waitress is of the highest order then a high standard of well-presented and garnished dishes can be used because the caterer can be sure that the prepared food will be correctly shown to the customer and that it will be transferred to the customer's plate in a neat, presentable fashion. If the waiters, waitresses or servers are untrained, care should be taken in the selection of dishes and only those dishes suitable for easy serving selected. Plate service, or the main food item plated and the vegetables and accompaniments placed on the table for the customer to help himself or herself, is becoming more popular.

Size and equipment of dining-room

The type of dining-room and its equipment have sometimes to be taken into account. The china in some catering establishments needs to be considered, particularly if it is coloured or highly patterned, as either factor can affect the appearance of certain foods.

The size of silver dishes needs to be considered when planning menus. This is important when a well-garnished dish such as Filet de bœuf bouquetière is planned which, if dressed on dishes that are too small, would risk losing its presentation value.

The amount of china and silver must be taken into account if a menu of several courses is required. Operating a five-course menu with only sufficient china and silver for three courses can raise problems which would upset the smooth running of the meal.

6 Price of menu

The proposed charge per head is obviously an important factor to consider when selecting food for any menu. When the caterer is asked to produce, for example, a meal to sell at a modest price per head he or she cannot consider expensive foods such as oysters, fresh salmon, game, asparagus, etc. Similarly, if asked to produce a meal for a good price then he or she should offer good value for the price charged. A useful working rule with regard to the cost of food for a meal is to see that the food cost does not exceed 40% of the selling price. For example, if the cost of the food per head for a meal is xp, then the suggested selling price would be

$$\frac{\text{xp} \times 100}{40} = £$$

This point is explained further in the section on costing.

7 Type of customers

The type of customers to be served, particularly at special parties, can sometimes affect the choice of foods. For example, the caterer may be asked to produce a dinner menu for a special party of:

Young people attending a twenty-first birthday party
Workers in heavy industry attending a celebration
Elderly people attending a conference
Old-age pensioners attending at Christmas
Football players after an international match
Visiting American students

In each of the above groups of people there are certain foods which would suit one group but not necessarily the others. See page 409 regarding social customs.

8 Supplies

When considering foods for a menu it is sound policy to think of any foods in season as they are usually plentiful, of good quality and reasonable price. At the same time the full range of part-prepared, ready prepared, and deep frozen food should also be borne in mind.

If large stocks of food are held, care should be taken, where possible, to use foods on the menu which are already in store, before ordering fresh supplies. This can help to avoid wastage of food and, of course, money.

The cold room or refrigerator should be inspected to see if there are any cooked foods that may be incorporated into menus. This is an important point and one that the experienced caterer makes full use of, as there are many ways in which cooked foods can be used to produce or help to produce attractive dishes. For example, a good hors d'œuvrier can use almost any item of cooked food.

Foods selected for a menu should be easy to obtain locally. Special foods are sometimes difficult to obtain at certain times of the year in some parts of the country. It is bad practice to offer customers particular dishes before checking that they are readily available.

If the customer has any special wishes, the good caterer should always do his best to comply with them.

9 Balance

This is particularly important when compiling special party menus, and the following points should be considered:

a) Repetition of ingredients

Never repeat the basic ingredients on one menu.

For example, if mushrooms, tomatoes, peas, bacon, etc, are used in one course of a menu then they should not reappear in any other course of the same menu. Examples:

Mushroom soup
Fillet of sole bonne-femme
Casseroled chicken grandmère
Mushroom and bacon savoury

(repetition of mushroom)

Tomato salad
Steak pie
Brussels sprouts; Sauté potatoes
Dutch Apple Tart

(repetition of pastry)

b) Repetition of colour

Where possible avoid repetition of colour. Examples:

Celery soup
Fricassée of veal
Buttered turnips; Creamed potatoes
Meringue and Vanilla ice cream

Tomato soup
Goulash of veal
Vichy carrots; Marquise potatoes
Peach Melba

Examples of a menu with repetition of both an ingredient and colour (peas and green):

Purée St-Germain
Darne de saumon grillée; sauce verte
Noisette d'agneau Clamart
"Gooseberry fool"

c) Repetition of words

Avoid repeating the same words on a menu. Examples:

Green pea soup
Grilled salmon steak, *Green* sauce
Neck of lamb with artichoke and *peas*
Small roast potatoes
Gooseberry fool

Crème portugaise
Carré d'agneau à la menthe
Purée d'epinards à la *crème*
Pommes boulangère
Fraises à la *créme*

Hors d'œuvre
Bœuf *braisé* aux nouilles
Endive *braisé*
Profiteroles au chocolat

Oysters
Fried *fillet* of sole
Grilled fillet steak
Parsley potatoes; Cauliflower
Apple pie

d) Overall balance of a menu

If many courses are to be served, care should be taken to see that they vary from dishes of a light nature to those of a more substantial nature and finish up with light dishes. In the case of a meal consisting of two or three meat courses, a rest may be made in the middle of the meal when a sorbet (a well-flavoured, lightly frozen water-ice designed to cleanse the palate) can be served; then a fresh start to the meal can be made with the roast. Examples:

Dîner

Melon frappé
Consommé royale
Délice de sole d'Antin
Escalope de ris de veau aux pointes d'asperges
Filet de bœuf bouquetière
Sorbet au Grand Marnier
Caneton d'Aylesbury rôti

Salade Mimosa
Soufflé à la vanille
Paillettes au parmesan

e) Texture of courses

Regard should be given to the texture of courses; some food should be soft, some should require chewing, crunching, biting, some should be swallowed and so on.

Example of varied texture menu:

Crème Dubarry
Escalope de veau viennoise
Petits pois au beurre; Pommes marquise
Flan aux pommes

Cream of cauliflower soup
Escalope of veal viennese style
Buttered peas; Marquis potatoes
Apple flan

Example of bad menus:

Cauliflower soup
Irish stew
Buttered peas; Mashed potatoes
Semolina pudding

Salami
Fried scallops
Veal escalope
Fried aubergine; Sauté potatoes
Apple fritters

f) Seasonings

If strong seasonings like onion, garlic or pungent herbs such as thyme, sage or bay leaf are used, ensure that they are not repeated in more than one course.

g) Sauces
If different sauces are served on one menu the foundation of the sauces should vary, eg demi-glace, velouté, tomato, butter thickened, arrowroot or cornflour thickened.

h) Garnishes
If traditional names are used for garnishes then the garnishes must be correct and they should not be repeated. Potatoes should only be named on a menu once, but they may be used in moderation on other courses; for example, Pommes duchesse may be piped around a dish of fish. Pommes parisienne, noisette or olivette may be used to help lightly garnish an entrée or light meat course on a menu where two or three meat courses are being served.

i) Food values
When a customer selects from an à la carte or table d'hôte menu the composition of the meal is the customer's own responsibility. When a set meal is offered for a special party or banquet the menu is usually more than adequate to fulfil the nutritional needs. Special attention should be paid, however, to the nutritional balance of meals for people engaged in light or heavy work. The manual worker will require more substantial food than the office worker. Meals served to school children, meals served in hospitals, hostels, homes for invalids or homes for the aged all need thought on nutritional balance. It is usual to prepare such menus a week or two in advance, and it necessary to see that the food provided over the period is satisfactory from the point of food value.

Further information on food value will be found in chapter 7, on Elementary nutrition.

10 Colour
This is a most important factor to consider in the presentation of food. The sensible use of colour will always help the eye-appeal of a dish. If it is sometimes necessary to use a little artificial colouring in order to finish certain foods correctly, care must be taken to see that these colours are used in moderation. Deep vivid colouring of any food should be avoided; the aim should be to use natural tints.

If a drab-looking main dish is being served, the careful use of a colourful garnish can greatly improve the presentation.

Plate appeal is a term which refers to the appearance of a course when it is served on a plate. The attractive appearance of each course is essential to the successful meal, and foods such as carrot, tomato, peas, watercress, parsley, truffle, oranges, cherries, etc, can give colour to the dullest coloured dishes. For example, a sprig of watercress with a dish of roast beef and Yorkshire pudding; chopped parsley on sauté potatoes; paprika on egg mayonnaise: neatly cut or turned vegetables and peas with a brown lamb stew; orange salad with roast duck – these are but a few examples of how important colour is in the presentation of a dish and in the planning of a menu.

11 Wording of menus

When all the previous points have been considered then the menu can be written. Many errors can occur in the writing of menus and the following are points which should be noted.

1 Consider the customer to be fed, and select language which will be easily understood.
2 If using French for menu writing give an English translation as a simple description under or by the side of each dish.
3 Having selected a language for the menu avoid using a mixture of languages on the same menu except in the case of certain national dishes which are best kept in the language of the country of origin. For example: in a first-class hotel or restaurant on a table d'hôte menu which may be of the highest order and written in French, dishes such as Irish stew, Lancashire hot pot, Steak and kidney pie should appear in the language of the country of origin not as Ragoût irlandaise or Irish stew aux légumes, Casserole Lancashire, Pie à la biftek et rognon.
4 When writing menus a decision has to be made regarding the use of capital letters for the names of dishes. In all cases consistency is the overriding principle. As a guide when writing in English, the writer of the menu decides to use capital letters according to style and layout. The type of establishment will, to a certain extent, affect the choice whether to use all capital letters, capital letters for major words or some other method.

   With menus written in French care needs to be taken with the use of capital letters as there are rules for their use. Trade practice, both in France and Britain, it is commonplace for all major words in the dish name to have capital letters. However, to be more accurate a capital letter would be used for the first word and for names of people eg Pêche Melba but names of places usually have a small letter eg Petits pois à la française. If follows that it is helpful to know the origin of the name of the dish and frequently this is hidden in history, a reason why so often in the trade, capital letters are used.

Certain menu terms such as hors d'œuvre, mayonnaise, hollandaise, bonne-femme, chasseur, need not be translated into English.

Where the place of origin of a food is known it may be used on the menu: Roast *Aylesbury* duckling; Grilled *Dover* sole; *Vale of Evesham* asparagus; Minted new *Jersey* potatoes; *Kentish* strawberries with *Devonshire* cream; *York* ham with *Belgian* chicory salad.

When writing a menu in French, care should be taken to see that the spelling is correct and that the accents are included.

A list of some of the names used in fish cookery to indicate certain garnishes or methods of cooking and serving

Name of garnish	Composition of garnish	Menu example
Bercy	poached with chopped shallots, parsley, fish stock and white wine; finished with white wine sauce and glazed	Filets de sole Bercy
bonne-femme	as for Bercy, with the addition of sliced mushrooms, glazed	Suprême de turbot bonne-femme
Bréval or d'Antin	as for bonne-femme, with the addition of chopped tomatoes, glazed	Suprême de barbue Bréval
Véronique	poached with white wine and fish stock; coated with white wine sauce; glazed and garnished with grapes	Sole Véronique
dieppoise	poached; coated with white wine sauce, garnish mussels, shrimps' tails and mushrooms	Suprême de carrelet dieppoise
Dugléré	poached with chopped shallots, parsley, tomatoes, fish stock; coated with a sauce made from the cooking liquor	Filet d'aigrefin Dugléré
Mornay	poached; coated with cheese sauce and glazed	Coquille de cabillaud Mornay
florentine	poached; dressed on a bed of leaf spinach; coated with Mornay sauce and glazed	Suprême de turbotin florentine
hollandaise	plain boiled; served with hollandaise sauce and plain boiled potatoes	Trouçon de turbot hollandaise
meunière	shallow fried on both sides; garnished with slice of lemon; coated with nutbrown butter, lemon juice and parsley	Truite de rivière meunière
belle meunière	as for meunière, with the addition of grilled mushroom, a slice of peeled tomato, and a soft herring-roe (passed through flour and shallow fried) (all neatly dressed on each portion of fish)	Sole belle meunière
Doria	as for meunière, with a sprinkling of small, turned, cooked pieces of cucumber	Suprême de flétan Doria
grenobloise	as for meunière, with the lemon in peeled segments and capers	Rouget grenobloise
bretonne	as for meunière, with picked shrimps and sliced mushrooms	Filet d'aigrefin bretonne
princesse	poached; coated with white wine sauce; garnished with asparagus heads and slices of truffle	Délices de sole princesse
Saint-Germain	filleted white fish dipped in butter; breadcrumbed; grilled; served with béarnaise sauce and noisette potatoes	Filets de plie St-Germain
Orly	deep fried and served with tomato sauce	Filet de cabillaud à l'Orly

Some meat and poultry garnishes with suitable menu examples

Name of garnish	Composition of garnish	Use of garnish	Menu term: example using garnish
boulangère	onions and potatoes	for roast lamb joints	Gigot d'agneau boulangère
bouquetière	artichoke bottoms, carrots, turnips, peas, French beans, cauliflower, château potatoes	for roast joints, usually beef	Filet de bœuf bouquetière
bourgeoise	carrots, onions, dice of bacon	for large joints	Bœuf braisé bourgeoise
bruxelloise	braised chicory, brussels sprouts, château potatoes	for large joints	Noix de veau bruxelloise
Clamart	artichoke bottoms filled with peas à la française or purée of peas and château potatoes	for large joints	Carré d'agneau Clamart
Dubarry	small balls of cauliflower Mornay, château potatoes	for large joints	Contrefilet de bœuf Dubarry
fermière	paysanne of carrot, turnips, onions, celery	for large joints	Poulet en casserole fermière
fleuriste	tomatoes filled with carrots, turnips, peas, French beans; château potatoes	for tournedos and noisettes	Noisette d'agneau fleuriste
Henri IV	pont-neuf potatoes and watercress	for tournedos and noisettes	Tournedos Henri IV
jardinière	carrots, turnips, French beans, flageolets, peas and cauliflower coated with Hollandaise sauce	for joints	Selle d'agneau jardinière
judic	stuffed tomatoes, braised lettuce, château potatoes	for joints	Longe d'agneau judic
mascotte	quarters of artichoke bottoms, cocotte potatoes, slices of truffle	for joints and poultry	Poulet sauté mascotte
Mercédès	grilled tomatoes, grilled mushrooms, braised lettuce, croquette potatoes	for joints	Selle de veau mercédès

Some meat and poultry garnishes with suitable menu examples *(contd)*

Name of garnish	Composition of garnish	Use of garnish	Menu term: example using garnish
Parmentier	1 cm dice of fried potatoes	for tournedos, noisettes and poultry	Poulet sauté Parmentier
portugaise	small stuffed tomatoes, château potatoes	for tournedos, noisettes and poultry	Tournedos à la portugaise
printanier	turned carrots and turnips, peas, dice of French beans	for entrées, stews	Navarin d'agneau printanier
Réforme	short batons of beetroot, white of egg, gherkin, mushroom, truffle and tongue	for lamb or mutton cutlets	Côtelette d'agneau Réforme
Sévigné	braised lettuce, grilled mushrooms, château potatoes	for tournedos, noisettes and joints	Carré de veau Sévigné
Soubise	onion purée or sauce	for cuts of mutton	Côte de mouton Soubise
tyrolienne	French fried onions, tomate concassé	for small cuts of meat	Noisette d'agneau tyrolienne
vert Pré	straw potatoes and watercress	for grilled meats	Entrecôte vert Pré
Vichy	with Vichy carrots	for entrées	Blanquette de veau Vichy
Washington	sweetcorn	for eggs and poultry	Poularde poêlé Washington
duxelle	finely chopped shallots and chopped mushrooms	this is a basic culinary preparation with many uses – for example, stuffed vegetables, sauces, etc	

Names used in sweet dishes to indicate a particular ingredient

	Composition	Menu example
Chantilly	whipped, sweetened, vanilla-flavoured cream	Meringue Chantilly
glacé	with ice cream	Meringue glacé Chantilly
meringuée	finished with piped meringue and glazed	Flan aux pommes meringuées
normande	with apple	Crêpes normande
Rubané	with different flavours and colours	Bavarois Rubané
Montmorency	with cherries	Coupe Montmorency
Condé	with rice	Poire Condé
Moka	coffee flavour	Glace moka
Suchard	chocolate flavour	Profiteroles, sauce Suchard
praliné	with chopped almond and hazel-nut toffee	Gâteau praliné
Melba	with vanilla ice cream and raspberry purée or sauce	Pêche Melba
cardinal	with strawberry ice cream and raspberry sauce, sliced almonds	Fraises cardinal
Hélène	vanilla ice cream, hot chocolate sauce	Poire Hélène
Jamaïque	rum flavour	Coupe Jamaïque

The words used to describe a dish on a menu must agree in gender and number (also see pages 308–9). For example:

A fillet of sole	Le filet de sole
Two fillets of sole	Deux filets de sole
Two fried fillets of sole	Deux filets de sole frits
Lamb	Agneau (m)
Lamb cutlet	La côtelette d'agneau
Two lamb cutlets	Deux côtelettes d'agneau
Grilled lamb cutlet	Côtelette d'agneau grillée
Two grilled lamb cutlets	Deux côtelettes d'agneau grillées

It will be noticed that the agreement is made with the cut or joint.

The definite article, *le* or *la*, if used with one dish should be used on all dishes; for example:

Le saumon fumé La selle d'agneau rôtie Le chou-fleur; sauce hollandaise Les pommes parisienne La pêche Melba	*and not*	Le saumon fumé Selle d'agneau rôtie Le chou-fleur; sauce hollandaise Les pommes parisienne Pêche Melba

Care should be taken over the use of the term *à la*, which means 'in the *or* after the style of'. This means that it can be used when a dish is prepared in the style of a certain place, town or country; for example:

Hors d'œuvre à la grecque	Poulet sauté à la portugaise
Filets de sole frits à la française	Petits pois à la flamande

When a dish of food is named after a person or place the name usually follows the food:

Pêche Melba, named after Madame Melba
Crème Dubarry, named after Madame Dubarry
Salade Waldorf, named after Waldorf Astoria Hotel

When using classical garnish names, see that any vegetable which is included in the garnish is not repeated as a vegetable. For example: Chou-fleur Mornay would not be put on the menu as a vegetable to be served with Selle d'agneau Dubarry.

The list which follows includes some of the terms used on a menu, which signify that a specific item of food will be served. For example: parmentier indicates that potatoes will be used; therefore Purée parmentier is potato soup; Omelette parmentier is an omelet garnished with potatoes.

Dubarry	Cauliflower	lyonnaise	Onions
bruxelloise	Brussels sprouts	florentine	Spinach
Clamart	Peas	portugaise	Tomatoes
Doria	Cucumber	princesse	Asparagus
provençale	Tomato and Garlic	Washington	Sweet corn

Breakfast menus

A breakfast menu can be compiled from the following foods:

1. *Fruits:* Grapefruit, orange, melon, apple.
   Fruit juices: Grapefruit, orange, tomato, pineapple.
   Stewed fruit: Prunes, figs, apples, pears.
2. *Cereals:* Cornflakes, shredded wheat, etc, porridge.
3. *Eggs:* Fried, boiled, poached, scrambled; omelets with bacon (streaky, back or gammon) or tomatoes or mushrooms or sauté potatoes.
4. *Fish:* Grilled herrings, kippers or bloaters; fried sole, plaice or whiting; fish cakes, smoked haddock, kedgeree.
5. *Meats (hot):* Fried or grilled bacon (streaky, back or gammon), sausages, kidneys, calves' liver, with tomatoes or mushrooms or sauté potatoes or potato cakes or bubble and squeak.
6. *Meats (cold):* Ham, bacon, pressed beef with sauté potatoes.
7. *Preserves:* Marmalade (orange, lemon, grapefruit, ginger), jams, honey.
8. *Fresh fruits:* Apple, pear, peach, grapes, etc.
9. *Beverages:* Tea, coffee, chocolate.
10. *Bread:* Rolls, croissants, brioche, toast.

Points to consider when compiling a breakfast menu:

a) It is usual to offer three of the above-mentioned courses. For example see menu below:

Fruit or Cereals
Fish, Eggs or Meat
Preserves, Bread, Coffee

b) As large a menu as possible should be offered, depending on the size of the establishment, bearing in mind that it is better to offer a smaller number of well-prepared dishes than a large number of hurriedly prepared ones.

c) A choice of plain foods such as boiled eggs or poached haddock should be available for the person who may not require a fried breakfast.

Breakfast menus may be table d'hôte or à la carte; a continental breakfast does not include any cooked dish. A typical continental breakfast would offer:

Rolls and Butter, Croissant, Toast
Preserves
Tea or Coffee

Examples of breakfast menus:

GOOD MORNING!

Breakfast is served from 7:00 a.m. until 11:00 a.m. daily

THE ENGLISH

Fresh Orange or Grapefruit Juice,
Pineapple, Tomato, Prune Juice or
Porridge or Cereal

Two Eggs, any style
with choice of
Ham, Bacon, Chipolata Sausages or Grilled Tomato

Croissants, Brioche, Breakfast Rolls or Toast
with Butter and Marmalade, Jam or Honey

Tea, Coffee, Sanka, Chocolate or Milk

THE CONTINENTAL

Fresh Orange or Grapefruit Juice,
Pineapple, Tomato or Prune Juice

A basket of Croissants, Brioche, Breakfast Rolls,
and Toast with
Butter and Marmalade, Jam or Honey

Tea, Coffee, Sanka, Chocolate or Milk

A LA CARTE

FRUITS & JUICES
Fresh Orange or Grapefruit Juice £. . . .Large £. . . .
Pineapple, Tomato or Prune Juice £. . . .Large £. . . .
Chilled Melon £. . . . Stewed Prunes £. . . .Half Grapefruit £. . . .
Stewed Figs £. . . . Fresh fruit in Season £. . . .

BREAKFAST FAVOURITES
Porridge or Cereal £. . . .
Eggs, any style: One £. . . . Two £. . . .
Ham, Bacon, Chipolata Sausages or Grilled Tomato £. . . .
Omelette, Plain £. . . . with Ham or Cheese £. . . .
Grilled Gammon Ham £. . . . Breakfast Sirloin Steak £. . . .
A Pair of Kippers £. . . . Smoked Haddock with a Poached Egg £. . . .
Pancakes with Maple Syrup £. . . .

FROM OUR BAKERY
Croissants or Breakfast Rolls £. . . . Brioche £. . . .
Assorted Danish Pastries £. . . . Toast £. . . .

BEVERAGES
Tea, Coffee, Sanka, Chocolate or Milk £. . . .

Service Charge 15%

Example of a hotel coffee shop menu:

Served throughout 24 hours

SANDWICH SELECTIONS

Club House: Sliced Turkey Breast, Bacon, Lettuce, Tomato and Mayonnaise £. . . .
Ham and Swiss Cheese £. . . . Tuna Fish Salad £. . . .
Cold Roast Beef £. . . . Turkey Breast £. . . .
Chicken Breast £. . . .

HOT SPECIALITIES

The following are served with French Fried Potatoes
Grilled Chopped Steak on a Toasted Bun £. . . .
Grilled Chopped Steak accompanied by a Fried Egg, Bacon, and Grilled Tomato £. . . .
Calf's Liver and Bacon £. . . . Omelettes to Choice £. . . .
Vegetables and Side Salads to Choice £. . . .

BREADS, PASTRIES AND ICE CREAMS

Gateaux £. . . . Toast and Butter £. . . .
Assorted Danish or French Pastries £. . . . Brown Bread and Butter £. . . .
Assorted Ice Creams £. . . . Assorted Biscuits £. . . .
Full Afternoon Tea £. . . .

BEVERAGES

Tea, Coffee, Sanka, Chocolate or Milk £. . . .

Luncheon menus

A luncheon table d'hôte menu may offer a choice of dishes or may be a set meal with little or no choice, depending on the type of establishment. If a special party luncheon menu is required, three or four courses are usually offered, for example:

Luncheon
Grapefruit cocktail
Fillets of sole bonne-femme
Roast Surrey chicken
French beans; Roast potatoes
Sherry trifle

Luncheon
Hors d'œuvre
Mixed grill
Brussels sprouts; Fried potatoes
Cream caramel

Almost all foods are suitable for serving at luncheon. In warm weather cold dishes are popular and a cold buffet should be available.

Only offer the number of courses and the number of dishes within each course that can be well prepared, cooked and served.

Luncheon menus can be compiled from the following foods:

First course

a) *Fruit cocktails:* Melon, grapefruit, orange, Florida.
b) *Fruits:* Melon, grapefruit, avocado pear.
   Fruit juices: Grapefruit, orange, pineapple, tomato.
c) *Shellfish, etc:* Potted shrimps, prawns, oysters, caviar, snails, crabmeat.
   Shellfish cocktails: Lobster, crab, prawn, shrimp.
d) *Smoked:* Salmon, trout, eel, sprats, buckling, mackerel, roe, ham, salami.
e) *Hors d'œuvre.*

Second course

Soups: Consommé with simple garnish. Cold in summer.
Purée: parmentier, St.-Germain, etc.
Crème: de volaille, de tomates, etc.
Potage: paysanne, minestrone, Scotch broth.

Third course

Farinaceous: Spaghetti, macaroni, ravioli, canneloni, gnocchi (italienne, romaine, parisienne), nouilles, risotto, pizza and quiche.

Fourth course

Eggs: Scrambled, poached, soft boiled, en cocotte, sur le plat, omelette.

When served for a luncheon menu egg dishes are usually garnished.

Fifth course

Fish. (Nearly all kinds of fish can be served, but without complicated garnishes. They are usually steamed, poached, boiled, grilled, deep or shallow fried.)

Mussels, scallops, herrings, skate, whiting, plaice, cod, turbot, brill, sole, scampi, trout, salmon trout, salmon, whitebait, kedgeree.

Sixth course

Entrees. (This is a dish of meat, game or poultry which is not roasted or grilled.)

Brown stews (ragoût de bœuf; navarin de mouton).
Braised steaks, braised beef, civet de lièvre.
Goulash de veau, braised oxtail, salmis of game.
Hot pot, Irish stew.
Meat pies, chicken pies.
Meat puddings.
Boiled meat (French and English style).
Fricassée, blanquette.
Calves' head, tripe, sautéd kidneys.
Vienna and Hamburg steaks, hamburgers.
Sausages, minced meat, chicken émincé.
Fried lamb, veal or pork cutlets or fillets.
Fried steaks (entrecôte, tournedos, fillets, etc).
Veal escalopes, sweetbreads.
Vol-au-vent of chicken or sweetbreads or both.
Sauerkraut, pilaff, kebab, chicken cutlets.
Chilli con-carne.

Seventh course

Roasts: Beef, pork, veal, lamb, mutton, chicken.

Eighth course

Grills: Steaks (chateaubriand, fillet, tournedos, point, rump).
Porterhouse, entrecôte.
Cutlets (single, double).
Chops (loin, chump).
Kidneys, mixed grill, chicken, chicken legs, kebabs.

Ninth course

Cold buffet: Salmon, lobster, crab.
Pâté or terrine.
Beef, ham, tongue, lamb.
Chicken, chicken pie, raised pies.
Pressed beef, duck, aspic dishes.
Ham mousse.

Tenth course

Vegetables: Cabbage, cauliflower, French beans, spinach, peas, carrots, tomatoes, etc.
Asparagus, globe artichoke, sea-kale (hot or cold with a suitable sauce).
Potatoes: Boiled, steamed, sautéd, fried, roast, creamed, croquette, lyonnaise, etc.

Eleventh course

Sweets: Steamed puddings (fruit and sponge).
Milk puddings.
Fruit (stewed, fools, flans, salad, pies, fritters).
Egg custard sweets (baked, bread and butter, cream, caramel, diplomat, cabinet).
Bavarois, savarin, baba.
Charlottes, profiteroles, gâteaux.
Pastries (mille-feuille, éclairs, etc).
Various ices and sorbets.

Twelfth course

Savouries: Simple savouries may also be served; for example, Welsh rarebit.

Thirteenth course

Cheese: A good selection of cheese. Biscuits, celery and radishes.

Fourteenth course

Dessert: Fresh fruit of all kinds and nuts.
Coffee.

Tea menus

These vary considerably, depending on the type of establishment. The high-class hotel will usually offer a dainty menu. For example:

Sandwiches (smoked salmon, ham, tongue, egg, tomato, cucumber) made with white or brown bread.
Bread and butter (white, brown, fruit loaf).
Jams, honey, lemon curd.
Small pastries, assorted gâteaux.
Fruit salad and cream, ices.
Tea (Indian, China, Russian, iced).

The commercial hotels, public restaurants, canteens, will offer simple snacks, cooked meals and high teas. For example:

Assorted sandwiches.
Buttered buns, scones, tea cakes, sally lunns, Scotch pancakes, waffles, sausage rolls, assorted bread and butter, various jams, toasted tea-cakes, scones, crumpets, buns.
Eggs (boiled, poached, fried, omelets).
Fried fish; grilled meats; roast poultry.
Cold meats and salads.
Assorted pastries; gâteaux.
Various ices, coupes, sundaes.
Tea; orange and lemon squash.

Dinner menus

A list of some of the foods suitable for dinner menus is given below, and both table d'hôte and à la carte menus should offer a sensible choice, depending upon the size of the establishment.

The number of courses on special party menus can vary from three to ten. The occasions for special dinner parties are often very important for the guest attending, therefore the compiling of such a menu is extremely important, calling for expert knowledge and wise judgement on the part of the caterer. The following are some of the foods used:

First course

a) *Cocktail:* Fruit and shellfish.
b) *Fruit:* Melon, fresh figs, avocado pear.
c) *Delicacies:* Caviar, oysters, snails, potted shrimps, prawns, foie gras.
d) *Smoked:* Salmon, trout, ham, salami, sausages, sprats, eel.
e) *Hors d'œuvre.*

Second course

Soup: Clear and consommé based, petite marmite, consommé de volaille, turtle soup, bisque de homard, crème de champignons, crème d'asperges, Germiny, velouté Agnès Sorel, soupe à l'oignon, Bortsch.

Cold soups: Vichyssoise, consommé, crème de pois.

Third course

Fish: Boiled salmon, turbot, trout (au bleu).

Shallow poached: sole, turbot, brill, halibut, with such classical garnishes as Newburg, américaine, Cubat, Véronique, bonne-femme.

Hot shellfish: lobster-Mornay, Thermidor, Newburg, cardinal; scampi, oysters.

Meunière: Sole, fillets of sole – grenobloise, belle meunière.

Fried: Sole, fillets of sole, goujons, scampi.

Grilled: Lobster, sole, salmon.

Cold: Salmon, salmon trout, trout, sole.

Fourth course

Entree: Light entrée dishes are used which are small and garnished.

Vegetables are not served with an entrée if it is followed by a rèlevé or roast with vegetables. Examples are:

Sweetbreads, hot mousse of ham or foie gras.
Sauté of chicken, tournedos, noisettes or cutlets of lamb.
Suprême de volaille sous-cloche.
Saddle of hare, filet mignon, vol-au-vent.

Fifth course

Releve: This is usually a joint which is carved and is cooked by braising or poêler such as:

Poularde poêlée aux champignons.
Poulet en casserole bonne-femme.
Poulet en cocotte grand'mère.
Selle d'agneau poêlée bouquetière.
Selle de veau poêlée, filet de bœuf poêlé, contrefilet de bœuf poêlé.
Braised ham, tongue, duck, pheasant, pigeon.

The relèvé is served with a good quality vegetable and potato.

Vegetable: French beans, broccoli, asparagus points, peas, broad beans, button brussels sprouts, aubergine, cauliflower, etc.

Potatoes: Parmentier, noisette, olivette, dauphine, nouvelles, rissolées, Mireille, duchesse, Byron.

Sixth course

Sorbet: This is a lightly frozen water ice sometimes flavoured with a liqueur or champagne and served with a wafer.

Seventh course

Roast: The roast course may be served with a salad, usually compound, such as lorette, niçoise, française.

Examples of roasts are:

Saddle of lamb or veal, fillet or sirloin of beef.
Poultry or game such as chicken, turkey, duck, goose, grouse,

partridge, pheasant, snipe, woodcock, guinea fowl, wild duck, plover, teal, venison, saddle of hare.

Eighth course
Cold dish: Such dishes as chicken in aspic or mousse de foie gras or jambon may be served.

Ninth course
Hot dish: A hot dish, usually a vegetable such as asparagus, globe artichoke, sea-kale, or a spinach, mushroom or ham soufflé, may be served.

Tenth course
Sweet: Hot soufflés or pancakes.

Cold, iced soufflé, bombes, coupes with fruit such as peaches, strawberries, raspberries, sorbet, posset, syllabub, mousse, bavarois, accompanied by petits fours.

Friandises, mignardises or frivolités (these are different names for very small pastries, sweets, biscuits, etc also known as Petit fours).

Eleventh course
Savoury: Any savouries such as canapé Diane, anges à cheval, quiche lorraine may be used on dinner menus.

Twelfth course
Cheese: All varieties may be offered.

Thirteenth course
Dessert: All dessert fruits and nuts may be served.
Coffee.

Examples of dinner menus (suitable for a medium-sized hotel):

A Table d'hôte

Dîner
Hors d'œuvre
Pamplemousse rafraîchi
Consommé Madrilène
Crème de volaille
Filet de sole bonne-femme
Truite de rivière meunière
Tournedos niçoise
Escalopes de ris de veau florentine
Suprême de volaille princesse
Petits pois Pommes rissolées
Épinards en purée Pommes parisienne
Céleri braisé Pommes maître d'hôtel
Poire Belle-Hélène
Meringue Chantilly
Bande aux cerises
Crêpes à l'orange

Hors d'œuvre
or
Cream of chicken
Shallow-fried river trout
Garnished sirloin steak
Braised ham, Sherry sauce
Garden peas Roast or
Spinach puree Creamed potatoes
Vanilla ice cream with meringue
and whipped cream
or
Cherry slice
or
Cheese board

B Special party

Dîner	Dinner
Crème de volaille	Chicken cream soup
Filets de sole cardinal	Fillets of sole, lobster sauce
Tournedos sauté au madère	Tournedos with madeira sauce
Petits pois fins au beurre; pommes parisienne	Buttered peas, baby roast potatoes
Soufflé arlequin	Vanilla and chocolate soufflé

C A la carte

Select from:

Hors d'œuvre

Escargots de Bourgogne
Crevettes roses
Coupe Floride
Saumon fumé
Honeydew melon
Potted shrimps

Potages

Bisque de homard
Soupe à l'oignon
Vichyssoise
Consommé Judic
Germiny
Velouté Agnès-Sorel

Poissons

Filets de sole caprice
Suprême de turbot Waleska
Homard à l'américaine
Saumon poché à la sauce hollandaise
Scampis frits à la sauce tartare
Truite aux amandes

Entrées

Suprême de volaille maréchale
Tournedos Rachel
Vol-au-vent de volaille
Escalope de veau au marsala
Côtelettes d'agneau Réforme
Poussin en casserole

Rôtis

Poulet de Surrey
Caneton d'Aylesbury
Selle d'agneau
Gibier (en saison)

Grillades

Chateaubriand
Filet de bœuf
Tournedos
Entrecôte
Chop d'agneau
Côtelettes d'agneau

Buffet froid

Mayonnaise de volaille
Jambon de York
Côte de bœuf
Langue de bœuf
Saumon d'Ecosse
Chicken pie

Salades

française, laitue, romaine, niçoise, lorette, japonaise

Légumes

Haricots verts, Endives au jus, Aubergines frites, Brocolis
Pommes: frites, rissolées, nouvelles, purée à la crème, olivette

Entremets

Pêche cardinal
Soufflé au Grand Marnier
Biscuit glacé praliné
Gâteau au chocolat
Sabayon au marsala
Crêpes Suzette
Ananas flambé
Sorbet au framboise

Canapés

Anges à cheval
Canapé Diane
Croûte Windsor
Canapé Baron

Les fromages au choix

Friandises

Café

Examples of contemporary menu items

Smoked trout mousse
Hot Stilton quiche
Layered fish pâté
Vegetable terrine
Royal of artichoke and mushrooms
Ham and cottage cheese cornets

Duck and orange pâté
Prawn and hazelnut salad
Pear with roquefort and watercress dressing
Avocado and fresh raspberries, raspberry vinaigrette
Fresh melon and kiwi fruit

Hot cream of apple and cucumber soup
Watercress and mushroom consommé
Stilton and celery soup

Iced cucumber and mint soup
Tomato and orange soup
Fish soup and chervil

Fricassée of turbot and scallops with basil and tomato
Fillets of monkfish with tomato and brandy sauce
Sea bass with crab and cucumber in cream sauce
Brioche of sole with oranges and green peppercorns

Suprême of fresh halibut, cream and watercress sauce
River trout pan fried with almonds and pine kernels
Sea bream with orange, lemon and bay leaves
Salmon with sorrel sauce

Noisettes of pork basquaise
Suprême of duck with juniper berry sauce
Roast quail with grape and Madeira sauce
Grilled lamb cutlets with herb butter
Breast of chicken stuffed with pâté, cream sauce and sherry
Grilled gammon steak, red wine sauce
Breast of chicken stuffed with Stilton and walnuts

Grilled guinea fowl with limes
Calf's liver with avocado
Loin of lamb with fresh basil and tomato
Fillet of beef on Stilton croûton, port wine sauce
Veal chop with tomatoes, garlic and fresh sage
Sweet and sour pork
Beef carbonnade with walnuts

Ratatouille
Carrots and diced bacon
Spiced beetroot
Sweetcorn pancakes

Braised red cabbage with apple
Spinach purée with pine kernels
Courgettes with tomatoes
Green sprouting broccoli

Strawberry Pavlova
Lemon sorbet
Pear in red wine
Coffee and praline ice cream
Ginger ice cream

Chocolate mint soufflé
Pineapple brioche
Honey and walnut tart
Lemon and lime sorbet
Passion fruit sorbet

Banquet menus

When compiling banquet menus certain points should be considered:

1 The food, which will possibly be for a large number of people, must be dressed in such a way that it can be served fairly quickly. Heavily garnished dishes should be avoided.
2 If a large number of dishes have to be dressed at the same time certain foods deteriorate quickly and do not stand storage, even for a short time in a hot plate – for example, deep-fried fish.

Banquet luncheon menus

A normal luncheon menu is used, bearing in mind the number of people involved. It is not usual to serve farinaceous dishes, eggs, stews or savouries. A luncheon menu could be drawn from the following courses:

First course
Soup, cocktail (fruit or shellfish), hors d'œuvre.

Second course
Fish.

Third course
Meat, hot or cold, but not a stew or made-up dish. Vegetables and potatoes or a salad should be served.

Fourth course
If the function is being held during the asparagus season, then either hot or cold asparagus with a suitable sauce may be served as a course on its own.

Fifth course
Sweet, hot or cold.
Cheese and biscuits.
Coffee.

Example of luncheon banquet menu

Hors d'œuvre
Fillets of sole bonne-femme
Grilled chicken and bacon
Minted garden peas   New Jersey potatoes
Dutch apple tart
Devon clotted cream
Cheese and biscuits

Banquet dinner menus
Here the caterer has the opportunity to excel, and two menu examples are given here:

a)
Consommé Alexandra
Délice de sole Newburg
Escalopes de ris de veau maréchale
Filet de bœuf bouquetière
Sorbet au champagne
Caneton rôti
Salade mimosa
Omelette surprise mylord
Quiche lorraine

b)
Clear oxtail soup with sherry
Poached Scotch salmon, Hollandaise sauce
Roast pheasant with celery
Broccoli with pine needles
Almond potatoes
Ice cream with cherries

Functions and banquets
It will be necessary to know the nature of function, as special facilities may be required. The type of function could be a wedding breakfast, silver or golden wedding anniversary, coming of age, retirement, presentation, conference, etc.

If there are any special diet requirements of guests, such as fish or eggs in place of meat, then the kitchen should know of them.

The type of meal for functions can vary from:

Buffets: Light, where the food is served or where the guests help themselves,
Buffets: Fork, hot or cold,
Buffets: Cold,
Buffets: Hot,

to a formal sit-down meal.

The cost of the menu is agreed with the organiser of the function, and if any of the following are required then they will be charged as extras:

Floral decorations	Place-name cards
Special menu printing	Table plan
Orchestra	Toastmaster or MC
Cabaret	Hire of rooms
Invitation cards	Special decorations

Light buffets (including cocktail parties)

Light buffets can include:

1 Hot savoury pastry patties of lobster, chicken, crab, salmon, mushroom, ham, etc.
2 Hot sausages (chipolatas), various fillings, such as chicken livers, prunes, mushrooms, tomatoes, etc, wrapped in bacon and skewered.
3 Bite sized items: quiche and pizza, hamburgers, meat balls with savoury sauce or dip, scampi, fried fish en goujons, tartare sauce.
4 Savoury finger toast to include any of the cold canapés. These may also be prepared on biscuits or shaped pieces of pastry. On the bases the following may be used: salami, ham, tongue, thinly sliced cooked meats, smoked salmon, caviar, mock caviar, sardine, eggs, etc.
5 Game chips, gaufrette potatoes, fried fish balls, celery stalks spread with cheese.
6 Sandwiches, bridge rolls open or closed but always small.
7 Sweets such as trifles, charlottes, jellies, bavarois, fruit salad, strawberries and raspberries with fresh cream, ice creams, pastries, gâteaux.
8 Beverages, coffee, tea, fruit-cup, punch-bowl, ice coffee.

Fork buffets

For these functions individual pieces of fish, meat and poultry are prepared so that they can be eaten by the guests standing up and balancing a plate in one hand. Salads should also be sensibly prepared so that they can be easily handled by the guest using only a fork; the lettuce should be shredded and kept in short lengths. Chicken or ham mousse, galantine, terrine, pâté, mayonnaise of salmon, lobster and chicken are all suitable dishes.

A menu example on a restaurant window

A buffet table

Full buffets

Speciality menus

a)

Coupe Hawaii
Selection of cold cuts of meat to include:
Roast Beef
York Ham
Chicken
Various Mixed Salads

Apple Strudel & Fresh Cream

Coffee

b)

Cocktail de fruits de mer
Saumon fumé
Truite de rivière fumée
Melon de la saison
Saumon d'Écosse bellevue
Côte de Bœuf Froide 'Belle Jardinière'
Suprême de Volaille 'Hawaienne'
Jambon et Langue 'Marguerite'

Salads

Russian, Tomato, Cucumber, Lorette, Waldorf
Hot Entrées (in chafing dish)
Petits kebabs de bœuf à l'oriental
Riz pilaw

ou

Goujonettes de sole frites à la sauce tartare

Sauces

Mayonnaise, Tartare, Cocktail, Vinaigrette
Plateau de fromages

Desserts

Macédoine de fruits
Babas au rhum
Mille feuilles
Friandises

Café

Self-service food counter

An example of a small country inn menu

Menu

Home-made vegetable soup	70p
Home-made Mallard and terrine	£1.15
Home-made pork, liver and brandy pate	£1.15
Avacado cream with prawns	80p
Rosey apple stuffed with cottage cheese and nuts	75p
Fresh melon, orange and grapefruit salad with yogurt	70p
Fresh calves liver with bacon and fried onions	£5.95
Supreme of chicken marinated in madeira and served with brandy and cream sauce	£5.75
Esk salmon baked in butter and served with Hollandaise sauce	£6.25
15-15oz sirloin steak and mushrooms	£6.45
Cider roast ham and mixed salad	£4.95
Parsnips and carrots in nutmeg	
Bubble and squeak	
Roast potatoes	
Leeks in cheese sauce	
Home-made apricot and brandy mousse	85p
Home-made bakewell tart and cream (original recipe)	75p
Duke of Cornwall flan (wholemeal pastry, bitter marmalade, butterscotch and cream)	75p
Fresh cream Drambuie meringue	95p
Cheeses	95p
Coffee pot	95p

Prices include a four course dinner
Value Added Tax and Service at 15% will be added.

First Dishes

SMOKED SALMON AND AVOCADO MOUSSE

TRADITIONAL GAME PATE
A forcemeat of hare and pigeon baked in an orange-flavoured pastry crust, served with Cumberland Sauce.

WALTONS' WINTER SALAD
Striplets of smoked goose tossed in a lime-flavoured dressing with pine kernels and a julienne of carrots and celeriac.

PATE OF SMOKED EEL
Fillets of smoked eel encased in a mousseline of salmon trout.

VEGETABLE TERRINE
WITH FRESH TOMATO SAUCE
Crisp young vegetables layered in a chicken forcemeat.

MUSHROOM SALAD
Sliced crisp white mushrooms tossed in a French dressing and topped with sour cream and chives.

Second Dishes

WALTONS' GLAZED PANCAKE
A wafer-thin pancake is stuffed with a mousseline of spinach and chicken, coated with Hollandaise, sprinkled with Parmesan cheese and glazed.

POACHED EGGS IN A CRUST
Each egg served on top of a puree of salsify and coated with a dill sauce.

STUFFED MUSHROOMS
Deep-cupped mushrooms are packed with Gruyere cheese and prawns, coated with breadcrumbs and fried.

HOT OR CHILLED SOUP OF THE DAY
A changing variety of soups is offered.

Principal Dishes

BROCHETTE OF CORNISH SCALLOPS, CHICKEN AND BACON
Served with a spicey risotto and saffron tomato sauce.

NOISETTES OF LAMB BEARNAISE
Each noisette is topped with lobster and coated with Bearnaise Sauce.

WALTONS' CHICKEN IN BEAUJOLAIS
The chicken is blazed in brandy, cooked in the red wine together with girolles and garnished with glazed onions and butter-fried croutons.

CREAMED FILLET OF BEEF WITH PEPPERS
Slivers of beef fillet, cooked in cream with paprika and a julienne of red peppers.

ESCALOPES OF VEAL SWISS STYLE
Thin pieces of veal are pan-fried, topped with mushrooms, tomatoes, Gruyere cheese and glazed.

BREAST OF PHEASANT DEVONSHIRE
Served with a port-enriched sauce, redcurrants and poached pears.

CALVES' LIVER WITH AVOCADO AND MANGO
Pan-fried and served with a buttery orange sauce.

SUPREME OF BRILL WALTONS' STYLE
Stuffed with a mousseline of salmon trout, the fish is poached, coated with lobster sauce and glazed.

OLD ENGLISH VENISON PIE
Cubes of venison, marinaded in red wine and juniper, are braised with chestnuts and baked with a puff-pastry crust.

MEDALLIONS OF BEEF A LA MODE DU CHEF
Two medallions cut from the fillet are cooked to your liking: one topped with asparagus and Hollandaise Sauce, the other with a rich dark mushroom sauce.

Puddings

CHOCOLATE MARQUISE
A rich dark chocolate mousse enclosed in thin slices of light brandy-soaked sponge cake.

STRAWBERRY BASKETS WITH DRAMBUIE

CHESTNUT AND RUM ICE CREAM

POACHED PEARS BALMORAL
Poached in vanilla syrup each pear is served on hazelnut ice-cream and coated with chocolate sauce.

SOUFFLE GLACE BENEDICTINE
An iced souffle aromatized with Benedictine.

LEMON AND LIME CHIFFON PIE WEST FLORIDA STYLE
A rich pastry case is filled with fruit puree and a light-textured lemon and lime mousse.

RASPBERRY VACHERIN
Hazelnut meringue shells filled with raspberries in Curacao and fresh raspberry sauce.

WALTONS' SPECIAL WATER ICES
Damson, passion fruit, redcurrant, kiwi fruit, mango.

STILTON CHEESE

COFFEE, MINT TEA

FRIANDISES

A contemporary English restaurant menu

LUNCHEON CLUB
MAIN RESTAURANT (Kitchen control copies)

Day	Fish	Nos prep	Nos sold	Salads and cold meats	Nos prep	Nos sold
Monday	Grilled Plaice with Lemon			Tongue Pâté		
Tuesday	Baked Bream Fillets			Tongue Pâté		
Wednesday	Fish Cakes & Shrimp Sauce			Gala Pie (No Egg) Chicken		
Thursday	Haddock Meuniére			Beef Cheese& Tomato Flan		
Friday	Grilled Cod Fried Cod*			Pork Smoked Mackerel		

The following items to be on counter daily:
1) Selection of Rolls, Bread, Ryvita, Vitawheat – with Butter. 2) Individual salad in lieu of Vegetable or Potato. 3) Cheese, Biscuits and Butter. 4) Custard Sauce wherever applicable. 5) Cheese, Salad

Industrial catering menus – pp 362–5

WEEK 1
DATE: 5 October

Main dish	Nos prep	Nos sold	Vegetables	Sweets (ice cream served with all sweets)
Roast Lamb and Onion Sauce Savoury Mince Vol-au-Vent Macaroni Cheese*			Croquette and Creamed Potatoes	Steamed Jam Sponge Apricot Crumble Fruit Salad Ice Cream & Choc. Sauce Semolina Pudding Fresh Fruit
Roast Chicken and Stuffing Steak and Mushroom Pie Savoury Pancakes*			Roast and Creamed Potatoes	Steamed Golden Sponge Baked Apples Stewed Cherries Lemon Jelly Whip Tapioca Pudding Fresh Fruit Assorted Yoghurts
Roast Beef and Yorkshire Pudding Moussaka Egg Mimosa*			Jacket and Creamed Potatoes	Blackcurrant Roly-Poly Coffee Gateau Stewed Golden Plums Rice Pudding Fresh Fruit
Roast Pork, Stuffing & Apple Sauce Braised Lamb Chop and Mushrooms Sausage Roll and Baked Beans*			Saute and Creamed Potatoes	Steamed Orange Sponge & Orange Sauce Black Cherry Cheesecake Ground Rice Pudding Fresh Fruit Assorted Yoghurts
Curried Beef Rice and Chutney Stuffed Potato			Chipped and Creamed Potatoes	Bakewell Tart Stewed Pineapple Rings Fruit Jelly Macaroni Pudding Fresh Fruit

LUNCHEON CLUB
SENIOR DINING ROOM

Day	Soup	Fish	Nos prep	Nos sold	Cold buffet	Nos prep	Nos sold
Monday							
	Kidney	Grilled Herrings in Oatmeal			Liver Pâté		
Tuesday							
	Cream of Celery	Fried Scampi & Tartare Sauces			Ham		
Wednesday							
	Chicken Noodle	Poached Cod Steak Portugaise			Gala Pie		
Thursday							
	White Onion	Grilled Whole Lemon Sole			Scotch Egg		
Friday							
	Champenoise	Grilled and Fried Cod			Tongue		

The following items to be on counter daily:
1) Selection of Rolls, Bread, Ryvita, Vitawheat – with Butter. 2) Individual salad in lieu of Vegetable or Potato. 3) Cheese, Biscuits and Butter. 4) Custard Sauce wherever applicable. 5) Cheese, Salad

WEEK 3
DATE: 19 October

Main dish	Nos prep	Nos sold	Grill	Nos prep	Nos sold	Vegetables	Sweets (ice cream served with all sweets)
Fried Chicken Leg & Mushroom Sauce Cottage Pie			Grilled Entrecote Steak and Tomato			Croquette and Creamed Potatoes	Almond Sliced Baked Pear Sponge Ground Rice Pudding Sherry Trifle Ice Cream and Meringue or Sardines on Toast
Braised Beef & Vegetables Bacon & Mushroom Vol-au-Vent			Grilled Pork Chop and Apple Sauce			Parmentier and Creamed Potatoes	Gooseberry Pie Mixed Fruit Sponge Lemon Cheesecake Tapioca Pudding Peach Melba or Cheese Tartlets
Roast Lamb & Redcurrant Jelly Calves Liver French Style			Grilled Rump Steak and Onions			Sauté and Creamed Potatoes	Bread and Butter Pudding Cherry Pie Macaroni Pudding Chocolate Eclairs Fruit Jelly or Canape Ivanhoe
Escalope of Veal Viennoise Curried Beef			Grilled Lamb Cutlets and Asparagus Tips			Roast and Creamed Potatoes	Treacle Tart Rhubarb Crumble Cream Caramel Stewed Oranges Semolina Pudding or Canape Fedora
Chilli Con Carne with Rice			Grilled Entrecote Steak and Mushrooms			Chipped and Creamed Potatoes	Baked Jam Roll Steamed Sultana Sponge Pineapple Flan Stewed Pears Rice Pudding or Welsh Rarebit

Industrial catering – special function menus

Fillet of sole bonne-femme

****

Roast duck, bigarrade sauce

****

New potatoes
Garden peas
Orange salad

****

Charlotte russe

****

Cheeseboard
Fruit

Avocado pear & prawns

****

Roast fillet of beef with

Creamed Horseradish Sauce

****

Château potatoes
Garden peas
Cauliflower and cream sauce

****

Strawberry gâteaux

****

Cheeseboard
Fruit

Sliced ham & tongue
Scotch eggs
Rollmops
Scotch sirloin
Sliced pâté
Hot sausages
Salmon pieces in aspic

****

Apple pie & cream

****

Cheeseboard
Fruit

Charentais melon

****

Poached Scotch salmon and
hollandaise sauce

****

New parsley potatoes
Cucumber salad
Garden peas

****

Lemon soufflé

****

Cheeseboard
Fruit

Melon
Grapefruit Cocktail
Prawn Cocktail
Tomato Soup
Chicken Noodle Soup

Roast Prime Rib of Beef, Yorkshire Pudding and Horseradish Sauce
Roast Leg of Lamb and Mint Sauce
Roast Leg of Pork, Seasoning and Apple Sauce
Cold Roast Joints

Roast Potatoes
Buttered Garden Peas
New Carrots
Salads in Season

Meringue Gateau
Fruit Salad
Strawberry Fool
Chocolate Gateau
Cheeseboard

Coffee

All prices are inclusive of service and V.A.T.

In the interest of hygiene, please refrain from smoking at the carving counters

PANCAKES

International Selection

Russian Style *(Meat)* Meat Blinza Pancakes Filled with Fresh Minced Beef Steak and Liver Filling ..

Russian Style *(Cheese)* Pancakes Filled with Sweet and Sour Double Cream Cheese and Lemon ..

Italian Style Pancake Filled with Chopped Veal Filling and topped with Creamy Cheese Sauce ..

Spanish Style Pancake Filled with Special Savoury Filling of Tomato, Rice and Mushrooms ..

★ ★ ★

American Style Savoury Pancakes

Showboat Pancake interlaced with a large helping of Sweetcorn and topped with Rasher of Bacon and Tomato ..

Kentucky Pancakes topped with Crispy Bacon, Sliced Tomatoes and Special Rich Tomato Sauce ..

New Orleans Pancakes topped with Chicken in Chef's White Sauce and Crushed Pineapple ..

Uncle Tom Pancakes Filled with Fresh Scrambled Egg and Crispy Rasher of Bacon ..

Plantation Pancakes with Sliced Frankfurter Sausages and Covered with a Rich Meat Sauce & Sliced Tomato

Mississippi Pancakes Layered with Ham, Tomato and Chef's Special Sauce ..

Florida Pancakes filled with Pure Beef Hamburger and Lettuce topped with Fresh Pancake and ring of Pineapple ..

American Pancakes topped with sliced Chicken and Cranberries ..

★SWEET PANCAKES ★

Regular Pancakes topped with Whipped Butter and Choice of Maple Syrup or Strawberry Melba or Chocolate Sauce and topped with Dairy Whipped Cream ..
Fresh Delicious Soft Ice Cream Extra.

Five Silver Dollar Pancakes topped with Whipped Cream and Morello Cherries and Melba Sauce ..

Hawaiian Style Pancakes topped with Pineapple, Melba Sauce and Whipped Dairy Cream ..
With Soft Ice Cream Extra

Cranberry Surprise Fresh Pancakes with a generous portion of Cranberry Sauce and Whipped Cream ..

Banana Pancake topped with Melba Sauce and Whipped Dairy Cream ..

Apple Surprise Pancake Chunks of Apple blended into a Honey Cinnamon Syrup and sandwiched between Pancakes ..
With Soft Ice Cream Extra

Special Crunch Pancake topped with Maple Syrup and Caramel Crunchnut and Whipped Dairy Cream ..
With Soft Ice Cream Extra.

★ ★ ★

American Pancake Pizza Selection

Pizza Neapolitan with Anchovies, Olives and Rich Neapolitan Sauce ..

Pizza Prosciutto topped with Ham

Pizza topped with Salami ..

Pizza Mississippi topped with Ham, Mushrooms, Salami, Olives and Anchovies

Toasted Sandwiches

Cheese and Tomato ..
Ham and Tomato ..
(Not served between 12 and 2 p.m.)

Double Decker American Club Sandwiches

Sliced Ham, Lettuce, Cheddar Cheese and Sweet Pickle ..
Double Cream Cheese, Bacon, Lettuce & Tomato ..
Sliced Chicken Ham, Cranberries and Lettuce ..
(All served with Potato Salad)

Waffles •

Freshly Baked Waffle topped with Whipped Butter and Choice of Maple Syrup or Strawberry Melba or Chocolate Sauce, and Whipped Cream
The Above with Soft Ice Cream Extra
Waffles served with Soft Sugar and Whipped Cream
Waffles topped with Ham & Fried Egg

• Beverages

Tea ..
Milk and Dash ..
Special Old Kentucky Coffee ..
Special Old Kentucky Coffee with Cream ..
Lemon Tea ..
New Orleans French Chocolate topped with Fresh Dairy Cream ..
Cola ..
Orange Squash ..

MINIMUM CHARGE AFTER 6.0 p.m. -
MINIMUM CHARGE 12 p.m. - 2.30 p.m. -

Speciality restaurant menu

An example of a children's menu

Appetisers

Pâté with hot toast........... 98p
Orange & Grapefruit cocktail.................. 65p
Chilled fruit juice........... 38p
Prawn cocktail............. 98p
Our own French Onion Soup............... 65p
Cream of Tomato Soup with Tarragon............. 60p
Chilled Melon............. 80p

Omelettes

Choice of Mushroom, Ham, Cheese, Tomato, Spanish, Chicken or Prawn & Tomato filling.................. £1.80

English Breakfast

until 11.30 am
Fruit juice. Fried egg, 2 rashers of back bacon, sausage & tomato. Toast & marmalade. Tea or coffee............. £2.15

Specialities

Grilled Gammon with Madeira sauce..... £2.30
Fillet Steak Garni (6oz uncooked weight)... £3.90
Fisherman's Platter....... £3.50
Scampi................... £3.50
Whole Grilled Plaice..... £2.45
Beef Stroganoff & Rice.. £2.95
Spaghetti Bolognaise..... £1.85
Lasagne................ £1.95
'Farmer Giles' Pie........ £2.95
Prawn or Chicken Curry with traditional accompaniments........ £2.65

Vegetables of the day .... Portion 30p
French Fried potatoes .... Portion 40p
Jacket potato with butter or soured cream & chives... 35p
Side salad................. 40p

Salads

Cold roast Chicken........ £2.35
Prawn.................... £2.80
Dressed Crab............. £2.60
Ham...................... £2.30
Quiche Lorraine........... £2.40
Cottage Cheese with mandarins & sultanas... £1.85

Ice Cream

Knickerbocker Glory........ 85p
Chocolate Nut Sundae..... 75p
Peach Melba................ 75p
Banana Split.............. 75p
Dairy Ice Cream... Portion 25p
Cornish, strawberry or chocolate

An example of a departmental store menu

Pastries Before 11.30 am and after 2.30 pm

Fresh cream pastries........42p
Toasted tea cake.............30p
Hot buttered toast...........20p
Danish pastries..............30p
Hot croissants with jam or honey..............40p
Scones & butter with jam or honey...............40p
Currant bun & butter........27p
Waffles with jam, honey or cottage cheese...........45p

Sweets & Cheese Board

Choice from trolley with cream.....................75p
(11.30 am to 2.30 pm)
Gateau.......................68p
Apple pie & cream..........65p
Lemon Meringue Pie.......65p
Selection from the cheese board................70p

Beverages

Coffee per cup with cream...32p
Your second cup.............25p
Rombouts filter coffee.......38p
Tea per pot per person with milk or lemon.........28p
Large chilled fruit juice.....45p
Jumbo milk shake with ice cream..................45p
Cola or orange...............25p
Milk........................20p

Wines Between 12.00 and 2.30 pm

Italian Chianti, Orvieto (white) or Rosé. French red or white Quarter Litre Bottle £1.20
French wine by the glass 5oz 50p 6⅔oz 60p

Cocktails Between 12.00 and 2.30 pm

'Bentalls Slinger' – white rum lemon juice, sugar, crushed ice. Shaken not stirred.....90p
'Singapore Sling' – gin, lemon juice, sugar, invigorated with soda and cherry brandy...98p
American – Campari and sweet vermouth topped with soda, cherry. Shaken not stirred.................98p
Liqueur coffee..............80p

Full bar available

All prices include V.A.T. Items subject to availability
Minimum charge between 12.00 until 2.30 pm of £1.50

All items featured on this menu are freshly prepared to your order

TO BEGIN

Large Fruit Juices
Served Chilled and Refreshing

1. **Orange** **38p**
2. **Tomato** **38p**

3. **Grapefruit Cocktail** **38p**
4. **Soup of the Day** **38p**
61. Enjoy a Bread Roll and Butter **22p**

CHEF'S CHOICE

11. **The Early Starter**
Griddled Egg, Pork Sausages, Rasher of Back Bacon, Tomato and Fried Bread **£1.75**

12. **The All Day Breakfast**
Large Fruit Juice or Grapefruit Cocktail, an Early Starter, Slice of Buttered Toast and a choice of Jam or Marmalade **£2.30**

14. **American-Style Breakfast**
Two Griddled Eggs, Two Rashers of Back Bacon and Two Hash Browns followed by Waffles with Mapleen Syrup **£2.10**

15. **Fillets of Plaice**
Two Fillets of Plaice, served with Half a Lemon and Chipped Potatoes **£2.35**

16. **Fillet of Cod**
Large Fillet of Fish, served with Half a Lemon and Chipped Potatoes **£2.20**

25. **Scampi Platter**
Golden-Crumbed Scampi, – deep fried – served with Half a Lemon and Chipped Potatoes **£3.45**

17. **Chef's Grill**
Bumper Burger, Rasher of Back Bacon, Pork Sausages, Griddled Egg, Tomato and Chipped Potatoes **£2.65**

18. **Liver and Bacon Grill**
Sliced Liver, Two Rashers of Back Bacon, Fried Onion Rings, Half a Tomato and Chipped Potatoes **£2.55**

26. **Gammon Steak Platter**
Traditional Cut Gammon Steak, griddled – served with Half a Peach, Griddled Egg, Half a Tomato and Chipped Potatoes **£2.95**

27. **Chopped Steak Platter**
Prime Chopped Steak, griddled – served on a Sesame Bap with Relish, Fried Onion Rings, Lettuce, Half a Tomato, Cucumber, a Bowl of Coleslaw and Chipped Potatoes **£2.85**

Waist Watcher's Salads
. . . for the Calorie Conscious . . .

19. **Gammon Ham with Fruit**
(370 Calories approx) **£2.35**

20. **Cottage Cheese with Fruit**
(340 Calories approx) **£2.35**

Served with Pineapple, Grapefruit, Half a Peach, Coleslaw, Lettuce, Tomato and Cucumber

As a Side Order . . .

23. **Bowl of Coleslaw**
(125 Calories approx) **25p**

24. **Waist Preserver**
Two Bumper Burgers, served with Cottage Cheese, a Bowl of Coleslaw, Lettuce, Half a Tomato and Cucumber **£2.35**

28. **Alabama Chicken Platter**
Deep Fried Breast of Chicken – cooked to Little Chef's own recipe – served with a Bowl of Coleslaw, Tomato and Chipped Potatoes **£2.95**

A popular catering menu

☆ BURGER RANGE ☆

Double Bumper Burgers

Two Bumper Burgers in a Sesame Bap with our Tasty Relish and your choice of filling – served with a Bowl of Coleslaw

36. **Cheeseburger**	£2.10
37. with Chipped Potatoes	£2.35
38. **Pineappleburger**	£2.10
39. with Chipped Potatoes	£2.35
40. **Eggburger**	£2.10
41. with Chipped Potatoes	£2.35

Pineappleburger
with Chipped Potatoes

Single Bumper Burgers

A Bumper Burger in a Sesame Bap with our Tasty Relish and your choice of filling – served with a Bowl of Coleslaw

42. **Cheese and Bacon Burger**	£2.10
43. with Chipped Potatoes	£2.35
44. **Egg and Bacon Burger**	£2.10
45. with Chipped Potatoes	£2.35

Or, if you're only feeling slightly peckish . . .

46. **A Single Bumper Burger** served on its own	£1.10
47. with Chipped Potatoes	£1.35

SNACKS

Super Snacks

54. Pork Sausage, Griddled Egg and Baked Beans on a Sesame Bap	98p
55. Slice of Cold Gammon Ham, served with a Griddled Egg, on Two Slices of Buttered Toast	98p
56. Double Egg on Toast	85p
57. Baked Beans on Toast	85p

58. **Black Forest Gateau** 80p

59. **American "Coffee Break" Cake**	43p
60. **Eccles Cake** – The "Tasty Pastry"	38p
61. Bread Roll and Butter	22p
62. Buttered Toast – Brown or White	16p
63. Buttered Bread – Brown or White	16p
64. Portion of Jam or Marmalade	16p

SWEETS

Pancake Specials

Our very special Pancakes served with your Choice of Filling and Vanilla Ice Cream*

68. **Jubilee Pancake** – with Red Cherries	85p
69. **Gooseberry Pancake**	85p
70. **Pineapple Pancake**	85p
71. **Raspberry Pancake**	85p

*Contains Non-Milk Fat

76. **Sugar and Lemon Pancakes** Two Pancakes topped with Sugar and Lemon	75p
77. **Mapleen Syrup Pancakes** Two Pancakes served with Mapleen Syrup and Vanilla Ice Cream*	75p
78. **Mapleen Syrup Waffles** Two Waffles served with Mapleen Syrup and Vanilla Ice Cream*	75p
81. **Apple Flan with Vanilla Ice Cream*** Enjoy our ***New*** Traditional Recipe	80p
82. **Aunt Mary's Apple Turnover** Served Hot with Vanilla Ice Cream*	80p

Speciality Ice Cream Bombes

Little Chef's selection of Dairy Ice Creams with a difference – choose from three exciting combinations . . .

85. **Chocolate and Raisin**	55p
86. **Coffee and Hazelnut**	55p
87. **Tutti-Frutti Vanilla**	55p

Go on-treat yourself!

TO DRINK

Our NEW LARGER cup of.....

89. Tea	22p
90. Lemon Tea	22p
91. Coffee	33p
94. Coca-Cola	35p
95. **Suncrush** Orange Squash	25p

Thick Milkshakes

96. Chocolate	40p
97. Raspberry	40p
98. Banana	40p
99. Chilled Milk	22p

4/82/10M

HOSPITAL MENU

2/5a	*BREAKFAST FOR SATURDAY*	*Diet Label*
	CHOICE OF FRUIT JUICE	
1	( ) GRAPEFRUIT JUICE	D.R.LF.LR.
2	( ) ORANGE JUICE	D.R.LF.LR.
	OR	
	CHOICE OF CEREAL	
3	( ) PORRIDGE	D.LF.
4	( ) CORNFLAKES	D.LF.LR.
5	( ) ALL BRAN	D.LF.
	CONTINENTAL BREAKFAST OF:-	
6	( ) ROLL, BUTTER, MARMALADE	LF.LR.
	OR	
7	( ) BOILED EGG	D.R.LR.
8	( ) WHITE BREAD	D.R.LF.LR.
9	( ) BROWN BREAD	D.R.LF.
10	( ) BUTTER	D.R.LF.LR.
11	( ) FLORA	D.R.LF.LR.
12	( ) MARMALADE	LF.

2/5a	*LUNCH FOR SATURDAY*	*Diet Label*
1	( ) STEAK & KIDNEY PIE	
2	( ) BRAISED HAM	D.R.LF.LR.
3	( ) MINCED LAMB	D.R.LF.LR.
4	( ) CHOPPED PORK SALAD	D.R.
5	( ) MIXED VEGETABLES	D.R.LF.
6	( ) CABBAGE	D.R.LF.
7	( ) NEW POTATOES	D.LF.LR.
8	( ) CREAMED POTATOES	D.LF.LR.
9	( ) CREAM CARAMEL	LR.
10	( ) CREAMED RICE PUDDING	D.LF.LR.
11	( ) ICE CREAM & WAFERS	D*.LR.
12	( ) DESSERT ORANGE	D.R.LF.
13	( ) CHEESE & BISCUITS	D.LR.
	( )SMALL PORTION	

2/5a	*SUPPER FOR SATURDAY*	*Diet Label*
1	( ) ASPARAGUS SOUP	LF.LR.
2	( ) GRILLED BEEF SAUSAGES	
3	( ) SAVOURY SCRAMBLED EGG	D.R.LR.
4	( ) EGG & TOMATO SANDWICH	
5	( ) CHICKEN SALAD	D.R.LF.
6	( ) GARDEN PEAS	D.R.LF.
7	( ) CREAMED POTATOES	D.LF.LR.
8	( ) FRUIT SALAD	D.R.LF.
9	( ) SEMOLINA PUDDING	D.LF.LR.
10	( ) DESSERT APPLE	D.R.LF.
11	( ) CHEESE & BISCUITS	D.LR.
	( )SMALL PORTION	

SPECIAL DIETS
If you are no a special diet, choose the dishes marked with your code

D = Diabetic
R = Reduction
LR = Low Residue
LF = Low Fat

D*. No Ice Cream Wafers

NAME .... NAME .... NAME ....

BED NO .... WARD .... BED NO .... WARD .... BED NO .... WARD ....

An example of a hospital menu

14 Buying, costing and control

Having read the chapter these learning objectives, both general and specific, should be achieved.

General objectives Understand the principles of portion and cost control. Be aware of the factors affecting purchase of food and know how to cost dishes and menus.

Specific objectives Specify the factors to consider when purchasing each commodity. Identify and use portion control equipment and state the amounts of food required per portion. Accurately cost dishes and menus. Be able to apply cost control.

The purchasing of commodities

The responsibility for the buying of commodities varies from company to company according to the size and the management policy. Buying may be the responsibility of the chef, manager, storekeeper, buyer or the buying department.

The following is a suggested list to assist efficient buying:

1 Acquire, and keep up to date, a sound knowledge of all commodities both fresh and convenience, to be purchased.
2 Be aware of the different types and qualities of each commodity that are available.
3 When buying fresh commodities be aware of part-prepared and ready-prepared items available in the market.
4 Keep a sharp eye on price variations. Buy at the best price to ensure the required quality and also an economic yield. The cheapest item may prove to be the most expensive if waste is excessive. When possible order by number and weight, eg:

   20kg plaice could be $80 \times \frac{1}{4}$ kg plaice
   $40 \times \frac{1}{2}$ kg plaice
   20×1 kg plaice

   It could also be 20 kg total weight of various sizes and this makes efficient portion control difficult.
5 Organise an efficient system of ordering with copies of all orders

kept for cross checking, whether orders are given in writing, verbally or by telephone.

6 Compare purchasing by retail, wholesale and contract procedures to ensure the best method is selected for your own particular organisation.

7 Explore all possible suppliers: local or markets; town or country; small or large.

8 Keep the number of suppliers to the minimum. At the same time have at least two suppliers for every group of commodities. The principle of having competition for the caterer's business is sound.

9 Issue all orders to suppliers fairly, allowing sufficient time for the order to be implemented efficiently.

10 Request price lists as frequently as possible and compare prices continually in order that you buy at a good market price.

11 Buy perishable goods when they are in full season in order to gain the best value at the cheapest price. To assist with the purchasing of the correct quantities it is useful to compile a purchasing chart for 100 covers from which items can be divided or multiplied according to requirement. Indication of quality standards can also be inserted in a chart of this kind.

12 Deliveries must all be checked against the orders given for quantity, quality and price. If any goods are delivered that are below an acceptable standard they must be returned either for replacement or credit.

13 Containers can account for large sums of money. Ensure that all the containers are correctly stored, returned to the suppliers and the proper credit given.

14 All invoices must be checked for quantities and prices.

15 All statements must be checked against invoices and passed swiftly to the office so that payment may be made in time to ensure maximum discount on the purchases.

16 Foster good relations with trade representatives because much useful up-to-date information can be gained from them.

17 Keep up-to-date trade catalogues, visit trade exhibitions, survey new equipment and continually review the space, services and systems in use in order to explore possible avenues of increased efficiency.

18 Organise a testing panel occasionally in order to keep abreast of new commodities and new products coming on to the market.

19 Consider whether computer application can assist the operation. See Chapter 15.

Portion control

Portion control means controlling the size or quantity of food to be served to each customer. The amount of food allowed depends on:

a) The type of customer or establishment
There will obviously be a difference in the size of portions served – for example, to heavy industrial workers and female clerical workers. In a restaurant offering a three-course table d'hôte menu for £x lunch including salmon, the size of the portion would naturally be smaller than in a luxury restaurant charging £x for the salmon on an a la carte menu.

b) The quality of the food
Better quality food usually yields a greater number of portions than poor quality food, eg cheap stewing beef often needs so much trimming that it is difficult to get six portions to the kilo, and the time and labour involved also means loss of money. On the other hand, good quality stewing beef will often yield eight portions to the kilogramme with much less time and labour required for preparation.

c) The buying price of the food
This should correspond to the quality of the food if the person responsible for buying has bought wisely. A good buyer will ensure that the price paid for any item of food is equivalent to the quality – in other words a good price should mean good quality. This should mean a good yield and so help to establish a sound portion control. If on the other hand an inefficient buyer has paid a high price for indifferent quality food then it will be difficult to get a fair number of portions, and the selling price necessary to make the required profit will be too high.

Portion control should be closely linked with the buying of the food; without a good knowledge of the food bought it is difficult to state fairly how many portions should be obtained from it. To evolve a sound system of portion control each establishment or type of establishment needs individual consideration. A golden rule should be 'a fair portion for a fair price.'

Portion control equipment
There are certain items of equipment which can assist in maintaining control of the size of the portions. For example:

Scoops – for ice cream or mashed potatoes.
Ladles – for soups and sauces.
Butter pat machines – butter pats can be regulated from 7 g onwards.
Fruit juice glasses – 75–150 g.
Soup plates – 14, 16, 17, 18 cm.
Milk dispensers and tea-measuring machines.
Individual pie dishes, pudding basins, moulds and coupes.

As an example of how portion control can save a great deal of money the following instance is true:

1 It was found that 0.007 litre of milk was being lost per cup by

spilling it from a jug.

32 000 cups = 214 litres of milk lost daily: resulted in a loss of hundreds of pounds per year.

2 When an extra pennyworth of meat is served on each plate it means a loss of £1000 over the year when 1000 meals are served daily.

The following list is of the approximate number of portions that are obtainable from various foods:

Soup: 2 – 3 portions to the $\frac{1}{2}$ litre.
Hors d'œuvre: 120–180 g per portion.
Smoked salmon: 16–20 portions to the kg when bought by the side; 20–24 portions to the kg when bought sliced.
Shellfish cocktail: 16–20 portions to the kg.
Melon: 2–8 portions per melon, depending on the type of melon.
Foie gras: 15–30 g per portion.
Caviar: 15–30 g per portion.

Fish

Plaice, cod, haddock fillet	8 portions to the kg
Cod and haddock on the bone	6 portions to the kg
Plaice, turbot, brill, halibut, on the bone	4 portions to the kg
Herring and trout	1 per portion (180 g–$\frac{1}{4}$ kg fish)
Mackerel and whiting	$\frac{1}{4}$ kg–360 g fish
Slip-sole	180 g–$\frac{1}{4}$ kg
Sole for main dish	300–360 g fish
Sole for filleting	$\frac{1}{2}$ –$\frac{3}{4}$ kg best size
Whitebait	8–10 portions to the kg
Salmon (gutted, but including head and bone)	4–6 portions to the kg
Crab or lobster	$\frac{1}{4}$ kg–360 g per portion

(A $\frac{1}{2}$ -kg lobster yields approx. 150 g meat; a 1-kg lobster yields approx. 360 g meat)

Sauces

8–12 portions to $\frac{1}{2}$ litre

Hollandaise
Béarnaise
Tomato
Any Demi-glace sauce
Custard
Apricot
Jam
Chocolate

10–14 portions to $\frac{1}{2}$ litre

Apple
Cranberry
Bread

15–20 portions to $\frac{1}{2}$ litre

Tartare
Vinaigrette
Mayonnaise

Meats

Beef:	
Roast on the bone	4–6 portions per kg
Roast boneless	6–8 portions per kg
Boiled or braised	6–8 portions per kg
Stews, puddings and pies	8–10 portions per kg
Steaks– Rump	120 g–$\frac{1}{4}$ kg per one portion
Sirloin	120 g–$\frac{1}{4}$ kg per one portion
Tournedos	90–120 g per one portion
Fillet	120–180 g per one portion
Offal:	
Ox-liver	8 portions to the kg
Sweetbreads	6–8 portions to the kg
Sheep's kidneys	2 per portion
Oxtail	4 portions per kg
Ox-tongue	4–6 portions per kg
Lamb:	
Leg	6–8 portions to the kg
Shoulder boned and stuffed	6–8 portions to the kg
Loin and best-end	6 portions to the kg
Stewing lamb	4–6 portions to the kg
Cutlet	90–120 g
Chop	120–180 g
Pork:	
Leg	8 portions to the kg
Shoulder	6–8 portions to the kg
Loin on the bone	6–8 portions to the kg
Pork chop	180 g–$\frac{1}{4}$ kg
Ham:	
Hot	8–10 portions to the kg
Cold	10–12 portions to the kg

Sausages are obtainable 12, 16 or 20 to the kg
Chiplotas yield approximately 32 or 48 to the kg

Cold meat	16 portions to the kg
Streaky bacon	32–40 rashers to the kg
Back bacon	24–32 rashers to the kg
Poultry:	
Poussin	1 portion 360 g (1 bird)
	2 portions $\frac{3}{4}$ kg (1 bird)
Ducks and chickens	360 g per portion
Geese and boiling fowl	360 g per portion
Turkey	$\frac{1}{4}$ kilo per portion

Vegetables

New potatoes	8 portions to the kg
Old potatoes	4–6 portions to the kg
Cabbage	
Turnip Parsnips Swedes Brussels sprouts Tomatoes French beans Cauliflower	6–8 portions to the kg
Spinach	
Peas	4–6 portions to the kg
Runner beans	

Cost control

It is important to know the exact cost of each process and every item produced. In order to be able to achieve this a system of cost analysis and cost information is essential.

The advantages of an efficient costing system are:

1 It discloses the net profit made by each section of the organisation and shows the cost of each meal produced.
2 It will reveal possible sources of economy and can result in a more effective use of stores, labour, materials, etc.
3 Costing provides information necessary for the formation of a sound price policy.
4 Cost records provide and facilitate the speedy quotations for all special functions, eg special parties, wedding receptions, etc.
5 Enables the caterer to keep to a budget.

No *one* costing system will automatically suit every catering business, but the following guide lines may be helpful.

a) The co-operation of all departments is necessary and essential.
b) The costing system should be adapted to the business and not vice versa. If the accepted procedure in an establishment is altered to fit a costing system then there is danger of causing resentment among the staff and as a result losing their co-operation.
c) Clear instructions in writing must be given to staff who are required to keep records. The system must be made as simple as possible so that the amount of clerical labour required is kept to the minimum. An efficient mechanical calculator should be provided to save time and labour.

To calculate the total cost of any one item or meal provided it is necessary to analyse the total expenditure under several headings.

Basically the total cost of each item consists of three main elements:

1 Food or materials cost.
2 Labour
3 Overheads (rent, rates, heating, lighting, equipment, repairs and maintenance)

(items 2 and 3: total cost)

a) Food or materials cost are known as variable costs because the level of cost will vary according to the volume of business. In an operation that uses part-time or extra staff for special occasions, the money paid to this category of staff also comes under variable costs. By comparison salaries and wages paid regularly to permanent staff come under fixed costs.
b) All cost of labour and overheads which are regular charges come under the heading of fixed costs.
c) Labour costs in the majority of operations fall into two categories: direct labour cost which is the cost of salaries and wages paid to staff such as chefs, waiters, barstaff, housekeepers, chambermaids; and where the cost can be allocated to income from food, drink and accomodation sales.

  Indirect labour cost would include salary and wages paid for example to managers, office staff and maintenance men who work for all departments and therefore their labour cost should be charged to all departments.

Cleaning materials

An important group of essential items that is often overlooked when costing is that of cleaning materials. There are over sixty different items that come under the heading of cleaning materials and cleaning utensils, and approximately 24 of these may be required for an average catering establishment. These may include: brooms, brushes, buckets, cloths, drain rods, dusters, mops, sponges, squeegees, scrubbing/polishing machines, suction/vacuum cleaners, wet and wet/dry suction cleaners, scouring pads, detergents, disinfectants, dustbin powder, washing-up liquids, fly sprays, sacks, scourers, steel wool, soap, soda etc.

It is important to understand the cost of these materials and to ensure that an allowance is made for them under the heading of overheads.

Profit

It is usual to express each element of cost as a percentage of the selling price. This enables the caterer to control his profits.

Gross profit or kitchen profit is the difference between the cost of the food and the selling price of the food.

Net profit is the difference between the selling price of the food (sales) and total cost (cost of food, labour and overheads).

Sales – Food cost = Gross profit (kitchen profit)
Sales – Total cost = Net profit
Food cost + Gross profit = Sales

Example: If food sales for 1 month = £4000
and food cost for 1 month = £1760
labour and overheads for 1 month = £1720
then total costs for 1 month = £3480
Gross profit (kitchen profit) = £2240
Net profit = £ 520

£4000 − £1760 = £2240
£4000 − £ 520 = £3480
£1760 + £2240 = £4000

Profit is always expressed as a percentage of the selling price.
∴ the percentage profit for the month was

$$\frac{\text{Net profit}}{\text{Sales}} \times 100 = £\frac{520 \times 100}{4000} = 13\%$$

A breakdown shows:

		Percentage of sales
Food cost	£1760	44%
Labour	£1000	25%
Overheads	£ 720	18%
	£3480	
Net profit	£ 520	13%
Sales	£4000	

If the restaurant served 4000 meals then the average spent by each customer would be

$$\frac{\text{Total sales £4000}}{\text{No of customers 4000}} = £1.00$$

As the percentage composition of sales for a month is now known, the average price of a meal for that period can be further analysed.

Average price of a meal = £1.00 = 100%
1p = 1%

which means that the customer's contribution towards:

Food cost	= 1p × 44 =	44p
Labour	= 1p × 25 =	25
Overheads	= 1p × 18 =	18
Net profit	= 1p × 13 =	13
Average price of meal	=	£1.00

A rule that can be applied to calculate the food cost price of a dish is: let the dish equal 40% and *fix the selling* price at 100%.

eg Cost of dish = 40p = 40%

$$\therefore \text{ Selling price } = \frac{40 \times 100}{40} = £1$$

Selling the dish at £1, making 60% gross profit above the cost price, would be known as 40% food cost. For example:

Sirloin steak (8 oz)
$\frac{1}{2}$ lb Entrecote steak at £3.60 a pound = £1.80p

$$\text{To fix the selling price at 40\% food cost } = \frac{1.80 \times 100}{40} = £4.50$$

The following will help with various food costings:

Food cost	To find the selling price multiply the cost price of the food by	If the cost price of food is £1 the selling price is	If cost price is 20p the selling price is	Gross profit
60%	$1\frac{2}{3}$	£1.66	32p	40%
55%	$1\frac{3}{4}$	£1.75	34p	45%
50%	2	£2	40p	50%
45%	$2\frac{2}{9}$	£2.22	44p	55%
40%	$2\frac{1}{2}$	£2.5	48p	60%
$33\frac{1}{3}$ %	3	£3	60p	$66\frac{2}{3}$ %

If food costing is controlled accurately the food cost of particular items on the menu and the total expenditure on food over a given period are found. Finding the food costs helps to control costs, prices and profits.

An efficient food cost system will disclose bad buying and inefficient storing and should tend to prevent waste and pilfering. This can assist the caterer to run an efficient business and help him to give the customer adequate value for money.

The caterer who gives the customer value for money together with the type of food that the customer wants is well on the way to being successful.

Further information: R.D. Boardman, *Hotel and Catering Costing and Budgets* (Heinemann); Michael Riley, *Understanding Food Cost Control* (Arnold); Hughes & Ireland, *Costing & Calculations for Caterers* (Stanley Thornes).

Decimalisation and metrication

What are decimals?

According to the dictionary, a decimal counting system is one which proceeds by tens: 1, 10, 100, 1000, etc. Below one, fractions are expressed in tenths, hundredths, thousandths, etc, and written as 0.1, 0.01, 0.001.

Simple examples of using decimals when adding, subtracting, multiplying and dividing:

Addition

Add 12.13, 2.157 and 7.934

```
       12.13
        2.157
        7.934
Ans.   22.221
```

Note that the decimal points are in a vertical line, including the decimal point in the total, and that the working is the same as ordinary addition.

Subtraction

Subtract 17.123 from 73.1

```
       73.1
       17.123
Ans.   55.977
```

Division

Divide 0.004 by 0.02

Make the divisor 0.02 into a whole number 2

To do this the point is moved two places to the right. The same must be done to the divident 0.004 which becomes 0.4

∴ 0.4 divided by 2 = .2

$$0.02\,\overline{)0.004} \text{ becomes } 2\,\overline{)0.4}^{\;0.2}$$

Multiplication

Multiply 1.345 by 4.38

```
         1.345
          4.38
       5380
         40350
         10760
Ans.   5.89110
```

Explanation: Multiply as with ordinary numbers, ignoring the decimal points for the time being, as shown.

1st line   1345 × 4(00), etc.
           starting figure (0) is placed *beneath the 4*

2nd line   1345 × 3(0)

starting figure (5) is placed *beneath the 3*

3rd line 1345 × 8

starting figure (0) is placed *beneath the 8*

4th line *a*) add lines 1, 2, 3 (these are called partial products) to give the figure 589110.

b) Count the number of figures after the decimal point in each of the two numbers being multiplied; thus 1.345 (3 fig. after d.p.); 4.38 (2 fig. after d.p.); giving a total of 5 figures after the decimal points.

c) Let the decimal point bounce 4 places from the right to left in our answer line, thus

5.8 9 1 1 0 = 5.89110 *Ans.*

Addition of pence

12p + 73p + 21p + 5p

```
  12
  73
  21
   5
----
111p = £1.11 Ans.
```

After adding, count two places from the right (dividing by 100) and insert the decimal point; the figures to the left of the point are pounds and those to the right are pence.

Addition of pounds and pence

£1.58 + £12.50 + 7p

```
£
  1.58
 12.50
 00.07
------
 14.15 = £14.15 Ans.
```

Note: The sum is headed by the £ sign only, usually situated over the first digit in the pounds column.

Subtraction

Deduct 22p from £1

```
1.00
0.22
----
0.78 = 0.78 or
           78p Ans.
```

Multiplication

£1.92 × 5

```
1.92
   5
----
9.60 = £9.60 Ans.
```

As there are two decimal figures after the point in the sum, there must be two decimal places in the answer.

Another example: $8\frac{1}{4}$ kg of cod fillet at £1.75 per kg ($8\frac{1}{4}$ = 8.25).

```
   8.25
   1.75
 -------
 825 00
 577 50
  41 25
 -------
1443 75
```

As there are four decimal places in the original numbers there must be four decimal places in the answer, which now reads 14.4375. The answer is in pounds and should be given correct to the second decimal place (correct to the nearest penny). *Ans.* £14.44.

Division

Make certain that the decimal point in the answer is immediately above the decimal point in the amount to be divided. *Example:* 24 people win £80.16 to be divided between them, how much does each receive?

```
      3.34
  24 |80.16
      72
      ----
       81
       72
      ----
        96
        96        Ans. £3.34 each
```

Examples of simple short cuts:

(i) To find the cost of 38 articles at $9\frac{1}{2}$ p each
(38 × 10p) minus (38 × $\frac{1}{2}$ p)
that is 38 × $9\frac{1}{2}$ p = 380 − 19 = = 361 = £3.61

(ii) To find the cost of 14 articles at 49p each
(14 × 50p) − 14p
that is 700 − 14 = 686 = £6.86

(iii) When considering discount for example:

1% discount of £1 equals 1p, therefore
5% discount of £110 equals 550p
550p = £5.50 deducted from £100 = £104.50

Changing to the metric system

The metric system is a decimal system applied to measurement. It was devised by the French after the Revolution as an attempt to rationalise their counting methods. Gradually a place in the system has been found for every aspect of life which requires to be measured, from quantities of flour to cosmic rays. The Systéme Internationale (SI), as it is now called, is being adopted by Britain.

There are six basic units

Quantity	*Basic unit*	*Symbol*
length	metre	m
mass	kilogram	kg
time	second	s
electric current	ampere	A
temperature	kelvin*	K
luminous intensity	candela	cd

and some of the supplementary units which concern the catering industry are

power	watt	W
area	square metre	m^2
volume	cubic metre (litre) m^3	
pressure	newton per sq metre	N/m^2
quantity of heat	joule	J
calorific value	joule	J/m^3
temperature	degree Celsius	deg C

As metric units are based on the number 10 their value must increase or decrease by tens. Whether measuring weight, capacity or length the multiples are expressed by the same prefix. Those which have relevance to the industry are

Prefix	*Symbol*	*Factor by which the unit is multiplied*
mega	M	1 000 000
kilo	K	1 000
hecto[1]	h	100
deca[1]	da	10
deci[1]	d	0.1
centi[1]	c	0.01
milli	m	0.001

* Do not be put off by the Kelvin. 1 degree Kelvin = 1 degree Celsius (centigrade) (0°C is equal to 273.15° on the Kelvin scale). Celsius is the preferred term to centigrade and will be used throughout; it is written as 75°C.

The following table shows these prefixes applied to the standard unit of length: the metre:

10 millimetres (mm)	=	1 centimetre (cm)
10 centimetres (cm)	=	1 decimetre (dm)
10 decimetres (dm)	=	1 metre (m)
10 *metres*	=	1 decametre (dam)
10 decametres (dam)	=	1 hectometre (hm)
10 hectometres (hm)	=	1 kilometre (km)

Thus there are 100 centimetres in 1 metre; 1000 metres in a kilometre, etc.

Working in metric units

Length

1 mm	=	0.039 inch
1 cm	=	0.394 inch
1 metre	=	39.37 inches or 1.09 yd
1 kilometre	=	0.62 or $\frac{5}{8}$ of a mile

Length

1 yard	=	0.914 m
1 foot	=	0.304 m
1 inch	=	2.54 cm or 25.4 mm
$\frac{3}{4}$ inch	=	1.9 cm or 19.05 mm
$\frac{1}{2}$ inch	=	1.27 cm or 12.7 mm
$\frac{1}{4}$ inch	=	0.635 cm or 6.35 mm
$\frac{1}{8}$ inch	=	0.317 cm or 3.17 mm
$\frac{1}{10}$ inch	=	0.254 cm or 2.54 mm

The basic unit for measuring length is the metre. For short lengths the centimetre (0.01 of a metre) and millimetre (0.001 of a metre) are used; and for long distances the kilometre (1000 metres).

Weight: the gram represents a very small mass and the kilogram therefore is the most used unit. For small quantities, however, the gram is suitable, for example 1 lb equals 453 g. Large weights are expressed in tonnes (t).

1 ton	=	1.016 tonne or 1016.05 kg
1 cwt	=	50.8 kg
1 stone	=	6.35 kg
1 lb	=	453 g
1 oz	=	28.35 g

1 kg	=	2.2 lb
28.35 g	=	1 oz

(It is worth noting that 1 g is the weight of 1 cubic centimetre (1 cm^3) of water.)

Capacity and volume: The cubic metre, which is the basic unit for measuring capacity and volume, is too large for everyday use and therefore the litre, which is 1/1000 or 0.001 of a cubic metre, is used.

1 centilitre	= 0.07 gill
1 decilitre	= 0.176 pint
1 litre	= 1.76 pints
1 decalitre	= 2.2 gallons
1 gallon	= 4.546 litres or 4.546 dm^3
1 quart	= 1.136 litres or 1.136 dm^3
1 pint	= 0.568 litre or 0.568 dm^3
1 gill	= 0.142 litre or 0.142 dm^3 or 1.42 decilitres
1 fluid ounce	= 28.4 cm^3 0.28 decilitre

Temperature: The Celsius (centigrade) scale is being used in place of the Fahrenheit scale. To change from Fahrenheit to Celsius subtract 32, multiply by 5 and divide by 9. For example

$$\begin{array}{r} 212 \\ -\ 32 \\ \hline 180 \\ \times\ 5 \\ \hline \div\ 9\ 900 = 100 \end{array} \qquad 121°F = 100°C.$$

For conversion of Celsius (centrigrade) to Fahrenheit multiply by 9, divide by 5 and add 32.

Pressure: Pressure has so far been measured in pounds per square inch; in the metric system the expression for pressure is newtons per square metre, which is abbreviated to N/m^2.

$1\ lb/in^2$
one pound per square inch = 6894.76 N/m^2

Newton (N) is the unit of force which when applied to 1 kilogramme of mass gives it acceleration of 1 metre per second.

Energy – heat: The terms Btu (British thermal unit), therm and Calorie will be replaced by the joule (J), which is the unit for measuring energy and quantity of heat.

1 therm = 105.506 MJ
1 Btu = 1.055 kJ

(For calorific value the term Jm^3 is used.)

Metric equivalent

	Approximate equivalent	*Exact equivalent*
1 oz	30 g	28.3 g
2 oz	60 g	56.6 g
4 oz	120 g	113.2 g
6 oz	180 g	169.8 g
8 oz	$\frac{1}{4}$ kg	227.0 g

12 oz	360 g	340.0 g
1 lb	$\frac{1}{2}$ kg	0.454 kg
1 stone	6 kg	6.35 kg
1 cwt	50 kg	50.8 kg

Metric equivalents

	Approximate equivalent	*Exact equivalent*
1 gill	$\frac{1}{12}$ litre	0.1065 litre
$\frac{1}{2}$ gill	$\frac{1}{16}$ litre	0.071 litre
$\frac{3}{4}$ pint	$\frac{3}{8}$ litre	0.426 litre
1 pint	$\frac{1}{2}$ litre	0.568 litre
1 quart	1 litre	1.136 litres
1 gallon	5 litres	4.543 litres
$\frac{1}{8}$ inch	3 mm	3.175 mm
$\frac{1}{4}$ inch	$\frac{1}{2}$ cm	0.635 cm
$\frac{1}{2}$ inch	1 cm	1.27 cm
1 foot	3 dm	3.05 dm

15

The computer in catering

Having read the chapter, these learning objectives, both general and specific, should be achieved.

General objectives Be aware of the increasing use of computers in the catering industry and understand their uses.

Specific objectives Explain how a computer can be used for control of stock and used for recipes and menus. Be able to correctly use the basic terminology associated with computers.

Introduction

In the computer industry they also have 'chips'. These are small plastic squares containing layers of minute electrical circuits which when pressed together are called silicon chips. They are very cheap to make, and scientists have found ways to link them together so that information can be stored as electrical signals in vast quantities.
No one living in such a technologically advanced age as ours, can escape the influence of computers. Not even caterers and hoteliers! The routine clerical jobs that hamper the efficient day to day workings of hoteliers and caterers can readily be solved with the thoughtful application of new computer technology.

For example, jobs concerned with information regarding menus, the recipes to use, the amounts of ingredients to order, depending upon stock in hand, can be instantly produced. This can be done accurately with both saving of time and effort.

Expense is no longer a problem. In the past, only large companies and organisations could afford the resources to devote large sums of money towards computer development. Today, a new generation of micro computers help to solve problems which hitherto could only be resolved on expensive machines. Hence, small businesses and institutions are now able to take advantage of the latest technological developments without an unacceptable initial capital outlay.

In this Chapter, we shall consider what makes the computer so helpful to caterers and how computers are being utilised in different sectors of the industry.

Computers still come in all shapes and sizes but are usually measured in terms of how much information they can store and 'digest' at any one time. The computers' duties include storing information (data) and using that data to provide fresh information or data. This requires the computer to process the information and gives rise to the term 'data processing' (or D.P.), around which a vast industry has developed over recent years.

Just as a cassette recorder requires the tape to be inserted before it can play the music or have sounds recorded on it, so the computer equipment (hardware) requires a tape or some medium containing instructions, before it can be brought into use. These instructions (software) are written in special computer language so that they can be understood by the hardware and are written in chapters just like a book, and these are called programs. Consequently, those who write programs are referred to as programmers.

It is a complicated subject and may be more difficult because the computer industry has invented new words and created a 'jargon' of its' own, without which, it is not possible to fully understand how the computer is able to help industry.

Hardware

This expression refers to the manufactured equipment we know as 'a computer'. It cannot work at all without instructions (programs or software) and usually requires several different pieces of equipment wired together to enable it to function.

If the computer is to help us in catering we have to be able to put data 'in' and take data 'out'. We also have to store it in the meantime until we need to use it. This will determine the main components of the computer. Each piece of equipment (or device) can be classified as an 'input device', an 'output device' or a 'storage' device.

The television screen with a keyboard like a typewriter is often the most distinctive feature of the computer. Data is input to storage by means of the keyboard and may be checked on the screen (Visual Display Unit or VDU) before it is stored away. In this way it becomes an 'input device'. It can also be used to take information out of storage and it is then being used as an 'output device'. The size and style of the VDU and keyboard vary from one manufacturer to another and some computers require no VDU at all and rely only on a printer as an output device. Others will allow the use of an ordinary television but these are usually small domestic computers and in general are less powerful than those used in industry.

The printer is like a typewriter without a keyboard. It types in different ways either using little dots joined together to form letters or like a traditional typewriter with characters on a wheel known as a 'daisy wheel'. Many popular printers type at speeds of 120-200 characters per second (CPS) and some can type at speeds of 3000 CPS or more. Generally speaking the price of the printer will depend on the speed and the quality of the typed print.

The silicon chip or a collection of many chips which makes all this possible is called the 'Central Processing Unit' (or CPU). They too come in different sizes and are measured by how many characters it will store at any given moment. These characters are counted in thousands (kilobytes or Kb) and are referred to as the 'Random Access Memory' (or RAM) of the computer. It is essential to know how big the RAM of the computer is, because this will determine what 'processing power' it has, and how clever the computer can be if programmed correctly. The bigger the CPU, the more expensive the computer. Domestic and other small computers need a 16Kb RAM, other larger computers can increase to 32Kb, 64Kb and 128Kb, all the way up to 1000Kb (or 1 Megabyte) and beyond.

Data is not always required to be instantly available, it can be unloaded from the RAM on other kinds of storage such as magnetic tape or disks. The magnetic tape acts just as the tape in a tape recorder and the various kinds of disks perform in just the same way but are a different shape. This 'off line' storage is also measured by the number of characters that can be stored and we measure them in 'kilobytes' (thousands/Kb) and 'megabytes' (millions/Mg). Most commonly used are the 'Floppy Disks' which are like small magnetic records similar to those used on a modern stereo record player. For larger volumes of

Computer operator types in data on to the visual display unit.

storage, these can become 'hard or fixed disks' and even small computers can have up to 10Mg of off-line (or secondary) storage for data.

Each piece of hardware is connected by wires to the CPU and with the use of British Telecom equipment connections between computers can be made from one side of the globe to another. As computer science progresses and computers become more widely used, so they become cheaper. Developments over the last decade have also made them smaller and smaller, and complicated computers costing hundreds of thousands of pounds which might have taken up a whole room in an office building at one time, can now be bought for just a hundredth of the price and sit neatly on a desk top.

Software

As we know, without instruction, the computer will not work and programs have to be devised which will make the hardware perform the required tasks. The same hardware can do different jobs at the same time, although in truth, it deals with each task in turn but does it so fast it appears to happen at the same time.

The instructions are written in special forms so that the computer will understand what should happen next. New 'high level' languages mean that programmers can write in English and use other programs to convert what they have written into machine language or 'Compiled Code'. The programmer must not only create the code, but he must understand what he wants to happen and how he wants the computer to react to his instructions.

In this way software can be made to suit different applications and some companies specialize only in providing software programs for one industry such as catering or hotels. It should not be necessary to understand all the internal processes of the computer before being able to use it, and if the software is helpful, most people can learn to use them after only a few hours training. Great care is needed when buying a computer to be sure that the software will perform all the tasks the user requires, and that it is easy to use, giving help to anyone who needs it. It is, therefore, a combination of reliable and dependable hardware, with helpful and dynamic software, which forms the basis of a good computer system.

The important information which is provided by the computer depends heavily upon how much information was put in, just as a recipe depends on the right ingredients. In the next section we shall consider how computers help those in catering.

Applications

Wherever meals are being provided in quantity, it is necessary to examine costs which occur and to ensure that some control exists to monitor payments to suppliers, and staff, and to ensure that these costs are accounted for. At the end of the day, profit levels can therefore be calculated. Additionally, institutional caterers must be in a position to be able to compare costs against allowances, so that overspending does not occur. Even the additional expense of 1p per meal, per person, becomes very significant in a factory feeding 10,000 people twice daily, five days a week. At the end of the year, this represents £52,000.

(10,000 people × 2 meals × 5 days × 52 weeks × £0.01 = £52,000)

In Chapter 14, we learnt how food costs have been calculated manually. The main element used in calculating what food costs have been incurred over a period of time, is by stocktaking. The simple formula for calculating the cost of each item is shown here:

No	Item	Opening Stock	+	Receipts of (Purchases)	–	Closing Stock	=	Consumption £
001	Bacon	244.00	+	105.00	–	165.00	=	184.00
002	Eggs	169.00	+	92.00	–	145.00	=	116.00
		413.00	+	197.00	–	310.00	=	300.00

From this simple example, we can see that at the beginning of the period we had an opening stock of £413.00 and during the period we bought £197.00 more bacon and eggs. On the last day of the period, we had only £310.00 worth of eggs and bacon and we were therefore able to

Computer operator inserts a floppy disc.

calculate exactly how much stock had been used. In this case it was £300.00.

Of course, the bacon and eggs had to be counted before the value of the closing stock could be worked out. The bacon would have been weighed and then valued at a cost per pound or kilogram. The eggs would have been counted and multiplied by their price. If this has to be done every week or every month for six hundred or a thousand stock items, it means a lot of paper-work. It is very time-consuming finding the right price of each item and multiplying it by the quantity which was left. It is likely, therefore, that the caterer would work the opening stock, receipts and closing stock by measurements of weight, volume or numbers and only express the consumption in monetary value.

No	Item	Opening Stock	–	Receipt of Purchases	–	Closing Stock	=	Consumption £
001	Bacon	225 lb 15 oz	+	97 lb 4 oz	–	152 lb 13 oz	=	170 lb 6 oz
002	Eggs	3755	+	2044	–	3221	=	2578 eggs

No	Item	Consumption	(times)	Unit Price	=	Value
001	Bacon	170 lb 6 oz	×	108p/lb	=	£184.00
002	Eggs	214doz 10/12th	×	54/doz	=	£116.00
						£300.00

Check the arithmetic of the bacon to see just how difficult it can be to get a precise result. Furthermore, these calculations would be far more complicated if the caterer wanted to know the value of the receipts or closing stock. To these figures might also be added more figures to make allowances for transfers from one kitchen to another, or to 'write-off' food items which have been spoilt or wasted through over-production.

So, it seems natural that these calculations should become the first job for computerisation. All the prices can be stored in the program and everything can be measured in terms of money and, if necessary, convert it to weights and measures. Conversely, if given the weights and measures, it could express it in monetary value. It does not relieve the caterer from the tedium of having to count every item at the end of the period, but it does reduce the amount of work, because the computer can prepare the papers with all the information printed on them and then it only requires the caterer to enter the amounts of the closing stock in figures, decimals or fractions.

Receipts of course, are gathered on a daily basis as goods are received. The invoices are entered into the system against each item, so that at the end of each period, not only can the value of receipts be calculated but also the caterer is in a position to make exact payments to each supplier. The price of each ingredient is collected from each invoice so that the system can value all the items at the correct price.

Those involved in new technology are now turning their attention to the collection of information for the closing stock. Small hand-held terminals with special light-sensitive pens can read the items' stock number and allow the stock-taker to enter the quantities on a small keyboard just like a calculator. This information can then be put straight into the system for processing, thus reducing even further, the amount of work to be done by the operator. Stocks which were manually calculated, required ten to fourteen days for completion, whereas, with the help of the computer the entire process can be completed in an hour or two.

Of course it is also possible to arrive at the value of stock consumed by another method. By dividing the quantities of the different ingredients, by the number of portions each recipe provides, and then multiplying that quantity by the number of portions prepared or sold. This will give a 'notional' or 'theoretical' stock consumption figure. To make these theoretical projections, the computer must not only have the recipes which have been used, but also, it must store details of the number of portions provided of each recipe.

Recipe report sheet

This then presents the second major area for the application of computers to caterers. Since details of all ingredients are already in the system for stocktaking purposes, these same records could be used as a source of data from which recipes could be made. Of course, we rarely use ingredients in the same measures as we buy them and it is therefore necessary to have a complete set of conversion tables, so that if an ingredient is bought by the pound or kilo, it can be used in recipes by the gram or by the ounce. Furthermore, these units of measure would have to be abbreviated to make the system easy to use.

This would then allow the operator to build up a group of ingredients into recipes and express each recipe as a cost per portion. The caterer could see the cost of each recipe, and, should the price of any ingredient change, so too would the cost of all the recipes using that ingredient. The application is generally referred to as 'production control' and presents very major advantages to the caterer.

Firstly, he can obtain from the computer comparisons on his **actual** consumption with his **notional** consumption and give the variance, or difference in terms of money and quantity. This enables him to see clearly, not only what he has actually used, but what he should be using if the recipes are followed correctly.

Of course explanations would have to be sought for large variances but the computer can indicate for the first time, which ingredients should be investigated.

Secondly, by putting in a series of menu for some days ahead, the caterer can ask the system exactly what ingredients are required for the next period.

This breaking down of menus and recipes into ingredients is called 'an explosion' and by creating such 'recipe explosions' the caterer can gauge his orders more accurately and so ensure a reasonable delivery system. Where food-cost-management and food-production have to be highly accurate, these tasks have been performed manually for many years.

In school meals and hospital catering, pupil and patient requirements can be analysed and converted into ingredients extremely quickly, by means of a card-reading device. This device can identify which dishes have been selected from a menu and give precise information to the caterer, on what must be produced for the next meal, and in the case of a central kitchen supplying various locations, to where the food should be despatched at the time of service.

It thus becomes only a short development, to say that, if a list of suppliers can be maintained alongside the recipes and ingredient files, then the computer could easily produce a purchase order list exactly to the specification required by the caterer.

In airline 'in-flight' catering, it may be necessary to write an invoice for meals supplied to passengers aboard a particular flight, or for one of the large industrial contract caterers, to bill each of the customers in turn, for what they have consumed. Yet more data, however, is needed for this function, since the system requires the selling prices of each portion of food supplied, on which an invoice can be based. Indeed, it may be necessary to have more than one selling price for each item, so that a complexed marketing policy can be followed

In the case of stock systems required for managing liquor stocks in bars, it is quite common for the same beer to be more expensive in one bar than another, even though it costs the same to purchase.

So, in this way the various applications expand to provide a wide range of information against which the caterer can make plans for the future and obtain results on the past activity with little or no delay. By making budgets for the future and keeping figures from the past, he can compare his performance with his objectives and base important management decisions from information on which he can rely.

Glossary

Central Processing Unit	Nerve centre of the computer.
Chips, Silicon	A combination of electronic circuits built on to a slice of silicon.
Compiled Code	Instructions for the computer translated into a machine language.
CPS	Characters per second.
Daisy Wheel	A metal wheel bearing characters for the typing of documents
Data	Information stored in the system which is not part of the program.
Data Processing	Usually applied to the manipulation of information and figures by the computer.
Domestic Computers	Small computers generally considered unsuitable for business use.

Fixed Disks	A solid, flat disk which is covered in magnetic material used for storing programs and data.
Floppy Disks	A flexible, flat disk also used for storing programs and data
Hard Disks	Same as a fixed disk but may be removable or interchangeable
Hardware	A collection of computer equipment.
Input Device	An equipment which allows information to be put into the computer.
Main Frames	Large installations of computers.
Megabyte	One million sets of binary digits usually used to measure the amount of storage.
Micro Computers	The smallest range of computers.
Mini Computers	Intermediate range of computers.
Off-line Storage	A means of keeping information for the computer which is filed away and not brought into the system until required.
Output Device	An equipment which can take information out of the system.
Processing	The manipulation of data by the computer.
Programmers	People who write instructions for computers.
Programs	Computer instructions.
RAM	Random access memory is the form of instant memory available without going to the storage facilities.
Software	Complete sets of programs required to make the system function.
Storage Device	Equipment used for storing data and programs usually in the form of fixed or floppy disks but also as tapes.

Books for reference

The Dictionary of Computers
by Anthony Chandor (Penguin)

Introducing Computers
by Peltu (National Computer Centre)

A Young Person's Guide to BBC Basic
by Milan (National Computer Centre)

Introducing Micro Processors
by Simons (National Computer Centre)

Introducing Computer Programming
by Collin (National Computer Centre)

Data Processing Made Simple
by Wooldridge (W.H. Allen)

In conjunction with *Practical Cookery* by V. Ceserani and R. Kinton (Edward Arnold) a computer program *Menumaster* is available from Shumwari Associates, 12 Marlin Court, Marlow, Bucks SC7 2AJ.

More information on the use of the micro computer in the catering industry can be obtained from:

XI DATA SYSTEMS
Ladywise House
Parkfield Street
Leeds LS11 5PH

Documentor computerised till being checked at a new Wimpy counter service fast food unit

16
Service of food

Having read the chapter, these learning objectives, both general and specific, should be achieved.

General objectives Appreciate the need for co-operation between kitchen and food service staff. Develop the attitude of good service through understanding from the point of view of the customer and of the food service personnel.

Specific objectives List the points that assist serving and food service staff to serve food supplied from the kitchen. Specify the various ways food may be served to the customer.

This chapter deals with the service of food from the kitchen, and there are several methods which are employed to enable the customer to receive the meal. The mode of service will depend on the type of establishment, and in some places more than one kind of service will be

Carvery

used, often from one kitchen. The cost of operating these methods varies, but the final objective is the same—that is, the food, when presented to the customer, should look attractive, should be of the right temperature, it should be as ordered and give value for money.

The kinds of service that are used are:

Waiter or waitress service,
Cafeteria service,
Hatch and counter service,
Snack bar, buffet service, take away

The kind of service to be provided will depend to some extent upon the type of people to be served, the number of people to be served, the dining accommodation, the number of sittings and the amount of time and money to be spent on the meal.

Waiter and waitress service

This method of service is the most costly to operate and it is used extensively in all kinds of restaurants. The food ordered by the customer to the waiter or waitress is collected from the hot plate by the waiting staff and served to the customer. (The details of this method are given in the chapter on Kitchen Organisation.)

From the kitchen point of view the following are of importance: the food ordered by the waiter should be ready when it is required. It should leave the kitchen at the correct temperature (hot if it should be hot, cold if it is meant to be cold), so that the food is presented correctly to the customer by the waiter. Orders received from the waiting staff should be dealt with in strict rotation and must go from the hot plate in the same order. Needless to say the waiters should not expect food to be on the hot plate before a reasonable period of time. The correct kitchen accompaniment to dishes should be sent from the hot plate with the appropriate dish, and the vegetables, as ordered, should be ready to go with the main dishes.

Co-operation between the kitchen and restaurant staff is essential for the successful service of food. The teaching of waiting to young people training to work in the kitchen is invaluable, since they are given an opportunity to see beyond the hot plate and to appreciate the waiters' problems. The need to serve portions of equal size becomes more obvious when a customer complains that the person opposite paying the same money has a portion larger than his or her own. The difficulty of sharing five potatoes between two people or trying to bone an in sufficiently cooked sole in front of the customer should cause the kitchen brigade to be careful in the way they serve the food.

Waiting staff will be used for a la carte and table d'hôte menus service (see chapter on Menu planning, page 328), also for banquets and for club service.

In serving food the kitchen staff should remember these points:

1 That it is necessary to dress the dish in such a way that the waiter may effectively transfer the food from the dish to the customer's plate.

2 To serve the food on the correct dish, which should be clean and of the right temperature.
3 The food should be arranged attractively; this is very important.
4 The dish should be clean after the food has been added to it and the correct amount should be placed on the dish.
5 Food which is served on a plate should be arranged neatly.
6 Hot foods (except those which are deep fried and would go limp and soft) should be covered with a lid.

Room service

Self-service food counter

Cafeteria service

Cafeteria service is used mainly in popular restaurants, industrial canteens and similar establishments.

The customer proceeds alongside the service counter, choosing food from the selection displayed. The counter contains bains-marie and has sections for hot and cold foods; usually one part is for beverages. The food is served and replenished by counter-hands.

The customer collects his or her own cutlery, and therefore staff are required only to clear tables. In some places the tables are laid; the customer then has only to choose his food and take it to the table.

When a large number of people have to be fed in a limited time, several counters would be provided, and the customers' meal times in industrial establishments would be staggered so as to reduce queuing.

The advantage of this method of service is speed, and it is therefore necessary for the kitchen to keep the counter supplied with food and not to cause a hold-up in the service by keeping the counter-hands waiting. The arrangement of the food should be such that the cold dishes are collected first and hot items placed towards the end of the service counter.

Hatch service

When high-speed service of meals is required this method may be used. Hatches connect the kitchen and dining-room. These hatches will be numbered and the numbers correspond to items numbered on the menu. The customer makes up his or her mind at the entrance and goes to the appropriate hatch and collects the meal.

Another form of this method of service is to provide the main course at the hatch and a selection of vegetables to be provided. This of course is a slower method of service because, as soon as a choice is provided, service slows down. This type of service is also referred to as counter service.

Staff restaurant at the Royal Lancaster Hotel, London

Snack bars

Snack bars usually provide beverages, sandwiches, salads, fruit, pastries and pies which are suitable for people requiring something quickly or something light. Seating accommodation is provided at the service counter as well as in the room.

Buffets

Full buffets, where all kinds of foods are used, some requiring carving or skilled service, are usually served from behind the buffet by members of the kitchen staff. The customer will help himself or herself to those foods not requiring the service of the staff. For such functions it is usual for the kitchen to replenish the buffet as required and to keep the buffet tables attractive.

For cocktail parties all the foods are dressed on dishes and the customers help themselves. When hot foods are served they are sent into the room while the customers are there so that they are still hot when waiters pass them round or when people help themselves.

Automatic vending

Because of the increased cost of labour there has been a considerable development in the use of vending machines: this is because their use does not require labour to serve, labour is only required to maintain and replenish the machines. Complete meal service can be achieved by using prepared meals held in a refrigerated compartment and used in conjunction with a microwave oven and a vending machine for drinks. This aspect of service can be of considerable importance to catering, particularly to replace or supplement existing catering services.

Some advantages of automation are:

1 Foods are available for 24 hours of the day.
2 Machines can relieve pressure of normal service at peak periods.
3 Staff are not required to serve from machines.
4 Clearing and washing-up are dispensed with as disposable cartons are used.
5 Refreshments are available on the job so reducing tea breaks.

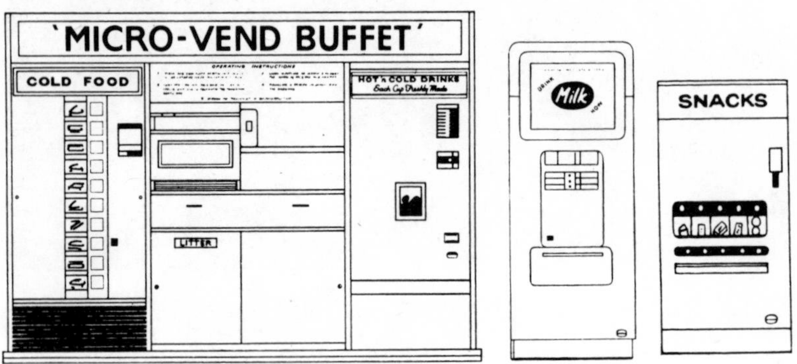

Speciality restaurants

Moderately priced speciality eating houses are in great demand and have seen a tremendous growth in recent years. In order to ensure a successful operation it is essential to assess the customers' requirements and to plan a menu that will attract sufficient customers to give adequate profit. A successful caterer is the one who gives the customer what he or she wants and not what the caterer thinks the customer wants. The most successful catering establishments are those which offer the type of food they can sell, which is not necessarily the type of food they would like to sell.

Well-cooked fish and chips has always been and will probably always be popular in Britain and it is interesting to note that one of the most successful speciality restaurant developments in the USA (the home of speciality restaurants) has been the English fish and chip shop complete with the food served in a bag made from a copy of *The Times* of the year 1778.

The Wimpy House, Texas Pancake House, the Health Food Restaurant, Kentucky Fried Chicken, the Sandwich Bar, Macdonalds plus numerous others are examples of speciality catering houses offering food and drink at moderate prices with menus carefully planned to fit their estimated markets (see pages 368-72).

Two methods of serving foods

Take away

Customer demand for prepared and cooked food to take-away has increased over the years and the enterprising caterer looking for the opportunity to increase his turn-over would be well advised to explore any potential market demand for his products. Suitable greaseproof containers need to be available for the service and carrying away of the food.

Ganymede dri-heat

This is a method of keeping foods either hot or cold. It is being used in some hospitals as it ensures that the food which reaches the patients is in the same fresh condition as it was when it left the kitchens.

A metal disc or pellet is electrically heated or cooled and placed in a special container under the plate. The container is designed to allow air to circulate round the pellet so that the food is maintained at the correct service temperature.

This is used in conjunction with conveyor belts and special service counters and helps to provide a better and quicker food service.

Further information: Lillicrap, *Food and Beverage Service* (Arnold).

The Lion Carvery–Excel Equipment Limited

17
The catering industry

Having read the chapter, these general objectives should be achieved.

General objectives Be aware of the various types of catering establishments and know of the particular characteristic peculiar to each.

Various types of catering establishments

Students need to be aware of the scope for employment in the industry, it is also desirable that they realise not only the social importance of the industry but its economic importance. The economic health of a nation is indicated by the food served in the home and in the eating establishments of the country. With full employment businesses boom; with overseas tourism expanding the effects are for the catering industry also to expand.

This country needs an industry capable of contributing to the national economy, therefore all aspects of the catering industry have an important part to play in the health of the economic state of the nation.

The provision of food for people of all ages, in all walks of life, at all times of the day or night, and in every situation shows the variety of scope which is to be found in the catering industry.

One thing is common to all – *the need for food to be cooked and served well.* Certain groups of people, however, have special food requirements; for example, some old people, due to poor digestion and to the fact that they may have dentures, require foods which are easily digested and need little chewing. Likewise when catering for young people it is particularly important to consider the nutritional needs of those who are still growing. An adequate supply of protein and calcium is essential.

Religious and national requirements are considerations of ever growing importance in a world which, due to increased travel, and better communications, demands an awareness of the social and religious requirements of others. Social customs involving the use of certain foods or dishes often originated because of religious requirements such as fasts, feasts, or other significant reasons. Many of the traditional observances are declining and the origins forgotten. Fish on Friday, pancakes on Shrove Tuesday have less significance today. This is affected not only by the changing influence of religion, social attitudes and customs, but increased use of technology. Perishable foods are now refrigerated. Fish does not have to be dried and salted so as to be available to meet religious demands.

The geographical situation dictates the physical reasons of what constitutes national diet. In certain areas rice will be commonplace; other areas yams or sweet potatoes, elsewhere wheat. Nationals from other countries, either visiting or working in the UK, should be considered so that their foods are available to them.

People restrict themselves to a vegetarian diet on religious grounds, humanistic reasons or because they are concerned with their physical well being. Those choosing not to eat meat or fish because they consider it to be morally wrong do so for humanitarian reasons. The provision of vegetarian foods should be available for those requiring them.

An awareness of people's food requirements and how to meet them is the responsibility of those employed in the catering industry.

The various types of catering establishments may be listed as follows:

Catering for industry (industrial catering)	Other aspects of catering
Chain-catering organisations	Residential establishments
Clubs	School meals service
Hospital catering	Transport catering
Hotels and restaurants	Welfare catering
Luncheon clubs	Wine bars, fast foods, take away

These may be grouped as *a*) hotels and restaurants; *b*) welfare and industrial catering; *c*) transport catering; *d*) other aspects of catering.

Hotels and restaurants

The great variety of hotels and restaurants is demonstrated when the palatial, first-class luxury hotel is compared with the small hotel owned and run as a family concern. With restaurants a similar comparison may be made between the exclusive top-class restaurant and the small restaurant which may serve just a few lunches.

Hotels are residential and most of them will provide breakfasts, lunches, teas, dinners and snacks. In some hotels banquets will be an important part of the business.

Restaurants will vary with the kind of meals they serve. Some will serve all types of meals whilst others will be restricted to the service of lunch and dinner or lunch and tea, etc. Again, banqueting may form an important part of the restaurant's service.

In some cases various types of meal service, such as grill rooms, or speciality restaurants may limit the type of foods served – eg smorrëbrod, steaks, eggs, etc – will be provided.

Wine bars, fast foods, take away

Customer demand has resulted in the rapid growth of a variety of units offering popular foods with a limited choice at a reasonable price, and with little or no waiting time, to be consumed either on the premises or taken away. See p 368 Speciality restaurants.

Clubs

These are usually administered by a secretary appointed by a management committee formed from club members. Good food and drink with an informal service in the old English style are the requirement in most clubs, particularly of the St James's area type.

Night clubs usually have the type of service associated with the restaurant trade.

Chain-catering organisations

There are many concerns having catering establishments spread over wide areas and in some cases over the whole country. Prospects for promotion and the opportunities are often considerable whether it is in a chain of hotels or restaurants. These are the well-known hotel companies and also the restaurant chains, the popular type of restaurant, chain stores and the shops with restaurants. These restaurants often serve lunches, teas and morning coffee and have snack bars and cafeterias.

Several nation-wide, moderately priced chain stores operate cafeterias. These organisations are backed by very progressive managements and one particular example has a very high reputation for cleanliness. In some cases considerable experimental work is carried out to standardise dishes throughout the company's concerns.

Further information: The Hotel, Catering and Institutional Management Association, 191 Trinity Road, London, SW17.

Welfare catering

The fundamental difference between welfare catering and the catering of hotels and restaurants is that the hotel or restaurant is run to make a profit and provide a service. The object of welfare catering is to provide a service without necessarily making a profit. The standards of cooking should be equally good, though the types of menu will be different.

Hospital catering

Hospital catering is classified as welfare catering, the object being to assist the nursing staff to get the patient well as soon as possible. To do this it is necessary to provide good quality food which has been carefully prepared and cooked to retain the maximum nutritional value, and presented to the patient in an appetising manner.

It is recognised that the provision of an adequate diet is just as much a part of the patients treatment as careful nursing and skilled medical attention.

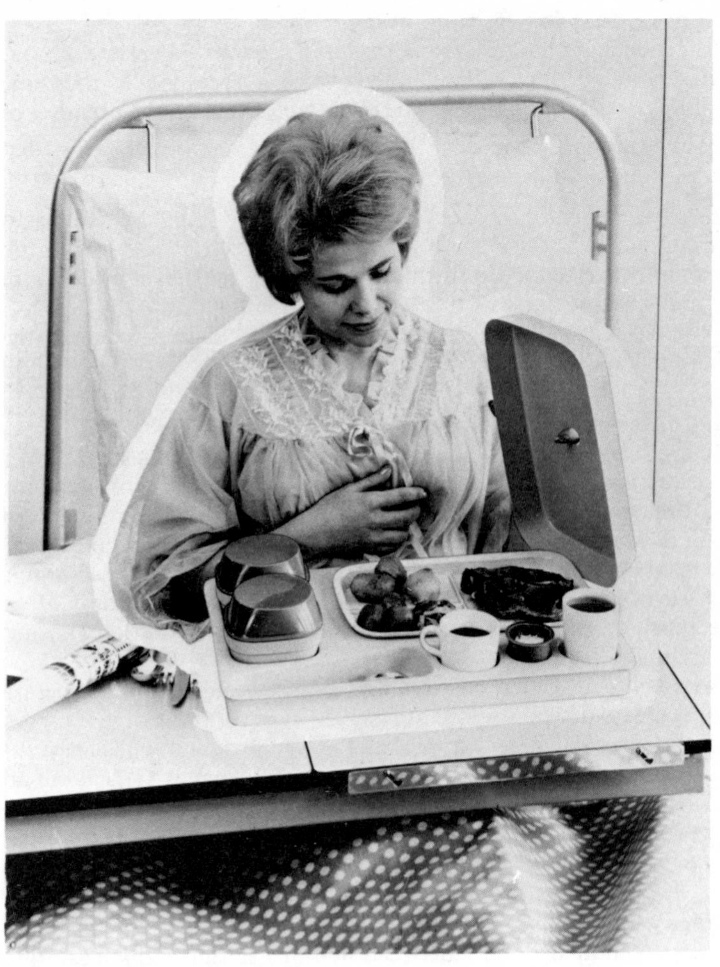

Hospital catering

Within the health service two million meals are served every 24 hours and the number served in one establishment can vary from 50 to 3000 people. In many hospitals patients are provided with a menu choice. The staff in the hospital catering service are organised as follows:

Catering managers plan menus, obtain supplies and supervise the preparation, cooking and service of the meals. They visit the wards to advise on the service of food to the patients, they also control the provision of the catering facilities for the doctors, nurses and other hospital employees.

Assistant catering managers assist and deputise for the catering managers with all or part of their duties, or they may be responsible for a small hospital.

Catering supervisors have similar responsibilities as do catering managers but this grade is only used in very small health service establishments.
Kitchen superintendents are responsible to the catering manager or the assistant catering manager for the running of one or more hospital kitchens.
Cooks are graded: 1 – assistant cook, 2 – assistant head cook, 3 – head cook. The head cook would be in charge of a kitchen under the control of the kitchen superintendent or catering manager.
Dining room supervisors are in charge of the staff during meal service and they are responsible to the catering manager.

People interested in being of service to the community and gaining job satisfaction could find this aspect of catering rewarding. Conditions, hours of work and pay as well as promotion prospects are factors which contribute to making this a worth while career.

Dietitians

In most hospitals a qualified dietitian is responsible for:

a) collaborating with the catering manager on the planning of meals;
b) drawing up and supervising special diets;
c) instructing diet cooks on the preparation of special dishes;
d) advising the catering manager and assisting in the training of cooks with regard to nutritional aspects;
e) advising patients.

In some hospitals the food will be prepared in a diet bay by diet cooks.

Diets

Information about the type of meal or diet(s) to be given to each patient is supplied by the ward sister to the kitchen daily. The information will give the number of full, light, fluid and special diets, and with each special diet will be given the name of the patient and the type of diet required.

The main hospital kitchen

All food except diets and food cooked in the ward kitchens is cooked here. In this kitchen all meals for patients, doctors, nurses, clerical and maintenance staffs are prepared. In hospitals where a canteen is provided for out-patients and visitors this will come under the control of the catering officer.

Hospital routine

Hospital catering has its own problems, which often make it very difficult to provide meals correctly served. Wards are sometimes spread over a wide area, or, in a large hospital, where there are long distances for the food to travel, provision of effective, silent trolleys is essential to keep the food hot.

The routine of a hospital is strictly timed and meals have to fit in with the duties of the nursing staff.

The amount of money the catering manager has to spend on food and drink is stated as so much per head. Good, wholesome varied meals can only be provided by careful buying and the elimination of waste.

The cost of providing the hospital catering service is periodically reviewed by the Government.

Further information: Hospital Caterers Association, John Ratcliffe Hospital, Oxford; and about dietetics from the British Dietetic Association, 103 Daimler House, Paradise Street, Birmingham.

School meals service

The provision of a mid-day meal at school for children up to and including secondary school age has been a statutory duty of local education authorities since 1941, and the policy governing this provision is laid down by the Department of Education and Science.

Over the years however, this policy has varied, generally due to the economic situation of the country at the time. At the present time local education authorities may decide for themselves:

a) their own free schools policy, subject to free meals being provided to pupils from families receiving supplementary benefit or family income supplement;
b) the kind of meals service they wish to provide;
c) the charge they wish to make for meals.

The school meals service is controlled by school meals organisers responsible for the catering service, including control of finance within the budget, menu planning, advice on food purchasing, kitchen planning and general administration. Supervision of the school meals kitchens is undertaken by catering officers or cook supervisors and in smaller units by a cook-in-charge. Many women are found in this area; they find that working in the school meals service can be fitted in with their responsibilities at home. Many work only for 2–2½ hours each day during the actual meal service, however kitchen staff at all levels receive training.

Many schools now offer a multi-choice menu with up to a dozen choices of both courses and operated on a cash cafeteria system as exists in most departmental stores. In recent years cook-freeze and cook chill systems have been introduced by some authorities to reduce labour costs.

School kitchens are well equipped. The highest standard of personal and kitchen hygiene is demanded.

Staff training 'in service' and by day and block release is the accepted method of ensuring a high standard of efficiency in the kitchens.

For up-to-date information on the Schools Meals Service obtain the current circulars issued by the Department of Education and Science, Elizabeth House, York Road, London, SE 1, or contact the Local Education Authority's School Meals Organiser.

Residential establishments

Under this heading are included schools, colleges, halls of residence, hostels, etc, where all the meals are provided. It is essential that in these establishments the nutritional balance of food is considered, as in all probability the people eating here will have no other food. It is important that food cooked for the residents should satisfy all the nutritional needs. Since many of these establishments cater for students and the age group which leads a very energetic life, these people usually have large appetites, and are growing fast. All the more reason that the food should be well cooked, plentiful, varied and attractive.

Further particulars of employment in this type of establishment can be obtained from the Hotel, Catering and Institutional Management Association, 191 Trinity Road, London, SW17 7HN.

Catering for industry (industrial catering)

The provision of canteens for industrial workers has given opportunities to many catering workers to be employed in first-class conditions. Apart from the main lunch meal, tea trolley rounds and/or vending machines may be part of the service. In some cases a 24-hour canteen service is necessary and it is usual to cater for the social activities of the workers. Not only are lunches provided for the manual workers but the clerical staff and managerial staff will in many cases have their meals from the same kitchen. There is ample scope for both men and women, and in this branch of the industry there are more top jobs for women than in any other.

Many industries have realised that output is related to the welfare of the employees. Well-fed workers produce more and better work and because of this a great deal of money is spent in providing first-class kitchens and dining-rooms and in subsidising the meals. This means that the workers receive their food at a price lower than its actual cost, the rest of the cost being borne by the company.

For an organisation chart of the Catering Division of the Office Management Department (London office) of the British Petroleum Company, see page 000.

Further information: Industrial Catering Association, 1 Victoria Parade-By-331 Sandy-combe Road, Richmond, Surrey.

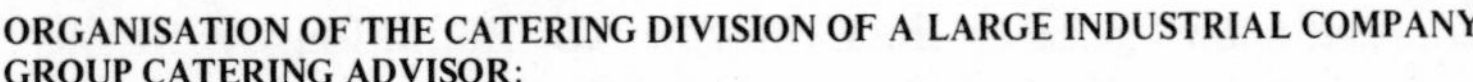

ORGANISATION OF THE CATERING DIVISION OF A LARGE INDUSTRIAL COMPANY

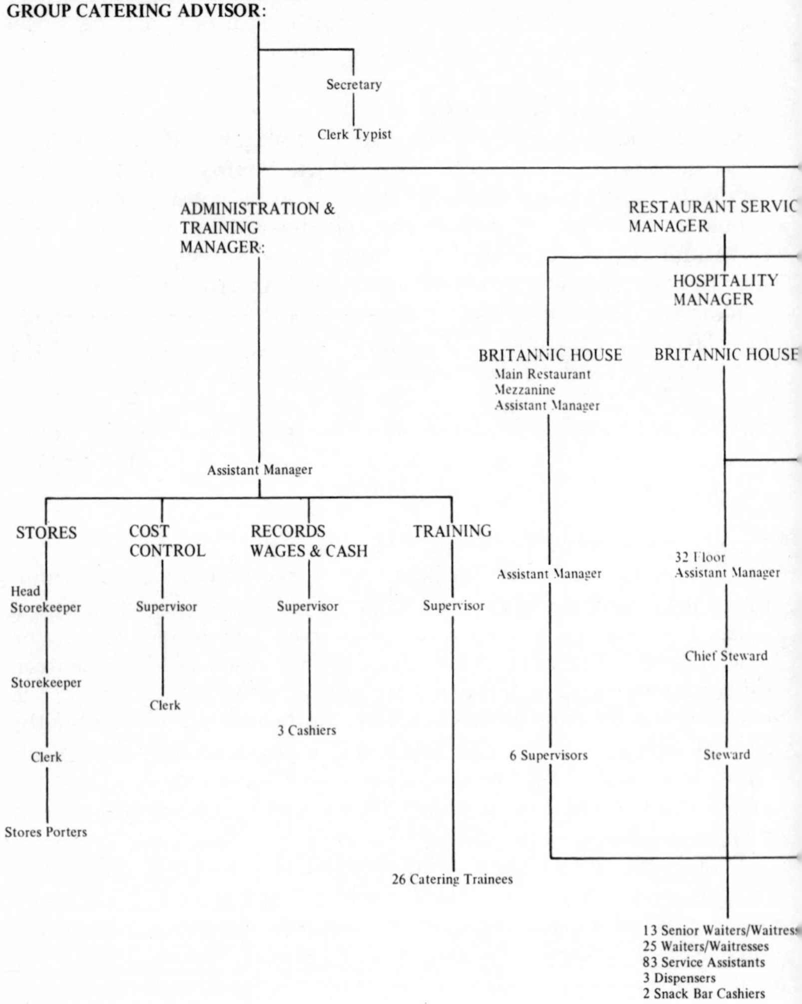

GROUP CATERING ADVISOR:

- Secretary
  - Clerk Typist
- ADMINISTRATION & TRAINING MANAGER:
  - Assistant Manager
    - STORES
      - Head Storekeeper
      - Storekeeper
      - Clerk
      - Stores Porters
    - COST CONTROL
      - Supervisor
      - Clerk
    - RECORDS WAGES & CASH
      - Supervisor
      - 3 Cashiers
    - TRAINING
      - Supervisor
      - 26 Catering Trainees
- RESTAURANT SERVIC[E] MANAGER
  - BRITANNIC HOUSE
    - Main Restaurant
    - Mezzanine
    - Assistant Manager
    - Assistant Manager
    - 6 Supervisors
  - HOSPITALITY MANAGER
    - BRITANNIC HOUSE
      - 32 Floor Assistant Manager
      - Chief Steward
      - Steward
  - 13 Senior Waiters/Waitress[es]
  - 25 Waiters/Waitresses
  - 83 Service Assistants
  - 3 Dispensers
  - 2 Snack Bar Cashiers

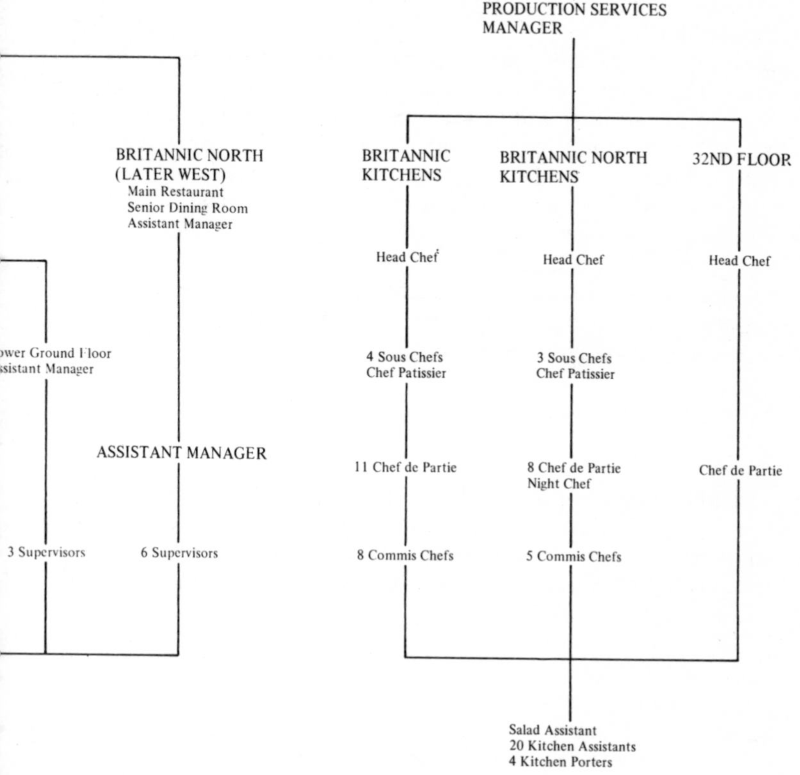
PRODUCTION SERVICES MANAGER
BRITANNIC NORTH (LATER WEST)
Main Restaurant
Senior Dining Room
Assistant Manager
BRITANNIC KITCHENS
BRITANNIC NORTH KITCHENS
32ND FLOOR
Head Chef
Head Chef
Head Chef
ower Ground Floor
ssistant Manager
4 Sous Chefs
Chef Patissier
3 Sous Chefs
Chef Patissier
ASSISTANT MANAGER
11 Chef de Partie
8 Chef de Partie
Night Chef
Chef de Partie
3 Supervisors
6 Supervisors
8 Commis Chefs
5 Commis Chefs
Salad Assistant
20 Kitchen Assistants
4 Kitchen Porters

Luncheon clubs

Clerical staffs in large offices are provided with lunching facilities, usually called a luncheon club or staff restaurant. These are usually subsidised and in some instances the meal may be supplied without charge. The catering is frequently of a very high standard and the kitchen or kitchens will provide meals for the directors, which will be of the best English fare or international cuisine.

Business lunches are served in small rooms so that there is privacy; the standard of food served will often be of the finest quality since the company will probably attach considerable importance to these functions. The senior clerical staff may have their own dining-room, whilst the rest of the staff will in some cases have a choice of an a la carte menu, a table d'hôte menu, waitress service or help-yourself and snack-bar facilities.

Luncheon clubs are provided by most large offices belonging to business firms, such as insurance head offices, petroleum companies, banks, etc. When luncheon facilities are not provided many firms provide their employees with luncheon vouchers.

Large stores also provide lunching arrangements for their staffs as well as the customers' restaurants.

Transport catering

Aircraft

Aircraft catering is concerned with the provision of meals during flights and this form of catering presents certain problems. Owing to very limited space on the plane special ovens are provided to heat the food, which is frozen. The food is prepared by the aircraft company or by an outside contract company.

Railway catering

The function of catering on trains in this country is carried out by British Rail. Meals may be served in restaurant cars, and snacks are served from buffet cars. The space in a restaurant car kitchen is very limited and there is considerable movement of the train, which causes difficulty for the staff.

Catering at sea

The large liners' catering is of similar standard to the large first-class hotels. Many shipping companies are noted for the excellence of their cuisine. The kitchens on board ship are usually oil-fired and extra precautions have to be taken in the kitchen in rough weather. Catering at sea includes the smaller ship, which has both cargo and passengers, and the cargo vessels and this includes the giant tankers of up to 100 000 tonnes.

Method of loading food on to aircraft (left)
The limited catering facilities and special equipment on an aircraft (right)

Other aspects of catering

The Services

Catering for the armed services is specialised and each branch has its own training centre; details of catering facilities and career opportunities can be obtained from Career Information Offices.

Contract catering

There are many catering concerns who are prepared to undertake the catering for businesses, schools, hospitals, etc, leaving these establishments free to concentrate on the business of educating or nursing or whatever may be their main concern. By employing contract caterers they are using the services of people who have specialised in catering, thus relieving themselves of the worry of entering a field outside their province. Contract catering is employed by nearly every branch of catering, including the armed forces. The arrangements made will vary. The contractor may meet certain operating costs or receive a payment from the company employing the contractor. Often the cost of food, wages and light equipment is the responsibility of the contractor, whilst the cost of fuel, heavy equipment maintenance is borne by the company.

The meal for the captive customer – in a prison (top) and in an aircraft

Outside catering

When functions are held where there is no catering set-up or where the function is not within the scope of the normal catering routine then certain firms will take over completely. Considerable variety is offered to people employed on these undertakings and often the standard will be of the very highest order. A certain amount of adaptability and ingenuity is required, especially for some outdoor jobs, but there is less chance of repetitive work. The types of function will include garden parties, agricultural and horticultural shows, the opening of new buildings, banquets, parties in private houses, etc.

Licensed house (pub) catering

There are approximately 70 000 licensed houses in England and Wales and almost all of them offer food in some form or other. To many people the food served in public houses is ideal in that it is the type of food they want, often simple, moderate in price and quickly served in a congenial atmosphere.

A great variation exists in public-house catering, from the ham and cheese roll operation to the exclusive à la carte restaurant. Public-house catering can be divided into four categories:

a) The luxury-type restaurant.
b) The speciality restaurant, eg steak bar, fish restaurant.
c) Fork dishes served from the bar counter where the food is consumed in the normal drinking areas.
d) Finger snacks, eg rolls, sandwiches, etc.

Further information: Pub Caterer magazine

18
Guide to study

Most students can be successful with their studies provided they wish to succeed in the catering industry and are determined to work reasonably hard throughout the course. A problem which is presented to many students is *how* to get the best out of their course, because it may be the first time they have entered an educational establishment where much of the responsibility for learning lies with them. Therefore some suggestions follow which may be helpful in guiding students how to study and how to cope with the various assessment techniques used in education.

The successful catering student is one who not only achieves the required standard of techniques but, equally important, has the right attitude to other students, teaching staff, customers and the industry that he has chosen.

Right attitude includes such qualities as reliability, conscientiousness, willingness, co-operativeness, honesty, punctuality and courtesy, and the student needs to develop these qualities if he is to be successful.

How to study

Catering is a very practical occupation, nevertheless there is a considerable amount of study needed for the learning of the theoretical aspects of catering courses and the following comments may be of assistance:

1 Learning requires effort and hard work, but thorough preparation gives confidence.
2 Studying in a quiet place is effective for most people, such as the local library's reading room.
3 Studying is more efficient when students are not tired.
4 The careful rewriting of lesson notes aids learning.
5 Do not leave revision until too near the examination, revise throughout the course, because spaced out learning is more effective than 'cramming'.
6 Study consistently throughout the course.
7 Organise regular times for both study and revision.
8 Practice in answering previous examination questions can be useful at appropriate stages of the course.

Most students need to know how they are progressing on the course of study both in relation to other members of the course and the standards required of the course. This information is obtained from teachers and various methods of evaluating students are used. It is essential that students know during the course as well as at the crucial stages, such as at the end of the year, if they are working in the best manner to reach the required standard.

What to study

The syllabus of most catering courses is written in objective terms, this enables the student to know precisely what should be known, and what the student should be able to do at the end of the course. An advantage of having the course content expressed in this way means that the student, and the teacher are both aware of the depth and breadth to which the course will be examined. Students towards the end of the course should check the syllabus for any omissions in their knowledge, the sensible student then accepts responsibility to remedy any gaps by further study and revision.

Evaluation procedures

Students are subjected to various techniques designed to assess what has been learned, therefore it is helpful to consider those methods in which catering students will be involved, although all students will not necessarily use all the methods. They include the evaluation of theoretical aspects by:

1 subjective-type examination; (brief answer and essay type)
2 objective-type examination;
3 continuous assessment.

and practical aspects by:

1 practical examination;
2 continuous assessment of practical work.

It is essential to understand that although the theoretical and practical aspects of the course may be assessed separately, the student should realise that theory and practice cannot be separated, because the understanding of the theory is necessary for achieving good results in the practical class.

Continuous assessment

One purpose of continuous assessment is to enable students to know how they are progressing during the course and this is achieved by periodic evaluations or assessments. The advantage of this system is that a person does not fail on one particular occasion as could occur with a single examination. Furthermore the student and the teacher can become aware of weaknesses early in the course and steps can be taken to remedy them.

Continuous assessment of practical work

This assessment should take place during a normal learning situation and should *not* take the form of an examination.

Students should receive details of the result of the assessment if possible at the end of the lesson.

The frequency of assessments will depend on the type of course, and the needs of the college and students. Some teachers will record marks for every practical class for every student in the group; others will assess a small number of students in certain situations. For City and Guilds of London Institute courses three aspects are assessed:

1 working methods;
2 practical skills;
3 professional practice.

It should be noted by students that it is in their interest to maintain a constantly good standard in these three areas during the whole of the course and continuous assessment should help in achieving this.

Continuous assessment of theory work

Some students may be expected to present work throughout the course in place of, or as well as, an examination. This work may include homework, work completed in the class, projects, etc and the marks awarded for these used in place of an examination mark. Again the student benefits; if the student is away or unwell on the day of the examination he or she has a problem, but with work spread over the course this problem does not occur.

Theory examinations

The object of a theory examination is to find out if the candidate knows the answers. The students have to convey to the examiner the required answers and this is usually done by objective-type questions (short answer) or subjective-type questions, which may range from a brief answer to a short essay or the use of both.

Objective-type questions

When answering objective-type questions it cannot be too strongly emphasised that students must *carefully* read the question, decide which is correct, then mark the paper accordingly. If the student is not *sure* of the answer it is wise to proceed to the next question. Having considered all the questions in this way then carefully re-read the questions which have not been answered and then mark the paper in the appropriate box.

For most students it will be unwise to review the questions they have answered straight away (provided the question has been carefully read); this is because with multiple-choice questions the distractors should be plausible and doubts might come into the student's mind.

Finally, more than adequate time is allowed for objective-type examinations, time is not a limiting factor, therefore there is no urgency to complete the paper quickly—accuracy is the key to success.

Examples of objective-type questions can be found in *Questions on*

Practical Cookery (Edward Arnold) and *Questions on Theory of Catering* (Edward Arnold).

Subjective-type questions

As with objective-type questions it is essential that care is taken when reading the question. With essay-type questions it is also necessary to take care in producing the answer and the following points should be considered.

1 *Writing:* This must be readable. The easier it is to read, the more favourable impression is given of the student's paper.

2 *Spelling:* The facts are what the examiner wants, but if there are many words spelt wrongly an unfavourable impression is given. French words should be correct. It is better to use correctly spelt English words than misspelt French words.
Extra care must be taken not to mis-spell words which are similar in French and English, eg filet and fillet, carotte and carrot.
Simple words are often spelt wrongly, such as gravy, plaice, gherkin, lettuce, etc.

3 *Answering questions:*

a) The meaning of the answer must be clear to the examiner.
b) The question should be understood and answered to the point.
c) The answer should be precise, with no padding.
d) No essential facts should be omitted from the answer.
e) All parts of a question should be answered.
f) Avoid using first person singular, eg 'Braising is a nice method of cooking, but I like fried steak.'
g) Avoid slang expressions and terms which are not good English.
h) Avoid vagueness and inaccurate phrases, eg: 'Spot of water'.
i) Use correct terms and words. eg: Sugar is not *diluted* , but dissolved. Gelignite does not go into bavarois!
j) Use accepted abbreviations for weights and measures, but avoid marge, veg, fridge.
k) Answer only the number of questions asked for. No extra marks will be gained if more are answered, but be certain to answer the required number of questions.
l) Answer questions in any order, but number them clearly.
m) Re-read the answers carefully when the paper is completed.

4 *Layout:* A paper laid out clearly creates a favourable impression. The following points may be helpful:

a) Leave a good margin.
b) Tabulate answers where suitable.
c) If an essay form answer is necessary give an introduction and have a good concluding paragraph.
d) Name the recipe, underline the heading.
e) Where diagrams can be used these should be drawn carefully.

Summary: It is necessary to know the facts to present them clearly. The writing must be legible, the English and spelling correct, the layout neat and the information to the point.

Examination advice

Hints on examinations

A problem which has to be faced by many people is nerves: these may only be controlled to a certain degree according to each individual. Confidence in one's own ability helps, and this confidence comes from knowledge; provided candidates have done their utmost to learn, there need be no excessive nervousness.

Preparation for examinations

1 Revise throughout the course, but in addition revise the whole course prior to the examination.
2 Determine when revision is most effective for you; some people prefer early morning, others late evening.
3 Near the examination, practise answering past examination questions in the same time allowed as in the actual examination.
4 Do *not rely only* on past examination questions.

Examination day prior to the examination

1 Allow plenty of time so as to arrive punctually.
2 Have an adequate number of pens.
3 If you wish to chew sweets select those without wrapping paper so that you do not disturb others.
4 Most examinations occur in the summer therefore wear suitable clothes.

The examination

If there is a choice of question then taken extra care reading them, so that you fully understand the question, you can then select those you can answer best.

Most candidates' nerves disappear as soon as they start writing in the theory examinations. Some people, however, find their minds go blank; in this case it may help to concentrate on one small part of the question, particularly questions on practical cookery. The student who may not have learnt a particular recipe by heart can often, by thinking hard of the practical preparation, arrive at a sufficiently sound answer to satisfy the examiner.

Practical examinations

Object

The object of the practical examination is to find out if the student can cook to a required standard, in a clean manner within a specified time.

The teacher will expect the candidate to work as if it was a normal working day and will be looking at the following:

1 The student,
2 The way the student works,
3 The manner in which the food is presented.

To assist the student to do well in practical examinations the following points may be considered:

Personal appearance

1 The student should appear professional, capable and confident (not over-confident or flustered).
2 Whites should be clean and worn correctly.
3 Knives should be clean and sharp.
4 Clean cloth or cloths and a clean swab are essential.

Some practical tests require an order of work to be completed before the practical part of the examination begins.

Order of work

The purpose of an order of work is to enable the student to sit down and think out the order in which the menu is to be prepared. The main thing to consider is that those items needing the longest time must be done first; for example, puff paste, meat pudding. The order of work for many menus will be a matter of opinion. The teacher has a copy of the order of work, and should the student wish to alter it the teacher must be informed.

Example of the order of work for the following menu:

Grapefruit Cocktail
Steak and Kidney Pudding
Parsley Potatoes Mixed Vegetables
Pancakes with Lemon

It is necessary to prepare the steak and kidney pudding first, therefore the making of the suet paste and preparation of the steak head the order of work, because of the time required to cook this dish. From an examination of the rest of the menu it is discovered that the pancake batter will need to stand, therefore it should not be left to the last. The grapefruit should be served cool, and once prepared can be put out of mind when in the cold room or refrigerator.

All the vegetables are best prepared at the same time and the table cleared of peelings. This leaves the small jobs such as chopping the parsley and cutting the lemons.

The order of work, then, may look like this:

1 Check ingredients and clear table.
2 Make suet paste and line basin.
3 Prepare the steak and kidney.
4 Finish the pudding and place in pan of boiling water (or steamer) to cook.

5 Make the pancake batter and allow to rest.
6 Peel the vegetables and potatoes.
7 Cut the vegetables and potatoes.
8 Cut the grapefruit, dress and place in refrigerator.
9 Cook the potatoes.
10 Chop the parsley and cut the lemons.
11 Cook the vegetables.
12 Serve the grapefruit.
13 Finish the potatoes and vegetables.
14 Serve the pudding.
15 Serve the vegetables and potatoes.
16 Cook and serve the pancakes.

The order of work should be brief and only the important points included. For most menus 16-20 well-thought-out points should be sufficient, because time spent doing this means there is less time available in the kitchen. The time of serving courses should be known before the commencement of the order of work.

Method of work

The way in which the student sets about the work is very important, and the following points can be borne in mind:

1 Perishable foods should be placed in the cold room.
2 Tables should be cleared ready to start work.
3 Only gases required at once should be lit.
4 Apart from the food which is being prepared the only other things on the table should be a triangle, board and knives.
5 Care should be taken to weigh foods accurately.
6 Trays or containers for rubbish, swill and trimmings should be kept.
7 Small equipment should be washed as soon as possible after use.
8 Correct knives should be used for the job and wiped after use.
9 The student should work cleanly, quickly and correctly.
10 Should something go wrong, for example the blind flan burn, and if there are sufficient ingredients available, the student should start again and if possible produce the dish.
11 Should something be spoiled at the last moment it does not mean that all is lost, because the teacher will have seen the work prior to the accident.

Service of food for tests

1 Food should be served on time.
2 Hot dishes should be hot, cold dishes cold.
3 All dishes should be clean.
4 Food must be correctly seasoned and
5 Served in right dishes.

6 The exact number of portions should be in the dish.
7 Appropriate accompaniment must be served.

Hygiene and safety

The student must work cleanly all the time. Stove, table and floor area should be kept clean, and at the end of the examination all pans used placed on the table for checking. It is essential that the student works in a manner that will not cause an accident to himself or herself or others working in the same situation.

Index